Wearable Sensors for Human Health Monitoring and Analysis

Wearable Sensors for Human Health Monitoring and Analysis

Guest Editors

Alessandro Scano
Alfonso Mastropietro
Rebecca Re
Paolo Perego

MDPI

Basel • Beijing • Wuhan • Barcelona • Belgrade • Novi Sad • Cluj • Manchester

Guest Editors
Alessandro Scano
Institute of Intelligent
Industrial Technologies and
Systems for Advanced
Manufacturing (STIIMA)
Italian National Research
Council (CNR)
Milano
Italy

Alfonso Mastropietro
Institute of Intelligent
Industrial Technologies and
Systems for Advanced
Manufacturing (STIIMA)
Italian National Research
Council (CNR)
Milano
Italy

Rebecca Re
Department of Physics
Polytechnic University of Milan
Milano
Italy

Paolo Perego
Department of Design
Polytechnic University of Milan
Milano
Italy

Editorial Office
MDPI AG
Grosspeteranlage 5
4052 Basel, Switzerland

This is a reprint of the Special Issue, published open access by the journal *Sensors* (ISSN 1424-8220), freely accessible at: https://www.mdpi.com/journal/sensors/special_issues/OC62G7APGK.

For citation purposes, cite each article independently as indicated on the article page online and as indicated below:

Lastname, A.A.; Lastname, B.B. Article Title. *Journal Name* **Year**, *Volume Number*, Page Range.

ISBN 978-3-7258-7060-8 (Hbk)
ISBN 978-3-7258-7061-5 (PDF)
https://doi.org/10.3390/books978-3-7258-7061-5

Cover image courtesy of Alessandro Scano

Contents

sensors

MDPI

Editorial

Wearable Sensors for Human Health Monitoring and Analysis

Alessandro Scano [1,*], Rebecca Re [2], Paolo Perego [3] and Alfonso Mastropietro [1]

1 Laboratory of Advanced Methods for Biomedical Signal and Image Processing, Institute of Intelligent Industrial Technologies and Systems for Advanced Manufacturing (STIIMA), Italian National Research Council (CNR), 20133 Milan, Italy; alfonso.mastropietro@cnr.it
2 Department of Physics of Politecnico di Milano, Institute of Photonics and Nanotechnologies, Italian National Research Council (CNR), 20133 Milan, Italy; rebecca.re@polimi.it
3 Department of Design, Politecnico di Milano, Via Candiani 72, 20158 Milan, Italy; paolo.perego@polimi.it
* Correspondence: alessandro.scano@cnr.it

1. Introduction

Many fields of research have recently made significant progress with the use of wearable sensors for the evaluation, monitoring, and analysis of health in several contexts, including healthcare [1,2], rehabilitation [3], industry [4], sports [5], and others [6]. Such sensors foster human-centered approaches to research and innovative applications and extend practical use, playing key roles in improving people's lives [7]. However, the adoption of wearable sensors poses several scientific challenges, including issues regarding data management, effective data analysis, calibration, the design of novel sensors (regarding wearability and power supply issues), and novel sensors' adherence to previously adopted standards. There are even logistical challenges to face, related, for example, to compatibility with and adherence to mandatory guidelines, especially in critical scenarios (such as those in medicine) [8].

In this Special Issue, "Wearable Sensors for Human Health Monitoring and Analysis," we aimed to foster scientific interest in these challenges and gathered a collection of 12 papers that describe the diverse applications of wearable biomedical sensors in the healthcare context pertaining to wearable sensors for human health monitoring and analysis. These contributions are grouped into key thematic areas, reflecting the breadth of research:

1. *Neurophysiology and brain signal analysis.* The authors of papers in this category utilized neurophysiological techniques such as electroencephalography (EEG) and functional near-infrared spectroscopy (fNIRS) to evaluate cerebral activity, pain intensity, and movement control, including studies on multilevel pain assessment using fNIRS and design suggestions for the optimal way of using wearable EEG equipment to evaluate motor imagery.
2. *Muscle and neuromotor analysis through physiological sensors.* This area of research investigates the monitoring or classification of neuromuscular and motor activity (EMG, sEMG, and muscle synergies), with applications in clinical, biomechanical, and sports contexts. It includes electromyographic comparisons of training equipment and multi-scale CNNs for sEMG-based hand gesture recognition for prosthetic devices.
3. *Wearable motion sensors and human motion evaluation.* These papers focus on the design, validation, and calibration of wearable sensors to analyze complex movements among frail individuals or those performing physical activities. Notable contributions address analyzing optimal wearable motion sensor placement for accurate classification of fall directions and the validity and test–retest reliability of spatiotemporal running parameter measurement. The utility of calibrating wearable sensors for quantifying infant leg movements is also explored.

4. *Design of and technologies used for wearable devices and digital health.* The studies in this category explore the design and development of novel hardware and software solutions for wearable systems, with emphasis on efficiency, energy, and data management. It includes studies on hybrid energy-efficient harvesting systems for healthcare wearables as well as the design and development of a smart fidget toy using blockchain technology to improve health data control.

Taken together, these papers highlight the key role of wearable sensors in understanding physiological mechanisms and advancing healthcare, medicine, industry, and sports. They further promote the adoption of these devices by providing insights for use, practical implementation strategies, software and hardware design considerations, and novel algorithms. A complete list of the contributors to this Special Issue is provided below:

- Zhu, W.; Lin, Y. Physiological Sensor Modality Sensitivity Test for Pain Intensity Classification in Quantitative Sensory Testing. Sensors 2025, 25, 2086. https://doi.org/10.3390/s25072086.
- Falivene, A.; Johnson, C.; Klingels, K.; Meyns, P.; Verbecque, E.; Hallemans, A.; Biffi, E.; Piazza, C.; Crippa, A. Time-Normalization Approach for fNIRS Data During Tasks with High Variability in Duration. Sensors 2025, 25, 1768. https://doi.org/10.3390/s25061768.
- Abelleira-Lamela, T.; Marcos-Pardo, P.J.; Abraldes, J.A.; González-Gálvez, N.; Espeso-García, A.; Esparza-Ros, F.; Vaquero-Cristóbal, R. Electromyographic Comparison of Traditional Fitness Machines, Outdoor Fitness Equipment Without Load Selectors, and Outdoor Fitness Equipment with Load Selectors in a Seated Chest Press Exercise in Trained Young Men. Sensors 2024, 24, 7740. https://doi.org/10.3390/s24237740.
- Fratti, R.; Marini, N.; Atzori, M.; Müller, H.; Tiengo, C.; Bassetto, F. A Multi-Scale CNN for Transfer Learning in sEMG-Based Hand Gesture Recognition for Prosthetic Devices. Sensors 2024, 24, 7147. https://doi.org/10.3390/s24227147.
- Bobrova, P.; Perego, P.; Boiano, R. Design and Development of a Smart Fidget Toy Using Blockchain Technology to Improve Health Data Control. Sensors 2024, 24, 6582. https://doi.org/10.3390/s24206582.
- Teng, S.; Kim, J.-Y.; Jeon, S.; Gil, H.-W.; Lyu, J.; Chung, E.H.; Kim, K.S.; Nam, Y. Analyzing Optimal Wearable Motion Sensor Placement for Accurate Classification of Fall Directions. Sensors 2024, 24, 6432. https://doi.org/10.3390/s24196432.
- Riglet, L.; Orliac, B.; Delphin, C.; Leonard, A.; Eby, N.; Ornetti, P.; Laroche, D.; Gueugnon, M. Validity and Test–Retest Reliability of Spatiotemporal Running Parameter Measurement Using Embedded Inertial Measurement Unit Insoles. Sensors 2024, 24, 5435. https://doi.org/10.3390/s24165435.
- Tohidinejad, Z.; Danyali, S.; Valizadeh, M.; Seepold, R.; TaheriNejad, N.; Haghi, M. Designing a Hybrid Energy-Efficient Harvesting System for Head- or Wrist-Worn Healthcare Wearable Devices. Sensors 2024, 24, 5219. https://doi.org/10.3390/s24165219.
- Carretero, A.; Araujo, A. Design Decisions for Wearable EEG to Detect Motor Imagery Movements. Sensors 2024, 24, 4763. https://doi.org/10.3390/s24154763.
- Khan, M.U.; Sousani, M.; Hirachan, N.; Joseph, C.; Ghahramani, M.; Chetty, G.; Goecke, R.; Fernandez-Rojas, R. Multilevel Pain Assessment with Functional Near-Infrared Spectroscopy: Evaluating ΔHBO2 and ΔHHB Measures for Comprehensive Analysis. Sensors 2024, 24, 458. https://doi.org/10.3390/s24020458.
- Oh, J.; Loeb, G.E.; Smith, B.A. The Utility of Calibrating Wearable Sensors before Quantifying Infant Leg Movements. Sensors 2024, 24, 5736. https://doi.org/10.3390/s24175736.

https://doi.org/10.3390/s26020575

- Scano, A.; Lanzani, V.; Brambilla, C.; d'Avella, A. Transferring Sensor-Based Assessments to Clinical Practice: The Case of Muscle Synergies. Sensors 2024, 24, 3934. https://doi.org/10.3390/s24123934.

A brief description of the novel contributions and main challenges faced in this Special Issue is provided in the following section.

2. Neurophysiology and Brain Signal Analysis

This section focuses on neurophysiology and brain signal analysis, sharing as a common topic the use of non-invasive neurophysiological techniques (EEG and fNIRS) to evaluate cerebral activity, pain intensity, or movement control.

"Time-Normalization Approach for fNIRS Data During Tasks with High Variability in Duration" presents a novel algorithm that aids in the comparison of multiple acquisitions during (fNIRS) recordings. The proposed algorithm performed well on a set of experimental data presented by the authors; it was distributed with the publication and has been made available to users.

"Multilevel Pain Assessment with Functional Near-Infrared Spectroscopy: Evaluating ΔHBO2 and ΔHHB Measures for Comprehensive Analysis" assesses pain in non-verbal patients. The study focuses on assessing pain via the analysis of fNIRS signals combined with machine learning, utilizing multiple fNIRS measures, including oxygenated and deoxygenated hemoglobin. The combination of these two measures demonstrated superior performance relative to when they were used independently in multilevel pain analysis.

Finally, "Design Decisions for Wearable EEG to Detect Motor Imagery Movements" investigated how one can make informed decisions regarding the design of wearable electroencephalography (wearable EEG) devices for the evaluation of motor imagery, drawing several conclusions regarding the optimal sampling frequency, the EEG system's comfort and portability, trials, the number of electrodes, algorithms, and their parameters, thus providing a set of practical suggestions for experimenters.

3. Muscle and Neuromotor Analysis Through Physiological Sensors

This section focuses on muscle and neuromotor analysis via physiological sensors, investigating the monitoring and classification of neuromuscular or motor activities (EMG, sEMG, muscle synergies, and movement) with clinical, biomechanical, or sports-related aims.

"Physiological Sensor Modality Sensitivity Test for Pain Intensity Classification in Quantitative Sensory Testing" deals with chronic pain and introduces a novel framework for objectively classifying pain intensity levels using physiological signals (blood volume pulse (BVP), galvanic skin response, EMG, respiration rate, skin temperature, and pupillometry) during Quantitative Sensory Testing sessions based on multiple wearable sensors, revealing that BVP is the most critical one for understanding pain levels.

Furthermore, "Electromyographic Comparison of Traditional Fitness Machines, Outdoor Fitness Equipment Without Load Selectors, and Outdoor Fitness Equipment with Load Selectors in a Seated Chest Press Exercise in Trained Young Men" describes an EMG assessment of the upper limbs of young men during their use of outdoor fitness equipment. The authors concluded that training with the outdoor seated chest press generated less EMG activity than traditional machine training. However, in general, an outdoor chest press equipped with a load selector system proved effective for strength training, yielding results comparable to those achieved the seated chest press.

"A Multi-Scale CNN for Transfer Learning in sEMG-Based Hand Gesture Recognition for Prosthetic Devices" describes how advancements in neural network approaches have enhanced the effectiveness of surface electromyography (sEMG)-based hand gesture

https://doi.org/10.3390/s26020575

recognition when measuring muscle activity. However, such methods hardly achieve high generalization and robustness, often requiring significant computational resources. In developing a robust model that can quickly adapt to new datasets using transfer learning, the authors highlight the effectiveness of transfer learning in creating adaptive, user-specific models for sEMG-based prosthetic hands.

Finally, "Transferring Sensor-Based Assessments to Clinical Practice: The Case of Muscle Synergies" expands on how advanced methods for analyzing motor control (muscle synergies) can be used to enhance the comprehension of the neuromotor system and adapt and predict the outcomes of therapies conceived to restore motor function. The paper also discusses the practical and technical challenges to be faced in adopting muscle synergies in clinical environments in research and in clinical practice.

4. Wearable Motion Sensors and Human Motion Evaluation

This section includes research that discusses wearable motion sensors and human motion evaluation, aiming at designing, validating, and calibrating wearable sensors to analyze complex movements or physical activity.

The authors of "Analyzing Optimal Wearable Motion Sensor Placement for Accurate Classification of Fall Directions" assessed inertial measurement unit (IMU) sensors placed at 12 distinct body locations to determine the most effective positions for capturing fall-related data. Statistical analyses of the results for the most effective classifier model demonstrated that the support vector machine is more effective than other classifiers across all sensor locations, with statistically significant differences in performance; the optimal sensor configuration for fall-direction classification involves strategically combining sensors placed on the pelvis, upper legs, and lower legs.

Furthermore, "Validity and Test–Retest Reliability of Spatiotemporal Running Parameter Measurement Using Embedded Inertial Measurement Unit Insoles" discusses the validation of the DSPro® insoles, developed to collect running parameters during tasks. The aim of this study was to assess the test–retest reliability and criterion validity of running gait parameters from DSPro® insoles compared to a motion-capture system. The test–retest reliability reflected moderate to excellent ICC values (ICC > 0.50).

Finally, in "The Utility of Calibrating Wearable Sensors before Quantifying Infant Leg Movements", the authors used wearable sensors to measure infant leg movement, showing that offset error in the measurement of gravitational acceleration is common among commercially available sensors. They demonstrate how offset and other errors can be measured using three wearable sensors available to professionals and how they affected a threshold-based movement detection algorithm for the quantification of infant leg movement. Additionally, they reveal how these offsets can be calibrated and corrected for to increase the reproducibility of results across sensors.

5. Design and Technologies for Wearable Devices and Digital Health

This section includes papers that describe design and technologies pertaining to wearable devices and digital health, focusing on the development of hardware/software wearable systems, with attention to efficiency, energy, and data management.

In "Designing a Hybrid Energy-Efficient Harvesting System for Head- or Wrist-Worn Healthcare Wearable Devices," the authors describe the architecture of a system based on highly efficient photovoltaic panels, compact thermoelectric modules, and two ultra-low-power BQ25504 DC-DC boost converters (Texas Instruments, Dallas, United States) which can increase battery life from 9.31 h to over 18 h, making it suitable for health-monitoring wearables worn on the head, face, or wrist region, specifically for outdoor workers.

https://doi.org/10.3390/s26020575

"Design and Development of a Smart Fidget Toy Using Blockchain Technology to Improve Health Data Control" explores the integration of blockchain technology in wearable health devices through the design and development of a SmartFidgetToy. Using an iterative user-centered design approach, the authors developed a mid-fidelity prototype of a physical fidget device with a blockchain-based web application. The study revealed high user interest (70%) in blockchain-based data control and sharing features and improved perceived security of data (among 90% of users) with blockchain integration.

6. Conclusions

This Special Issue, "Wearable Sensors for Human Health Monitoring and Analysis," presents a comprehensive collection of cutting-edge research in the field of wearable sensors. The 12 papers featured in the collection emphasize the rapid evolution and interdisciplinary nature of this field, offering significant contributions to both fundamental understanding and practical applications in human-health monitoring. Spanning novel approaches to pain *assessment using* fNIRS and sEMG-based prosthetic device control, advancements in fall detection and energy-efficient power solutions, and the pioneering use of blockchain for health data management, the collection highlights the enormous potential of wearable technology in healthcare. These advancements in the use of wearable technologies pave the way for improved practices, novel design approaches and paradigms, and advanced research programs. As biomedical sensor technology continues to evolve, the findings from these studies hold significant promise in revolutionizing medical practices and addressing complex health-related challenges, ultimately leading to better human health and well-being.

Author Contributions: Conceptualization, A.S., R.R., P.P., and A.M.; methodology, A.S., R.R., P.P., and A.M.; investigation, A.S., R.R., P.P., and A.M.; resources, A.S., R.R., P.P., and A.M.; writing—original draft preparation, A.S., R.R., P.P., and A.M.; writing—review and editing, A.S., R.R., P.P., and A.M.; visualization, A.S., R.R., P.P., and A.M.; project administration, A.S., R.R., P.P., and A.M. All authors have read and agreed to the published version of the manuscript.

Funding: This research received no external funding.

Acknowledgments: We would like to extend our appreciation to all the authors and scholars who contributed to this Special Issue. The success of this collection, exemplified in this editorial, is proof of the dedication, expertise, and hard work of each author in regard to pursuing scientific objectives. These valuable insights, innovative methodologies, and groundbreaking findings can enrich the field of wearable sensors in various applications. We extend our gratitude to the editorial and technical teams for their valuable and continuous support throughout the Special Issue's timespan, ensuring a smooth and efficient publication journey. Finally, we express our gratitude to the readers and the broader scientific community for their interest and engagement in this Special Issue, as it is their enthusiasm that drives the advancement of knowledge and fosters progress in wearable sensors in healthcare applications.

Conflicts of Interest: The authors declare no conflicts of interest.

References

1. Ajami, S.; Teimouri, F. Features and application of wearable biosensors in medical care. *J. Res. Med. Sci.* **2015**, *20*, 1208–1215. [CrossRef] [PubMed]
2. Dunn, J.; Runge, R.; Snyder, M. Wearables and the medical revolution. *Pers. Med.* **2018**, *15*, 429–448. [CrossRef] [PubMed]
3. Patel, S.; Park, H.; Bonato, P.; Chan, L.; Rodgers, M. A review of wearable sensors and systems with application in rehabilitation. *J. Neuroeng. Rehabil.* **2012**, *9*, 21. [CrossRef] [PubMed]
4. Schall, M.C., Jr.; Sesek, R.F.; Cavuoto, L.A. Barriers to the adoption of wearable sensors in the workplace: A survey of occupational safety and health professionals. *Hum. Factors* **2018**, *60*, 351–362. [CrossRef] [PubMed]

https://doi.org/10.3390/s26020575

5. Rana, M.; Mittal, V. Wearable sensors for real-time kinematics analysis in sports: A review. *IEEE Sens. J.* **2020**, *21*, 1187–1207. [CrossRef]
6. Heikenfeld, J.; Jajack, A.; Rogers, J.; Gutruf, P.; Tian, L.; Pan, T.; Li, R.; Khine, M.; Kim, J.; Wang, J.; et al. Wearable sensors: Modalities, challenges, and prospects. *Lab Chip* **2018**, *18*, 217–248. [CrossRef] [PubMed]
7. Park, S.; Jayaraman, S. Enhancing the quality of life through wearable technology. *IEEE Eng. Med. Biol. Mag.* **2003**, *22*, 41–48. [CrossRef] [PubMed]
8. Fletcher, R.R.; Poh, M.Z.; Eydgahi, H. Wearable sensors: Opportunities and challenges for low-cost health care. In Proceedings of the 2010 Annual International Conference of the IEEE Engineering in Medicine and Biology, Buenos Aires, Argentina, 31 August–4 September 2010; IEEE: Piscataway, NJ, USA, 2010; pp. 1763–1766.

https://doi.org/10.3390/s26020575

sensors MDPI

Perspective

Transferring Sensor-Based Assessments to Clinical Practice: The Case of Muscle Synergies

Alessandro Scano [1,*], Valentina Lanzani [1], Cristina Brambilla [1] and Andrea d'Avella [2,3]

[1] Institute of Intelligent Industrial Systems and Technologies for Advanced Manufacturing (STIIMA), Italian Council of National Research (CNR), 20133 Milan, Italy; valentina.lanzani@stiima.cnr.it (V.L.); cristina.brambilla@stiima.cnr.it (C.B.)
[2] Laboratory of Neuromotor Physiology, IRCCS Fondazione Santa Lucia, Via Ardeatina 306-354, 00179 Rome, Italy; a.davella@hsantalucia.it
[3] Department of Biology, University of Rome Tor Vergata, Via della Ricerca Scientifica, 00133 Rome, Italy
* Correspondence: alessandro.scano@stiima.cnr.it

Abstract: Sensor-based assessments in medical practice and rehabilitation include the measurement of physiological signals such as EEG, EMG, ECG, heart rate, and NIRS, and the recording of movement kinematics and interaction forces. Such measurements are commonly employed in clinics with the aim of assessing patients' pathologies, but so far some of them have found full exploitation mainly for research purposes. In fact, even though the data they allow to gather may shed light on physiopathology and mechanisms underlying motor recovery in rehabilitation, their practical use in the clinical environment is mainly devoted to research studies, with a very reduced impact on clinical practice. This is especially the case for muscle synergies, a well-known method for the evaluation of motor control in neuroscience based on multichannel EMG recordings. In this paper, considering neuromotor rehabilitation as one of the most important scenarios for exploiting novel methods to assess motor control, the main challenges and future perspectives for the standard clinical adoption of muscle synergy analysis are reported and critically discussed.

Keywords: muscle synergies; kinematic synergies; functional synergies; clinical assessment; instrumental assessments

1. Introduction

Sensor-based assessments in medical practice and rehabilitation include the use of measurement of physiological signals such as electroencephalography (EEG), electromyography (EMG), electrocardiography (ECG), and near-infrared spectroscopy (NIRS), and the recording of movement kinematics and interaction forces [1]. Even though these measurements are commonly employed in clinics for assessing patients, some of them have not been fully exploited [2]. In fact, while instrumental assessments of patients may shed light on mechanisms underlying physiopathology and motor recovery, they have been used mostly in research studies, with a reduced impact on clinical practice, at least for some of these techniques [3]. This is especially the case for muscle synergies, a multichannel EMG-based analysis technique method for the evaluation of motor control commonly used in neuroscience. Muscle synergy refers to the coordinated activation of groups of muscles working together to produce specific movements or perform a particular task [4]. The concept of muscle synergies is based on the idea that the nervous system organizes and controls complex movements by recruiting and coordinating the activity of multiple muscles as functional units or modules [5]. In this way, rather than controlling each muscle involved in a movement individually, the nervous system simplifies the control process by organizing muscles into synergistic groups and flexibly combining their activation. These muscle synergies help streamline motor control and contribute to the efficiency of movements. Thus, muscle synergies are often studied in motor neuroscience to understand how the nervous system orchestrates movements. Researchers use muscle synergy-based approaches to analyze muscle activity and movement patterns during various tasks, finding applications in many clinical scenarios [6], including post-stroke patients [7–9], Parkinson's disease [10], cerebral palsy [11], and others. Understanding muscle synergies may have implications for many fields such as rehabilitation [12], sports science [13], and robotics [14], as it provides insights into how the nervous system controls and coordinates movements [15]. Muscle synergy approaches have been used in many clinical applications. First, they have been used to investigate different tasks. In fact, muscle synergies have been identified as biomarkers of cortical damage in stroke patients during walking [16], showing that patients usually have a lower complexity in motor control. Other studies expanded synergistic assessment to upper limb movements, finding that shoulder muscle synergies are altered in hemiplegic patients [9,17]. Secondly, muscle synergies have been embedded into multidomain evaluations. In fact, muscle synergy assessment has been coupled with other signals, such as kinematics and EEG, to provide a comprehensive evaluation of patients [18,19]. Multiparameter approaches also include muscle synergies with spinal map activities, connecting EMG recorded at muscle level to the spatiotemporal motoneuronal activity in patients with spinal cord injuries [20]. Muscle synergies have also been employed to describe device-assisted rehabilitation. In fact, a field that has benefited from synergistic approaches is robotic rehabilitation, in which muscle synergies have been used to assess motor control improvement, showing that robotic treatment has beneficial effects on upper limbs [21,22] and hands [23]. Rehabilitation in virtual reality environments has also been assessed with muscle synergies in post-stroke patients [24] and children with cerebral palsy [25]. Lastly, muscle synergies have been used to guide functional electrical stimulation (FES) controller during lower limb rehabilitation [26] and to control myoelectric hand prosthesis [27].

However, despite their increasing diffusion in motor neuroscience research, the impact of synergies is still limited. Indeed, research findings and innovations reached in this field have had few effects on clinical scenarios or social and economic aspects of society, since the practical application of synergies is still very limited. From the analysis of the current literature, which is very wide and whose systematic report is beyond the scope of this contribution, fundamental issues emerge, showing that, although the method of muscle synergy analysis is spreading, it is far from being fully exploited in clinical scenarios. Some limitations are more evident, such as the limited number of participants in the studies; other limitations are subtler and include the use made of synergies merely as an evaluation metric. In addition, although some reviews that summarize the findings are available for many sub-fields of applications of muscle synergies, quantitative meta-analyses have not been provided. Therefore, with this contribution, we try to address the main issues that emerged from our screening by discussing several challenges that have only partially been met so far by the scientific and clinical communities.

In this study, referring mainly to neuromotor rehabilitation as the most promising scenario for applying synergistic approaches, the main challenges for the standard clinical adoption of muscle synergies are reported and critically discussed. Thus, the aim of the study is to summarize the main challenges needed to transfer muscle synergies for major exploitation in the clinical field.

2. Challenges

The literature that is available regarding muscle synergies has been screened to identify the most representative work published so far about muscle synergies, especially regarding possible application to clinical scenarios. The corpus of available studies is very wide. Muscle synergies have become a major topic in neuroscience, as portrayed in Figure 1, which shows the number of English papers published in indexed journals from 1999 available in Scopus (updated until 30 April 2024). The subject area is limited to medicine, neuroscience, engineering, and computer science; the document type is limited to articles and reviews; and the language is limited to English.

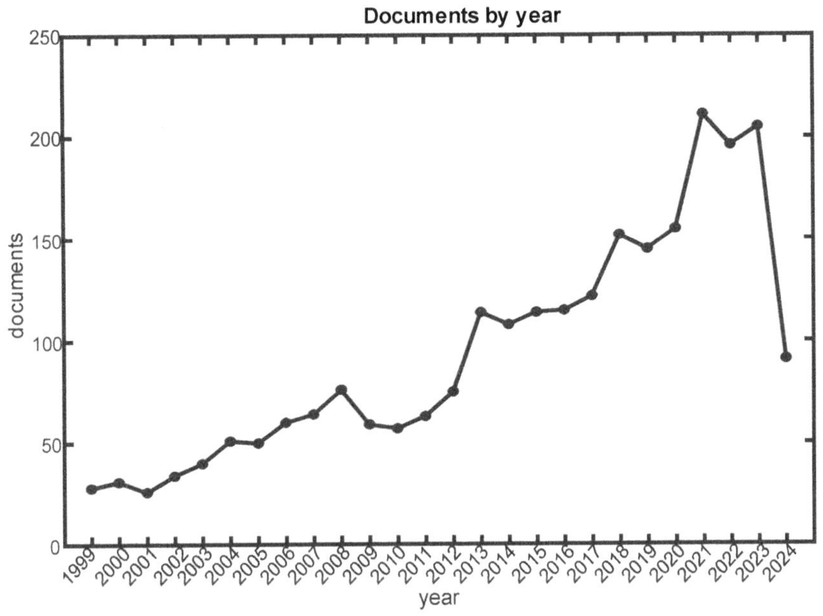

Figure 1. Number of papers found on Scopus when using "muscle AND synergy OR synergies" as keyword for the query and searching in abstract, keywords, and title, updated in April 2024.

The main added value of muscle synergies regards the comprehension and quantification of neuro-motor variables in the domain in which they are generated. In fact, standard clinical scales and kinematic analysis only investigate the output of the non-linear neuro–muscle–skeletal system dynamics, disregarding the neural variables and processes that account for such output. This is reflected in the consideration that most of the rehabilitative approaches focus on improving task performance and/or joint performance to indirectly improve neural activation, while a more direct approach should focus on assessing the patient's neural commands and observe which kinematic coordination will result [7,28]. Whether muscle synergy analysis is more useful than conventional clinical metrics for assessing neurological patients' severity remains unclear; however, further support for the need of muscle synergies comes from the motor abundance principle [29], which states that similar motor outputs are possibly generated by multiple manifolds of muscle activations [30]. In this way, the investigation of higher levels (with respect to kinematics) of the hierarchical and modular organization of the neuro-motor system is required to fully understand the system.

Surprisingly, despite their increasing diffusion in motor neuroscience research, the impact that synergies have reached (intended as the influence, significance, or effect of research findings) is still poor and is limited mainly to basic research. It follows that clinical, practical, real world, social, and economic impacts of the method are still very limited. Thus, a great challenge for the field is to transfer the synergistic approaches from basic research to clinical use, which could be one of the most valuable applications, if not the most attractive, of this method.

The screening of the literature allows to identify three main groups of challenges connected to the adoption of muscle synergies in clinical scenarios: logistic challenges, technical challenges, and research challenges (Figure 2).

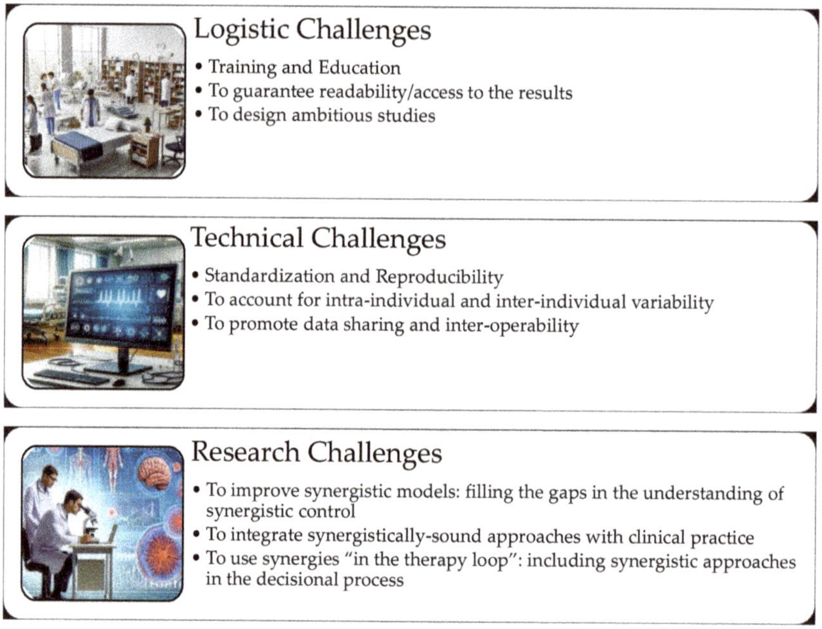

Figure 2. A summary of the challenges to be faced as summarized in this work. Logistic challenges include training and education, to design ambitious studies, to guarantee readability of the results. Technical challenges include to guarantee standardization and reproducibility, to account for intra-individual and inter-individual variability, to promote data sharing and inter-operability. Research challenges include to improve synergistic models, filling the gaps in the understanding of synergistic control, to integrate synergistically sound approaches with clinical practice, to use synergies "in the therapy loop", including synergistic approaches in the decisional process.

Logistic Challenges. They include the resources needed and organizational issues to be solved to make the method compatible with clinical scenarios. The following logistic challenges have been identified:

- Training and education;
- Design of ambitious studies;
- Guarantee of readability/access to the results.

Technical Challenges. They include the technical application of the method and the scientific activities connected to the method. The following technical challenges have been identified:

- To guarantee standardization and reproducibility;
- To account for intra-individual and inter-individual variability;
- To promote data sharing and inter-operability.

Research Challenges. They include wider scoped questions that relate to the improvement and the full understanding of the method itself and its application to clinical scenarios. The following research challenges have been identified:

- To improve synergistic models, filling the gaps in the understanding of synergistic control;
- To integrate synergistically sound approaches with clinical practice;
- To use synergies "in the therapy loop", including synergistic approaches in the decisional process.

2.1. Logistic Challenges

2.1.1. Training and Education

Synergy analysis is a highly multidisciplinary approach to understanding neuromotor control, which requires knowledge in many fields. A general rule for success in multidisciplinary fields is to gather multidisciplinary scientific teams able to cover all the areas of knowledge needed to gather data, perform the analysis, and interpret the results [31]. To correctly acquire multichannel EMG signals, the first step is to guarantee accurate synergistic assessment. It has been shown that specific professional figures might be instructed to perform signal recordings rather than relying only on bioengineers or on medical personnel without a specific training [32,33]. Furthermore, there is a need for bioengineers who are instructed and updated on the variety of steps for synergy extraction and have a wide view of the available literature, including signal processing, algorithms for extraction, and synergistic models. Experienced physicians are also needed to drive scientific questions and to support the selection of the best pipelines and algorithms to adopt for synergy analysis, as they encompass a physiological interpretation of the findings [34]. It is of paramount importance that all these figures work as one to improve the impact of synergy-based studies, while much work in the field is carried on by a subset of them.

2.1.2. To Design Ambitious Studies

In terms of statistical significance, most of the synergy-based studies cannot compare with standard clinical studies that use clinical scales for patients' assessments, e.g., to evaluate the outcome of standard occupational therapy or the effect of drugs. This is probably due to their intrinsic design based on the nature of research, highly relying on pilot and concept work. Indeed, a recent review highlighted that clinical studies with muscle synergies involve only about 15 patients on average [6]. Other factors may include the relatively long time needed for preparing the set-up and instrumenting patients with sensors, and the complexity and effort needed for analyzing the data. A specific field that well exemplifies this limitation is robotic rehabilitation, where, according to the authors' best knowledge, the study based on synergistic evaluation with the largest number of enrolled patients is from Lencioni et al. [21], originally featuring 40 post-stroke patients (32 of whom completed the protocol and were analyzed). All the other studies on robotic rehabilitation enrolled a lower number of patients, thus limiting the significance of the results, even when they were based on very refined designs and algorithms. Consequently, systematic reviews conducted on muscle synergy studies are usually based on proof of concept studies rather than on structured clinical trials.

Moreover, in many clinical contexts, multichannel EMG (eight or more channels) might not be available, since this kind of equipment can be very expensive for clinics [35]. It is indeed necessary that many clinical structures decide to equip themselves with such instruments, with the consideration of muscle synergies as a cost-effective technique to be employed and supported.

2.1.3. To Guarantee Readability/Access to the Results

Synergistic approaches generate output data that are hard for engineers to read and interpret, as they lack neurological or physiology knowledge, and for physicians, as they are not always fully aware of the details of each synergy extraction method and how it impacts the interpretation of the outputs. For example, the level of impairment and the side of lesion in stroke patients may influence the synergy structure [36,37]; therefore, understanding of the pathological state is needed for a correct interpretation of synergy analysis. However, in a few works, the data regarding the type of lesion are available. A fundamental tool to fill this gap may come from the adoption of detailed reports that summarize the main findings achieved in the study. Synthetic reports should be comprehensive of explanations and physiological interpretation of the modifications of synergies induced by treatments or when comparing conditions, coupling clinical and synergistic findings. Such reports might also involve synthesis of the outputs at different time frames, for example, in longitudinal studies. They might also exploit artificial intelligence or reasoning systems to provide accurate reports, shared across studies if based on the same evaluation systems.

2.2. Technical Challenges

2.2.1. Standardization and Reproducibility

One of the most crucial challenges regards the establishment of standardized methods and pipelines to extract synergies from EMG signals. This process is made of many steps and includes electrode positioning, signal filtering (high-pass and low-pass cut-off frequencies, whose effects have been studied systematically, e.g., in [38]), EMG normalization (according to various techniques), data pooling (i.e., whether data should be averaged, concatenated, or analyzed separately [39]), and other preprocessing steps. Two further analysis steps are particularly crucial: the choice of the extraction algorithm, which may vary depending on the aim and on the study design, and the choice of the criteria for selecting the number of synergies. Both these design selections intrinsically impact in a fundamental way the interpretation of the data and the clinical application of synergistic models. There are many algorithms that can be used, such as the synchronous or spatial model [40,41], the temporal model [42,43], the spatiotemporal model [44], the space-by-time model [45], and the autoencoder [46]. Interestingly, each model may capture features that help describe synergistic control under specific assumptions, and thus intrinsically incorporate a specific knowledge and understanding of the physiology underlying neural control of movement. A clear example of this can be found in Berger et al. [47,48], who showed that the spatial model could not identify specific impairments in the motor organization of patients with cerebellar damage. It was instead shown that temporal and spatiotemporal synergies could capture relevant modifications. Despite the variety of algorithms and approaches available, almost no other study has compared synergy extraction approaches to understand the range of applicability and efficacy of each algorithm on clinical data, limiting the analysis to one synergistic model per work. Similarly, no guideline is available regarding which models should be coupled to specific conditions or pathologies. It follows that one major challenge would be to couple the most appropriate algorithm(s) for the investigated domain.

Another still unresolved challenge is the selection of the optimal number of synergies. Even though a large variety of methods have been proposed and tested, including standard VAF/R^2 thresholds, minimum VAF variation, a priori selection, linear fit, multi-VAF resolution analysis [49], and others, it is still a matter of debate whether an optimal number of synergies can really be defined, and how [50]. Indeed, since synergies are expected to be organized at the neural level, one would expect that it would be possible to define univocally the "right" number of synergies, even though recent work suggests that a multiresolution analysis might be more suitable and accurate for interpreting synergy data [51]. How these problems relate to observed findings such as synergy merging and fractionation [9] is also a matter of investigation for a better understanding of motor control to foster the clinical application of synergistic approaches.

Some attempts to improve reproducibility and standardization have been made by proposing reference evaluation protocols [51] or by distributing toolboxes for synergistic control in some applications, including upper limb [52], general purpose and lower limb [53], hand [54], and exoskeleton walking [55]. However, these attempts of standardizing synergy analysis are mostly from a single group or a few research groups rather than from a wide consortium of synergy users.

2.2.2. To Account for Intra-Individual and Inter-Individual Variability

Movement shows intrinsic variability that can be observed both at the intra-individual level, due to the differences in the multiple executions of the same task by the same subject [56], and at the inter-individual level, related to the variation of motor patterns when multiple subjects perform the same task [57]. Movement variability is reflected in synergy variability [58]. A very important need is to build a definitive highly reliable quantification of the intra-individual and inter-individual variability of synergies in order to define normative datasets that allow to evaluate, in a comparative sense, the results with referable data. Generating such datasets would implicitly constrain many protocol choices to a standard procedure (e.g., number and positioning of the EMG electrodes) and allow to quantify at what extent differences and modifications are due to inherent modifications of the neural system and which are due to intrinsic physiological variability between trials and subjects. Movement variability may affect the correct interpretation of synergies, capturing changes that are not related to neurological improvements [59]. This limitation afflicts many longitudinal studies and has rarely been quantified in previous works, even though it is a fundamental step toward an aware use of the method.

2.2.3. To Promote Data Sharing and Inter-Operability

As clinical scales show some limitations such as poor resolution in evaluating motor capabilities and potential inter-operator biases [60], they are in general completely inter-operable between clinical centers and standardized in their formulation. The same standardization is not found with muscle synergies [61]. Thus, transferring synergies to clinical practice would require individuating common and shared platforms for data sharing; individuating common and shared data formats; and building publicly interconnected repositories. All these aspects are at the moment completely missing and contribute to making the results of many studies difficult to compare. Lastly, there is also a lack of shared databases that can be exploited by the scientific community; in fact, the vast majority of the data underlying synergistic studies are not published in shared repositories (probably also due to the limited number of subjects available).

2.3. Research Challenges

2.3.1. To Improve Synergistic Models, Filling the Gaps in the Understanding of Synergistic Control

Improving synergistic models would facilitate clinical practice as it would simplify the choice of methods and algorithms to be employed to study each clinical condition. Several communications and reviews have pointed out specific directions to follow, which are summarized below.

The neural basis of synergistic control. More work should be carried out to understand whether synergistic control reflect neural implementation and what are the neural bases underlying it [62]. Much effort has been made to understand the roles of the spinal cord, brain stem, and motor cortical areas in the activation, organization, and fine-tuning of muscle synergies. A recent review suggested to constrain models with neurophysiological knowledge [63]. Discerning the applicability of spatial, temporal, and spatiotemporal models should also be further considered. Moreover, the alteration in muscle synergies might be dependent on the site of lesions, the severity of impairment, and the stage of disease [64], and these data are often not considered in the available studies.

Synergy models. It has also been suggested to encourage the concurrent use of various synergy models–spatial, temporal, and spatiotemporal synergies. The number of synergies represents either the dimension of the spatial structure or the number of independent temporal patterns, and it has been observed that these two aspects are often mixed in the analysis. To select a number, criteria based on noise estimates through bootstrap methods, reliability of analysis results, or functional outcomes of the synergies provide interesting substitutes to criteria solely based on variance thresholds [65].

Link to the task space. A major step toward the conception of synergy-driven approaches may also come from the linking of neural synergies to the task space variables [14], which has been addressed with recent novel algorithms that extend the concept of synergies by incorporating task variables to the neural drive [66–68] or factorizations based on multidomain approaches [18]. In such approaches, muscle weights are flanked with task weights (typically kinematic or force) that show how the neural synergy results in a task output.

The role of non-linearities. Most of the algorithms used so far naturally employ a linear combination of synergies to produce purposeful movement. While this is compatible with most of the observations made on humans and on animals, it is possible that future work will have to consider more complex models, and thus they may require to adopt non-linear models, such as the recently used autoencoder-based approaches [46], kinematic–muscular synergies [66], or information and network theory to measure the task relevance of the synergies [67].

2.3.2. To Integrate Synergistically Sound Approaches with Clinical Practice

One complex challenge to face is to find a proper trade-off between standard clinical assessments and synergistic approaches by harmoniously integrating the two frameworks. In fact, finding a common ground between structured protocols typical of research laboratories and clinical demands poses a challenge that has not yet been solved. In laboratory studies, volunteers are typically requested to follow structured protocols (e.g., standard multitarget reaching movements exploring many directionalities throughout many repetitions [69] or, in any case, a wide variety of movements and task repetitions). In a clinical environment, a sufficient level of motion variability should be persevered (i.e., when possible, a reasonable number of tasks and repetitions should be performed to guarantee that synergistic extraction is adequately representing the repertoire of modules available to subjects) keeping in mind the limitations due to patients' motor capability and cooperation [6]. A reasonable trade-off might be an approach combining synergies with functional movements resembling clinical scales [70,71], even though more should be carried out to find a consensus.

2.3.3. To Use Synergies "in the Therapy Loop", including Synergistic Approaches in the Decisional Process

So far, synergies have been used only as an evaluation tool in the vast majority of studies [7]. Typically, synergies are used as a mere evaluation method to compare two conditions (e.g., pre- or post-therapy synergies in longitudinal studies; robot-assisted synergies, and free movement in single-session studies). However, even though such approaches are valuable in principle, they do not fully exploit the potential of a synergistic approach, as they evaluate synergies as an output of an intervention conceived in another domain. In fact, there is a lack of synergy-based rehabilitation protocols, i.e., protocols conceived to promote, restore, or reshape specific synergies or synergy combinations. So, the neural frame is used only to evaluate the effects of therapies and interventions that are not conceived in the same domain but rather in the tridimensional space or in the joint domain inspired to human kinematics. Few studies have focused on developing effective ways to use muscle synergies as a training target or a control input signal for intervention systems; however, changes in muscle synergies would maximize the motor function improvement [7]. Creating synergistic-based rehabilitation protocols might be an example on how to foster such a view, for example, by changing the destination and use of the robot due to synergistic assessment, to customize therapies on the basis of synergy clustering in order to restore specific synergies, or to reshape temporal commands in case synergies cannot be changed (as in chronic patients). Another option would be, in the robotic field, to create control-based algorithms based on synergistic approaches [72] or to tune therapies or assistance depending on the recorded synergies.

A summary of the main needs and lessons learned to transfer muscle synergy to a standard suitable for clinical practice are shown in Table 1.

Table 1. Needs and lessons learned regarding the application of synergistic approaches in clinical environments.

Type of Challenge	Needs	Lessons Learned/Possible Solutions
Logistic	Training and Education	• To correctly acquire multichannel EMG signals with personnel devoted to data acquisition. • To enroll bioengineers to support analysis. They should be updated on synergistic models and pipelines. • Physicians should drive clinical questions and needs.
Logistic	To design ambitious studies	• To increase the number of structured studies. • To enroll a large number of patients/healthy controls to increase study significance. • To consider expensive equipment as cost-effective.
Logistic	To guarantee readability of the results	• To provide synthetic reports for clinicians, including physiological explanations, multisession comparisons, etc., readable by technical and medical personnel.
Technical	To guarantee standardization and reproducibility	• To generate consensus guidelines for processing and protocols. • To provide freely usable synergy-based toolboxes. • To limit custom choices in research by defining technical standards. • To associate algorithms and pipelines to specific research questions.
Technical	To account for intra-individual and inter-individual variability	• To build shared and public reference databases. • To investigate intra-session, inter-session, intra-subject, and inter-subject variability.
Technical	To promote data sharing and inter-operability	• To individuate common platforms for data sharing. • To individuate common and shared data formats. • To create publicly interconnected repositories.
Research	To improve synergistic models, filling the gaps in the understanding of synergistic control	• To improve the understanding of the neural basis of synergistic control. • To concurrently implement multiple synergy models. • To create a link to the task space. • To evaluate the role of non-linearities.
Research	To integrate synergistically sound approaches with clinical practice	• To generate consensus protocols, at least for some classes of experiments. • To harmonize the requirements of the clinics with the requirements of synergistic approaches.
Research	To use synergies "in the therapy loop", including synergistic approaches in the decisional process	• To create synergy-based rehabilitation protocols. • To change the destination and use of the therapy/robot/treatment due to synergistic assessments. • To create control-based algorithms based on synergistic approaches.

3. Perspectives

3.1. Logistic Challenges

Logistic challenges that have to be addressed include the need for professional resources to make the method compatible with clinical scenarios. A multidisciplinary team is necessary to apply synergy analysis and to understand the neuro-motor control. The knowledge needed for data gathering, performing the analysis, and interpreting the results should be covered by qualified experts in the field. Therefore, expert professionals are needed to correctly acquire multichannel EMG signals, as a fundamental preliminary step for the data analyses aiming at synergy extraction in order to reduce signal noise, cross-talk, artefacts, and inaccuracies. Then, the analyses should be supported by bioengineers who are updated on synergistic models and data processing and able to select the pipelines that best fit the aims of the work. An appropriate physiological interpretation and a selection of relevant clinical implications should instead be led by physicians, providing insights from a medical point of view. Only a constant interaction with all these professional figures may guarantee the exploitation of the full potential of the method. From this perspective, readability of the results of synergistic analysis becomes a crucial issue. Indeed, the outputs of the analysis can be non-trivial to read and interpret for both engineers and physicians. Therefore, the main findings should be summarized in detailed reports that provide comprehensive explanations and physiological interpretation of the modifications of synergies induced by treatments or when comparing conditions, adding clinical interpretation to synergistic outputs according to systematic criteria.

Moreover, comprehensive studies applying muscle synergy analysis are not available in the literature and are usually limited to pilot research studies. Therefore, more structured studies in which the longitudinal course of therapy is investigated should be designed and a large number of participants and patients should be enrolled so that the significance and generalizability power of the studies increase. A last barrier to overcome refers to the non-negligible cost of the equipment. For a systematic adoption of muscle synergy analysis, clinical structures need to equip themselves with such instruments and consider muscle synergies as a cost-effective technique to be employed and supported.

3.2. Technical Challenges

The standardization and the reproducibility of the methodological approach is fundamental for increasing the applicability of synergy analysis in clinical scenarios. To overcome this challenge, it is necessary to generate consensus guidelines for processing and protocols, including the pre-processing steps, the choice of the synergy model and the corresponding factorization algorithm, the criteria for selecting the number of synergies, and others. These guidelines may be associated with specific research questions in which different methodological steps are needed, so that guidelines are specified for each investigation. Defining technical standards and reducing custom choices will help to spread the use of these techniques, and especially increase the possibility of comparisons between studies. The adoption of freely usable and distributed synergy-based toolboxes [52] should be encouraged to guarantee easy-to-use methods for all researchers and clinicians and promote common background between researchers.

Another important technical aspect poorly assessed in the literature is the need to quantify the intra-individual and inter-individual variability of synergies, to avoid the misinterpretation of synergistic output by assuming that changes or modifications to synergies are due to neurological improvements rather than physiological differences between experimental sessions or experimental biases (e.g., imperfect inter-session electrode positioning). Therefore, normative datasets need to be created to investigate intra-session, inter-session, intra-subject, and inter-subject variability and publicly shared in order to build normative databases of muscle synergies that can be used as reference. The limited data sharing and inter-operability are also technical challenges to face, that add further need of standardization to the previously described limitations. To promote data sharing, it is fundamental to identify common data formats that can be used by anyone and to identify common platforms in which data can be shared. Publicly interconnected repositories can be created for this purpose; indeed, this will help to exploit research findings and to compare studies.

3.3. Research Challenges

From a research point of view, synergistic models need to be improved to facilitate their use in clinical practice. First of all, the role of neural structures, such as the spinal cord, the brain stem, and motor cortical areas, in implementing synergistic control should be better understood so that synergy models can be linked more directly to neuromotor physiology. Moreover, since different synergy models provide different insights into motor control, multiple synergy models should be implemented concurrently to provide comprehensive perspectives when describing motor control in a specific scenario. Another important step to improve the interpretation of the outcomes of synergy models is the inclusion in the model of a direct link to the task space, showing how neural synergies are mapped into task outputs. Some models have been developed [66], but more research needs to be performed in future to overcome limitations such as linearity of the models.

One complex challenge to face is to integrate synergistically sound approaches with clinical practice. Indeed, a trade-off between standard clinical assessments and synergistic approaches should be found. In fact, a sufficient variety and quantity of movements is necessary for extracting reliable synergies; however, this need should consider the limitations due to patients' motor capability and cooperation. Therefore, consensus protocols should be generated to harmonize the requirements of clinical treatments with the requirements of synergistic approaches. One possible option is to use assessments based on functional movements, resembling those performed in clinical scales [71], that are already administered to patients when they are evaluated in clinics with standard protocols.

Finally, the potential of the synergistic approach has not been fully exploited, as synergies have been used only as an assessment method. Indeed, the synergistic approach should be included in the decisional process performed by clinicians in selecting specific rehabilitative programs. Synergy-based rehabilitation protocols, conceived to promote, restore, or reshape synergies, should be encouraged and their efficacy assessed. For example, the destination and use of a robotic operator for rehabilitation may be decided on the basis of synergistic assessment and control-based algorithms based on synergistic approaches. In this way, the effects of therapies and interventions are evaluated in the muscle synergy domain, with the aim of maximizing motor function improvement.

4. Conclusions

Many challenges arise when aiming at the systematic application of muscle synergies to clinical practice. However, to date, the challenges and issues presented in this work have not been addressed and solved all together, leaving a wide margin for improving the adoption of the method in clinical practice. There is a need for comprehensive and ambitious future studies addressing and solving these challenges. In fact, some of the challenges have been partially addressed but never with a comprehensive uniform vision, which is ultimately needed for transferring muscle synergy assessment to standard clinical practice.

Author Contributions: Conceptualization, A.S. and A.d.; methodology, A.S.; validation, A.S.; formal analysis, A.S., V.L. and C.B.; investigation, A.S.; resources, A.S.; original draft preparation, A.S., V.L., C.B. and A.d.; writing—review and editing, A.S., V.L., C.B. and A.d.; visualization, A.S., V.L. and C.B.; project administration, A.S.; funding acquisition, A.S. All authors have read and agreed to the published version of the manuscript.

Funding: This work was supported by the project "Fit4MedRob—Fit for Medical Robotics", "Piano Nazionale Complementare (PNC)—PNC0000007". This work was funded by the Italian Ministry of University and Research, under the complementary actions to the Plan of National Recovery and Resilience (PNRR) "Fit4MedRob-Fit for Medical Robotics" Grant (# PNC0000007).

Institutional Review Board Statement: Not applicable.

Informed Consent Statement: Not applicable.

Data Availability Statement: Data are contained within the article.

Acknowledgments: The authors wish to acknowledge the project Fit4MedRob—"Fit for Medical Robotics", Piano Nazionale Complementare (PNC)—PNC0000007.

Conflicts of Interest: The authors declare no conflicts of interest.

References

1. Scano, A.; Guanziroli, E.; Brambilla, C.; Amendola, C.; Pirovano, I.; Gasperini, G.; Molteni, F.; Spinelli, L.; Molinari Tosatti, L.; Rizzo, G.; et al. A Narrative Review on Multi-Domain Instrumental Approaches to Evaluate Neuromotor Function in Rehabilitation. *Healthcare* **2023**, *11*, 2282. [CrossRef] [PubMed]
2. Maura, R.M.; Rueda Parra, S.; Stevens, R.E.; Weeks, D.L.; Wolbrecht, E.T.; Perry, J.C. Literature Review of Stroke Assessment for Upper-Extremity Physical Function via EEG, EMG, Kinematic, and Kinetic Measurements and Their Reliability. *J. Neuroeng. Rehabil.* **2023**, *20*, 21. [CrossRef] [PubMed]
3. Campanini, I.; Disselhorst-Klug, C.; Rymer, W.Z.; Merletti, R. Surface EMG in Clinical Assessment and Neurorehabilitation: Barriers Limiting Its Use. *Front. Neurol.* **2020**, *11*, 934. [CrossRef] [PubMed]
4. Tresch, M.C.; Jarc, A. The Case for and against Muscle Synergies. *Curr. Opin. Neurobiol.* **2009**, *19*, 601. [CrossRef] [PubMed]
5. Bizzi, E.; Cheung, V.C.K.; d'Avella, A.; Saltiel, P.; Tresch, M. Combining Modules for Movement. *Brain Res. Rev.* **2008**, *57*, 125–133. [CrossRef] [PubMed]
6. Zhao, K.; Zhang, Z.; Wen, H.; Liu, B.; Li, J.; d'Avella, A.; Scano, A. Muscle Synergies for Evaluating Upper Limb in Clinical Applications: A Systematic Review. *Heliyon* **2023**, *9*, e16202. [CrossRef] [PubMed]
7. Hong, Y.N.G.; Ballekere, A.N.; Fregly, B.J.; Roh, J. Are Muscle Synergies Useful for Stroke Rehabilitation? *Curr. Opin. Biomed. Eng.* **2021**, *19*, 100315. [CrossRef]
8. Cheung, V.C.K.; Piron, L.; Agostini, M.; Silvoni, S.; Turolla, A.; Bizzi, E. Stability of Muscle Synergies for Voluntary Actions after Cortical Stroke in Humans. *Proc. Natl. Acad. Sci. USA* **2009**, *106*, 19563–19568. [CrossRef] [PubMed]
9. Cheung, V.C.K.; Turolla, A.; Agostini, M.; Silvoni, S.; Bennis, C.; Kasi, P.; Paganoni, S.; Bonato, P.; Bizzi, E. Muscle Synergy Patterns as Physiological Markers of Motor Cortical Damage. *Proc. Natl. Acad. Sci. USA* **2012**, *109*, 14652–14656. [CrossRef]
10. Mileti, I.; Zampogna, A.; Santuz, A.; Asci, F.; Del Prete, Z.; Arampatzis, A.; Palermo, E.; Suppa, A. Muscle Synergies in Parkinson's Disease. *Sensors* **2020**, *20*, 3209. [CrossRef]
11. Steele, K.M.; Rozumalski, A.; Schwartz, M.H. Muscle Synergies and Complexity of Neuromuscular Control during Gait in Cerebral Palsy. *Dev. Med. Child Neurol.* **2015**, *57*, 1176–1182. [CrossRef] [PubMed]
12. Dipietro, L.; Krebs, H.I.; Fasoli, S.E.; Volpe, B.T.; Stein, J.; Bever, C.; Hogan, N. Changing Motor Synergies in Chronic Stroke. *J. Neurophysiol.* **2007**, *98*, 757–768. [CrossRef] [PubMed]
13. Frère, J.; Hug, F. Between-Subject Variability of Muscle Synergies during a Complex Motor Skill. *Front. Comput. Neurosci.* **2012**, *6*, 99. [CrossRef] [PubMed]
14. Alessandro, C.; Delis, I.; Nori, F.; Panzeri, S.; Berret, B. Muscle Synergies in Neuroscience and Robotics: From Input-Space to Task-Space Perspectives. *Front. Comput. Neurosci.* **2013**, *7*, 43. [CrossRef] [PubMed]
15. Taborri, J.; Agostini, V.; Artemiadis, P.K.; Ghislieri, M.; Jacobs, D.A.; Roh, J.; Rossi, S. Feasibility of Muscle Synergy Outcomes in Clinics, Robotics, and Sports: A Systematic Review. *Appl. Bionics. Biomech.* **2018**, *2018*, 3934698. [CrossRef] [PubMed]
16. Clark, D.J.; Ting, L.H.; Zajac, F.E.; Neptune, R.R.; Kautz, S.A. Merging of Healthy Motor Modules Predicts Reduced Locomotor Performance and Muscle Coordination Complexity Post-Stroke. *J. Neurophysiol.* **2010**, *103*, 844–857. [CrossRef] [PubMed]
17. Roh, J.; Rymer, W.Z.; Perreault, E.J.; Yoo, S.B.; Beer, R.F. Alterations in Upper Limb Muscle Synergy Structure in Chronic Stroke Survivors. *J. Neurophysiol.* **2013**, *109*, 768–781. [CrossRef] [PubMed]
18. Pierella, C.; Pirondini, E.; Kinany, N.; Coscia, M.; Giang, C.; Miehlbradt, J.; Magnin, C.; Nicolo, P.; Dalise, S.; Sgherri, G.; et al. A Multimodal Approach to Capture Post-Stroke Temporal Dynamics of Recovery. *J. Neural. Eng.* **2020**, *17*, 045002. [CrossRef] [PubMed]
19. Belfatto, A.; Scano, A.; Chiavenna, A.; Mastropietro, A.; Mrakic-Sposta, S.; Pittaccio, S.; Molinari Tosatti, L.; Molteni, F.; Rizzo, G. A Multiparameter Approach to Evaluate Post-Stroke Patients: An Application on Robotic Rehabilitation. *Appl. Sci.* **2018**, *8*, 2248. [CrossRef]
20. Bellitto, A.; De Luca, A.; Gamba, S.; Losio, L.; Massone, A.; Casadio, M.; Pierella, C. Clinical, Kinematic and Muscle Assessment of Bilateral Coordinated Upper-Limb Movements Following Cervical Spinal Cord Injury. *IEEE Trans. Neural. Syst. Rehabil. Eng.* **2023**, *31*, 3607–3618. [CrossRef]
21. Lencioni, T.; Fornia, L.; Bowman, T.; Marzegan, A.; Caronni, A.; Turolla, A.; Jonsdottir, J.; Carpinella, I.; Ferrarin, M. A Randomized Controlled Trial on the Effects Induced by Robot-Assisted and Usual-Care Rehabilitation on Upper Limb Muscle Synergies in Post-Stroke Subjects. *Sci. Rep.* **2021**, *11*, 5323. [CrossRef] [PubMed]
22. Tropea, P.; Monaco, V.; Coscia, M.; Posteraro, F.; Micera, S. Effects of Early and Intensive Neuro-Rehabilitative Treatment on Muscle Synergies in Acute Post-Stroke Patients: A Pilot Study. *J. Neuroeng. Rehabil.* **2013**, *10*, 103. [CrossRef] [PubMed]
23. Scotto di Luzio, F.; Cordella, F.; Bravi, M.; Santacaterina, F.; Bressi, F.; Sterzi, S.; Zollo, L. Modification of Hand Muscular Synergies in Stroke Patients after Robot-Aided Rehabilitation. *Appl. Sci.* **2022**, *12*, 3146. [CrossRef]

24. Maistrello, L.; Rimini, D.; Cheung, V.C.K.; Pregnolato, G.; Turolla, A. Muscle Synergies and Clinical Outcome Measures Describe Different Factors of Upper Limb Motor Function in Stroke Survivors Undergoing Rehabilitation in a Virtual Reality Environment. *Sensors* **2021**, *21*, 8002. [CrossRef] [PubMed]
25. Booth, A.T.C.; van der Krogt, M.M.; Harlaar, J.; Dominici, N.; Buizer, A.I. Muscle Synergies in Response to Biofeedback-Driven Gait Adaptations in Children With Cerebral Palsy. *Front. Physiol.* **2019**, *10*, 437973. [CrossRef] [PubMed]
26. Ferrante, S.; Chia Bejarano, N.; Ambrosini, E.; Nardone, A.; Turcato, A.M.; Monticone, M.; Ferrigno, G.; Pedrocchi, A. A Personalized Multi-Channel FES Controller Based on Muscle Synergies to Support Gait Rehabilitation after Stroke. *Front. Neurosci.* **2016**, *10*, 425. [CrossRef] [PubMed]
27. Furui, A.; Eto, S.; Nakagaki, K.; Shimada, K.; Nakamura, G.; Masuda, A.; Chin, T.; Tsuji, T. A Myoelectric Prosthetic Hand with Muscle Synergy-Based Motion Determination and Impedance Model-Based Biomimetic Control. *Sci. Robot.* **2019**, *4*, eaaw6339. [CrossRef] [PubMed]
28. Patel, V.; Craig, J.; Schumacher, M.; Burns, M.K.; Florescu, I.; Vinjamuri, R. Synergy Repetition Training versus Task Repetition Training in Acquiring New Skill. *Front. Bioeng. Biotechnol.* **2017**, *5*, 9. [CrossRef] [PubMed]
29. Latash, M.L. The Bliss (Not the Problem) of Motor Abundance (Not Redundancy). *Exp. Brain Res.* **2012**, *217*, 1–5. [CrossRef]
30. Kang, N.; Shinohara, M.; Zatsiorsky, V.M.; Latash, M.L. Learning Multi-Finger Synergies: An Uncontrolled Manifold Analysis. *Exp. Brain Res.* **2004**, *157*, 336–350. [CrossRef]
31. Manca, A.; Cereatti, A.; Bar-On, L.; Botter, A.; Della Croce, U.; Knaflitz, M.; Maffiuletti, N.A.; Mazzoli, D.; Merlo, A.; Roatta, S.; et al. A Survey on the Use and Barriers of Surface Electromyography in Neurorehabilitation. *Front. Neurol.* **2020**, *11*, 573616. [CrossRef]
32. Merletti, R.; Temporiti, F.; Gatti, R.; Gupta, S.; Sandrini, G.; Serrao, M. Translation of Surface Electromyography to Clinical and Motor Rehabilitation Applications: The Need for New Clinical Figures. *Transl. Neurosci.* **2023**, *14*, 20220279. [CrossRef]
33. Merletti, R. Metrology in sEMG and Movement Analysis: The Need for Training New Figures in Clinical Rehabilitation. *Front. Rehabil. Sci.* **2024**, *5*, 1353374. [CrossRef]
34. McManus, L.; De Vito, G.; Lowery, M.M. Analysis and Biophysics of Surface EMG for Physiotherapists and Kinesiologists: Toward a Common Language With Rehabilitation Engineers. *Front. Neurol.* **2020**, *11*, 576729. [CrossRef]
35. Feldner, H.A.; Howell, D.; Kelly, V.E.; McCoy, S.W.; Steele, K.M. "Look, Your Muscles Are Firing!": A Qualitative Study of Clinician Perspectives on the Use of Surface Electromyography in Neurorehabilitation. *Arch. Phys. Med. Rehabil.* **2019**, *100*, 663–675. [CrossRef]
36. Pellegrino, L.; Coscia, M.; Pierella, C.; Giannoni, P.; Cherif, A.; Mugnosso, M.; Marinelli, L.; Casadio, M. Effects of Hemispheric Stroke Localization on the Reorganization of Arm Movements within Different Mechanical Environments. *Life* **2021**, *11*, 383. [CrossRef]
37. Roh, J.; Rymer, W.Z.; Beer, R.F. Evidence for Altered Upper Extremity Muscle Synergies in Chronic Stroke Survivors with Mild and Moderate Impairment. *Front. Hum. Neurosci.* **2015**, *9*, 6. [CrossRef]
38. Kieliba, P.; Tropea, P.; Pirondini, E.; Coscia, M.; Micera, S.; Artoni, F. How Are Muscle Synergies Affected by Electromyography Pre-Processing? *IEEE Trans. Neural Syst. Rehabil. Eng.* **2018**, *26*, 882–893. [CrossRef]
39. Oliveira, A.S.; Gizzi, L.; Farina, D.; Kersting, U.G. Motor Modules of Human Locomotion: Influence of EMG Averaging, Concatenation, and Number of Step Cycles. *Front. Hum. Neurosci.* **2014**, *8*, 335. [CrossRef]
40. Tresch, M.C.; Saltiel, P.; Bizzi, E. The Construction of Movement by the Spinal Cord. *Nat. Neurosci.* **1999**, *2*, 162–167. [CrossRef]
41. Ting, L.H.; Macpherson, J.M. A Limited Set of Muscle Synergies for Force Control during a Postural Task. *J. Neurophysiol.* **2005**, *93*, 609–613. [CrossRef]
42. Ivanenko, Y.P.; Poppele, R.E.; Lacquaniti, F. Five Basic Muscle Activation Patterns Account for Muscle Activity during Human Locomotion. *J. Physiol.* **2004**, *556*, 267–282. [CrossRef]
43. Ivanenko, Y.P.; Cappellini, G.; Dominici, N.; Poppele, R.E.; Lacquaniti, F. Coordination of Locomotion with Voluntary Movements in Humans. *J. Neurosci.* **2005**, *25*, 7238–7253. [CrossRef]
44. d'Avella, A.; Saltiel, P.; Bizzi, E. Combinations of Muscle Synergies in the Construction of a Natural Motor Behavior. *Nat. Neurosci.* **2003**, *6*, 300–308. [CrossRef]
45. Delis, I.; Panzeri, S.; Pozzo, T.; Berret, B. A Unifying Model of Concurrent Spatial and Temporal Modularity in Muscle Activity. *J. Neurophysiol.* **2014**, *111*, 675–693. [CrossRef]
46. Buongiorno, D.; Cascarano, G.D.; Camardella, C.; De Feudis, I.; Frisoli, A.; Bevilacqua, V. Task-Oriented Muscle Synergy Extraction Using An Autoencoder-Based Neural Model. *Information* **2020**, *11*, 219. [CrossRef]
47. Berger, D.J.; Masciullo, M.; Molinari, M.; Lacquaniti, F.; d'Avella, A. Does the Cerebellum Shape the Spatiotemporal Organization of Muscle Patterns? Insights from Subjects with Cerebellar Ataxias. *J. Neurophysiol.* **2020**, *123*, 1691–1710. [CrossRef]
48. Berger, D.J.; Ferrari, F.; Esposito, A.; Masciullo, M.; Molinari, M.; Lacquaniti, F.; d'Avella, A. Changes in Muscle Synergy Organization after Neurological Lesions. In *Converging Clinical and Engineering Research on Neurorehabilitation II*; Ibáñez, J., González-Vargas, J., Azorín, J.M., Akay, M., Pons, J.L., Eds.; Springer International Publishing: Cham, Switzerland, 2017; pp. 939–943.
49. Pale, U.; Atzori, M.; Müller, H.; Scano, A. Variability of Muscle Synergies in Hand Grasps: Analysis of Intra- and Inter-Session Data. *Sensors* **2020**, *20*, 4297. [CrossRef]

50. Ranaldi, S.; De Marchis, C.; Severini, G.; Conforto, S. An Objective, Information-Based Approach for Selecting the Number of Muscle Synergies to Be Extracted via Non-Negative Matrix Factorization. *IEEE Trans. Neural Syst. Rehabil. Eng.* **2021**, *29*, 2676–2683. [CrossRef]
51. Longatelli, V.; Torricelli, D.; Tornero, J.; Pedrocchi, A.; Molteni, F.; Pons, J.L.; Gandolla, M. A Unified Scheme for the Benchmarking of Upper Limb Functions in Neurological Disorders. *J. Neuroeng. Rehabil.* **2022**, *19*, 102. [CrossRef]
52. Russo, M.; Scano, A.; Brambilla, C.; d'Avella, A. SynergyAnalyzer: A Matlab Toolbox Implementing Mixed-Matrix Factorization to Identify Kinematic-Muscular Synergies. *Comput. Methods Programs Biomed.* **2024**, *251*, 108217. [CrossRef] [PubMed]
53. Santuz, A. musclesyneRgies: Factorization of Electromyographic Data in R with Sensible Defaults. *J. Open Source Softw.* **2022**, *7*, 4439. [CrossRef]
54. Lapresa, M.; Zollo, L.; Cordella, F. A User-Friendly Automatic Toolbox for Hand Kinematic Analysis, Clinical Assessment and Postural Synergies Extraction. *Front. Bioeng. Biotechnol.* **2022**, *10*, 1010073. [CrossRef]
55. Zhvansky, D.S.; Sylos-Labini, F.; Dewolf, A.; Cappellini, G.; d'Avella, A.; Lacquaniti, F.; Ivanenko, Y. Evaluation of Spatiotemporal Patterns of the Spinal Muscle Coordination Output during Walking in the Exoskeleton. *Sensors* **2022**, *22*, 5708. [CrossRef] [PubMed]
56. Rimini, D.; Agostini, V.; Knaflitz, M. Intra-Subject Consistency during Locomotion: Similarity in Shared and Subject-Specific Muscle Synergies. *Front. Hum. Neurosci.* **2017**, *11*, 586. [CrossRef] [PubMed]
57. De Marchis, C.; Schmid, M.; Bibbo, D.; Bernabucci, I.; Conforto, S. Inter-Individual Variability of Forces and Modular Muscle Coordination in Cycling: A Study on Untrained Subjects. *Hum. Mov. Sci.* **2013**, *32*, 1480–1494. [CrossRef]
58. Zhao, K.; Zhang, Z.; Wen, H.; Scano, A. Intra-Subject and Inter-Subject Movement Variability Quantified with Muscle Synergies in Upper-Limb Reaching Movements. *Biomimetics* **2021**, *6*, 63. [CrossRef] [PubMed]
59. Shuman, B.; Goudriaan, M.; Bar-On, L.; Schwartz, M.H.; Desloovere, K.; Steele, K.M. Repeatability of Muscle Synergies within and between Days for Typically Developing Children and Children with Cerebral Palsy. *Gait Posture* **2016**, *45*, 127–132. [CrossRef]
60. Maceira-Elvira, P.; Popa, T.; Schmid, A.-C.; Hummel, F.C. Wearable Technology in Stroke Rehabilitation: Towards Improved Diagnosis and Treatment of Upper-Limb Motor Impairment. *J. Neuroeng. Rehabil.* **2019**, *16*, 142. [CrossRef]
61. Lang, C.E.; Barth, J.; Holleran, C.L.; Konrad, J.D.; Bland, M.D. Implementation of Wearable Sensing Technology for Movement: Pushing Forward into the Routine Physical Rehabilitation Care Field. *Sensors* **2020**, *20*, 5744. [CrossRef]
62. Bruton, M.; O'Dwyer, N. Synergies in Coordination: A Comprehensive Overview of Neural, Computational, and Behavioral Approaches. *J. Neurophysiol.* **2018**, *120*, 2761–2774. [CrossRef]
63. Cheung, V.C.K.; Seki, K. Approaches to Revealing the Neural Basis of Muscle Synergies: A Review and a Critique. *J. Neurophysiol.* **2021**, *125*, 1580–1597. [CrossRef] [PubMed]
64. Singh, R.E.; Iqbal, K.; White, G.; Hutchinson, T.E. A Systematic Review on Muscle Synergies: From Building Blocks of Motor Behavior to a Neurorehabilitation Tool. *Appl. Bionics. Biomech.* **2018**, *2018*, 3615368. [CrossRef] [PubMed]
65. Turpin, N.A.; Uriac, S.; Dalleau, G. How to Improve the Muscle Synergy Analysis Methodology? *Eur. J. Appl. Physiol.* **2021**, *121*, 1009–1025. [CrossRef] [PubMed]
66. Scano, A.; Mira, R.M.; d'Avella, A. Mixed Matrix Factorization: A Novel Algorithm for the Extraction of Kinematic-Muscular Synergies. *J. Neurophysiol.* **2022**, *127*, 529–547. [CrossRef]
67. O'Reilly, D.; Delis, I. Dissecting Muscle Synergies in the Task Space. *eLife* **2024**, *12*, RP87651. [CrossRef] [PubMed]
68. Lapresa, M.; Corradini, V.; Iacca, A.; Scotto di Luzio, F.; Zollo, L.; Cordella, F. A Comprehensive Analysis of Task-Specific Hand Kinematic, Muscle and Force Synergies. *Biocybern. Biomed. Eng.* **2024**, *44*, 218–230. [CrossRef]
69. Scano, A.; Dardari, L.; Molteni, F.; Giberti, H.; Tosatti, L.M.; d'Avella, A. A Comprehensive Spatial Mapping of Muscle Synergies in Highly Variable Upper-Limb Movements of Healthy Subjects. *Front. Physiol.* **2019**, *10*, 1231. [CrossRef] [PubMed]
70. Irastorza-Landa, N.; García-Cossio, E.; Sarasola-Sanz, A.; Brötz, D.; Birbaumer, N.; Ramos-Murguialday, A. Functional Synergy Recruitment Index as a Reliable Biomarker of Motor Function and Recovery in Chronic Stroke Patients. *J. Neural. Eng.* **2021**, *18*, 046061. [CrossRef]
71. Funato, T.; Hattori, N.; Yozu, A.; An, Q.; Oya, T.; Shirafuji, S.; Jino, A.; Miura, K.; Martino, G.; Berger, D.; et al. Muscle Synergy Analysis Yields an Efficient and Physiologically Relevant Method of Assessing Stroke. *Brain Commun.* **2022**, *4*, fcac200. [CrossRef]
72. Camardella, C.; Barsotti, M.; Buongiorno, D.; Frisoli, A.; Bevilacqua, V. Towards Online Myoelectric Control Based on Muscle Synergies-to-Force Mapping for Robotic Applications. *Neurocomputing* **2021**, *452*, 768–778. [CrossRef]

sensors

Brief Report

The Utility of Calibrating Wearable Sensors before Quantifying Infant Leg Movements

Jinseok Oh [1], Gerald E. Loeb [2] and Beth A. Smith [1,3,4,*]

1 Division of Developmental-Behavioral Pediatrics, Children's Hospital Los Angeles, Los Angeles, CA 90027, USA; joh@chla.usc.edu
2 Alfred E. Mann Department of Biomedical Engineering, Viterbi School of Engineering, University of Southern California, Los Angeles, CA 90089, USA; gloeb@usc.edu
3 Department of Pediatrics, Keck School of Medicine, University of Southern California, Los Angeles, CA 90033, USA
4 Developmental Neuroscience and Neurogenetics Program, The Saban Research Institute, Children's Hospital Los Angeles, Los Angeles, CA 90027, USA
* Correspondence: bsmith@chla.usc.edu

Abstract: While interest in using wearable sensors to measure infant leg movement is increasing, attention should be paid to the characteristics of the sensors. Specifically, offset error in the measurement of gravitational acceleration (g) is common among commercially available sensors. In this brief report, we demonstrate how we measured the offset and other errors in three different off-the-shelf wearable sensors available to professionals and how they affected a threshold-based movement detection algorithm for the quantification of infant leg movement. We describe how to calibrate and correct for these offsets and how conducting this improves the reproducibility of results across sensors.

Keywords: calibration; movement quantification; offset error; inertial measurement unit (IMU); wearable sensors; reproducibility

1. Introduction

The consistent quantification of infant leg movements has potential clinical applications. For example, some researchers found a significant correlation between the frequency of kicking movements in infancy and the onset of walking [1]. Others have shown that leg movement characteristics are altered in infants born preterm [2], with low birth weight [3], or with Down's syndrome [1,4]. These findings suggest the diagnostic potential of infant leg movement kinematics, particularly when measured at a very early age.

While infant leg movements can be studied with three-dimensional motion capture systems, these systems are only useful for measuring movement over a short period (minutes) of observation. The same is true of video-based movement analysis. Because it currently requires visual annotation by experienced coders, in our experience, analyzing 5 min of infant leg movement data takes a few hours. In both cases, quantification is thus limited to minutes of behavior and does not capture the full range of behaviors that infants produce across one or more days. Alternatively, interest is growing in measuring the full-day movements of infants in their natural environments using wearable sensors [5–7]. Commercially available wearable sensors are relatively inexpensive and easy to operate, and some sensors can even be shipped to the homes of participants. With the currently available battery life, such sensors can record infant leg movements continuously for days. It is important to note, however, that while collecting wearable sensor data with off-the-shelf sensors is relatively easy, analyzing the data is not. Sensor-based metrics need to be created and validated for infants, as off-the-shelf analyses are not available for analyzing infant leg movements.

Any measurement technique requires decisions about what aspects of behavior are useful to measure and what methods for data analysis are needed to quantify those aspects. Movement duration is one such aspect that may help distinguish infant-generated movements from any other movement such as cuddling of caregivers or vibrations from different sources. A previous study reported that infant spontaneous kicks are typically less than 500 milliseconds long [8]. This translates to some number of data points depending on the sampling frequency (f_s) of the sensor. Movement intensity is another aspect to consider. Commercially available inertial measurement units (IMUs) generally transduce and quantify all three axes of translational acceleration plus all three axes of gyroscopic rotational velocity. Infant leg movements can be defined according to a minimal acceleration magnitude and counted according to the number of times the magnitude crosses a predefined threshold [7].

Researchers interested in using wearable sensors to measure the frequency of infant leg movements and estimate physical activity intensity, particularly those newly introduced to the methodology, need to understand the kinds of errors to which multi-axis accelerometers are prone and how they might affect the results of movement detection and counting algorithms. This knowledge will facilitate not only the collection but also the comparison of data from multiple studies and IMU devices. Specifically, errors of gain (i.e., the static effects of gravity reported are proportional to but not identical with $1\ g = 9.8\ \text{m/s}^2$), non-orthogonality of axes (i.e., there is no orientation in which all the effects of gravity appear on only one axis), offsets (i.e., when oriented orthogonally to gravity, the signal is non-zero), and noise (fluctuations around the mean when the sensor is at rest) need to be considered.

This study aims to raise awareness on the need for calibration adjustments for IMU data analysis by (1) reporting the gain, misalignment (non-orthogonality of axes), offset, and noise of measurement axes for IMUs used to collect movement data, (2) providing a tutorial on a simple calibration and correction method [9], which has been noted by researchers over time [10,11], and (3) demonstrating its ability to improve the comparability of movement counts among the different IMUs. Three samples from each of three different models of IMUs for each of these types of error were tested. The results demonstrate that only offset errors were significant and that they had significant effects on movement detection, which were mitigated by the calibration procedure described herein.

2. Background

Identifying discrete, countable movements from continuous analog signals generated by wearable sensors necessarily requires threshold-based algorithms such as those developed and validated previously [7] and that employed in this study. Briefly, the time-stamped Euclidian norm of acceleration ($\| a \|$) measured along three axes is calculated (square root of the sum of the squares of each axis of acceleration). Translational accelerometers respond to the static effects of gravity, so $\| a \|$ is detrended ($\| a \|_d$) by subtracting its median (based on the assumption that the infant is at rest at least 50% of the time) so that the values of the detrended norm are centered around the baseline of $0\ g$. Positive and negative threshold values are dynamically set near $\pm 0.1\ g$ to detect movements that start with or against the static effects of gravity, respectively. A movement starts with a value of $\| a \|_d$ above the positive threshold or below the negative threshold. The movement ends after consecutive values cross the baseline twice from different directions within 1.5 s (indicating a movement that is accelerated from and decelerated to a resting posture), with at least one of the values crossing the threshold opposite to the one already crossed. During the entire duration of the movement, the angular velocity norm with its median subtracted should be greater than 0. The portion of $\| a \|_d$ (connected pink dots) marked by a green line in Figure 1B showcases the described movement.

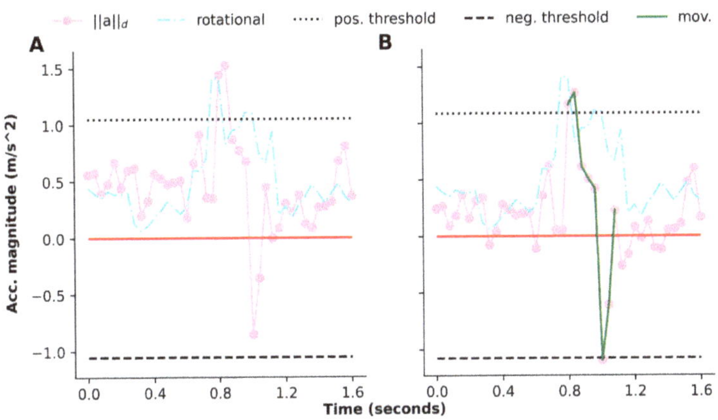

Figure 1. Movement detection algorithm's output for raw and calibrated wearable data analysis. (**A**) Magnitude of the detrended acceleration norm ($\| a \|_d$) in a window of 1.6 s when gain and/or offset is not corrected. The positive threshold (pos. threshold) is crossed once, but the negative threshold (neg. threshold) is not crossed. This is because, overall, $\| a \|_d$ is shifted upward within this window. No movement is defined within this window. (**B**) Same data after offset and gain error corrected; $\| a \|_d$ values are shifted downward, and the trajectory crosses both the positive and the negative threshold, allowing a movement (mov.) to be detected after calibration. Median subtracted angular velocity norm (rotational) is greater than 0 during the duration of the movement.

A movement may be missed when accelerometer analyses are not adjusted for calibration. Figure 1A showcases where uncalibrated infant leg movement data are provided to the algorithm. The $\| a \|_d$ values during this 1.6 s period are shifted upward from the baseline of 0 g (red solid line) because of the uncorrected errors in the offset. The trajectory of $\| a \|_d$ rises and crosses the positive threshold but does not cross the negative threshold. The algorithm thus decides that this trajectory is not a movement. When errors are corrected, the missed movement is counted (Figure 1B). For asymmetrical movements, failure to correct offset errors can also result in the false-positive identification of movements.

3. Materials and Methods

3.1. Selected Wearable Sensors

Three tri-axial IMU models in common use were tested to compare the presence of gain and offset error: Opal version 2 (APDM Inc., Portland, OR, USA), Ax6 (Axivity Ltd., Newcastle, UK), and Movesense Active HR2 (Movesense Ltd., Vantaa, Finland). Opal has been a popular choice that has been validated for recording various aspects of human movement [12–14] and even used as a reference when validating other sensors [15]. The Ax6 and Active sensors are also often used to measure or recognize human movement [15–19].

Specifications of the sensors are provided in Table 1. The sensors had distinct shapes. Opal and Ax6 were cuboids with round corners, having orthogonal faces aligned with the axis measuring g (gravity axis). In contrast, Active had a button-like shape with top and bottom surfaces well aligned with gravity. The other two axes were found to have limited accuracy, because the orientation of the axes was not indicated on the circular shape of the sensors.

Table 1. Specifications of the wearable sensors.

Name	Dimension (mm) [1]	Sampling Frequency (Hz) [2]	Range [3]	Resolution [3]	Noise [3]
Opal V2	43.7 × 39.7 × 13.7 (L × W × H)	20	±16 g (A), ±2000 deg/s (G)	14 bits (A), 16 bits (G)	120 μg/$\sqrt{\text{Hz}}$ (A), 0.0025 deg/s/$\sqrt{\text{Hz}}$
Ax6	23 × 32.5 × 8.9 (L × W × H)	25		16 bits (A, G)	N/A
Movesense Active HR2	36.6 × 10.6 (D × H)	52		16 bits (A, G)	N/A

[1] Opal V2 and Ax6 are cuboid-shaped, while Movesense Active has a button-like shape (L: length, W: width, H: height, D: diameter). [2] Frequency used to record data is reported. [3] Measures for an accelerometer (A) or a gyroscope (G), respectively; N/A: not provided from data sheets.

3.2. Preparation of Datasets

Two types of datasets were prepared. Calibration datasets were generated from 60 s recordings of each of the three IMU models. Each sensor was placed on a level floor in two different orientations (up and down) per axis for 10 s (10 × 6 = 60 s of recording per sensor). This allowed one of the three axes to be parallel to the direction of gravity and measure g along that specific axis. These measurements were later compared with the expected g value (−1 or +1) to estimate the gain and offset [9,20]. Three calibration datasets were prepared for each of the three IMU models using three different sensors of each model. An iOS application, *Movesense Showcase* (Suunto Oy, version 1.1.0), was used to find the approximate x and y axes when preparing the calibration dataset of the Active sensors. The sensors were connected through Bluetooth, and the app streamed the g value measured along the axes. While observing the streamed values, the sensor was placed in orientations that maximized the measured g (−1 or 1) along a single axis (x or y) while minimizing the amount along the other two axes (i.e., measuring values close to 0 g).

A movement dataset was prepared from 5 min of recording using all three sensors wrapped together and placed on the right dorsal forearm of an adult (JO). The specific location of the sensor placement was 5 cm from the wrist (a line connecting the styloid processes of the ulnar and radius bones). The combined mass and bulk of all three sensors precludes carrying this out on an infant limb. The sensors recorded 200 linear forearm movements (100 thrusts and 100 pull-backs) that mimicked the movement of the lower leg during typical supine kicking [21] by an infant. Supine kicking is one of the earliest and most prominent movements observed among infants between 0–6 months. It involves the synchronized movements (flexion or extension) of an infant's hip and knee joint. In previous studies, leg movements including kicking were recorded using wearable sensors placed near the ankles of infant legs [7]. To imitate kicking recorded by sensors at the ankle, sensors were placed near the wrist of the adult, and a movement started with the shoulder abducted at around 45°. The internal rotation of the shoulder joint and the extension of the elbow joint (thrust) imitated an infant's extension of the hip and the knee joints. The external rotation of the shoulder and the flexion of the elbow (pull-back) corresponded to the synchronized flexion of the joints of an infant. During movements, the arm was in the air, not touching any surface.

3.3. Estimation of Errors and Measurement Noise

Prior to gain estimation, offsets for each axis were determined and corrected. If a non-gravity axis produced the same non-zero value when the IMU was flipped in the gravity axis, then that was indicative of a simple offset. If the non-zero values were different in the two orientations, this was indicative of an alignment error. Misalignment is not important for movement detection because it is based on the non-dimensional norm, but it must be considered to determine the magnitude of the offset, which is the mean of these two signed values. Consequently, the mean of each axis' values when it was a non-gravity axis (expected value: 0 g) was the axis offset. Axis misalignment was the mean of the absolute values of the axis after the axis offset was removed. Finally, offset-removed measurements of ± 1 g were arranged. Windowed mean of values near 1 g whose standard deviation was less than 0.01 represented a 1 g measurement of the axis (m_{1g}). The same was true for values near -1 g (m_{-1g}). Ultimately, the gain (G_{axis}) of an axis was estimated using the following equation: $G_{axis} = (m_{1g}) - (m_{-1g})/2$.

Measurement noise, the spread around the mean of a steady-state signal, was also measured along each axis. A noisy sensor will automatically generate a larger acceleration norm and possibly more movement counts than a noise-free sensor. To examine if this was true with our algorithm, the measurement noise of each axis was defined as twice the standard deviation of the values used to derive m_{1g}.

3.4. Error-Correction Procedure

The movement dataset for each sensor was first corrected for the offset of the corresponding sensor. The axis-specific offset error was subtracted from the recorded linear acceleration values of the matching measurement axis. The values were then each divided by the gain value of the corresponding axis. Misalignment error and measurement noise were not addressed. The former was not considered because it did not have any influence on the movement detection algorithm based on the dimensionless norm. The latter was not corrected based on a post hoc assessment of its influence on the work of the algorithm.

3.5. Filtering High Frequency Components

The three different IMU models had different sampling rates, reflective of the different frequency responses available from their sensors. To be certain that the same movement parameters were detected by all three, it was necessary to filter out the higher-frequency components of the datasets from Ax6 and Active to be equivalent to the lower-frequency Opal. A first-order Finite Impulse Response (FIR) low-pass filter with the Hamming window and a cut-off frequency set at 8 Hz was applied to all three sensor types. This was below the Nyquist frequency of the Opal sensor (10 Hz) and above the maximum frequency of the recorded movements (~4 Hz). The digital filter was designed using the *firwin* function of Python's SciPy module (Python 3.10.6; SciPy 1.11.0).

4. Results

4.1. Offset, Misalignment, Gain, and Noise of Measurement Axes

All three sensors exhibited varying levels of offset, misalignment, and gain errors and noise. When the sensors were in various orientations, the acceleration in the non-gravitational Y axis was not consistently zero, as depicted in Figure 2. When the X axis was the gravity axis, measured g values (solid lines) were near -0.07, even when the sensor was flipped. When the Z axis was the gravity axis, one orientation returned Y values near -0.065, while the other generated values near -0.095. This is indicative of axis misalignment as well as offset. The calculated offset and misalignment of the axis were -0.074 g and 0.016 g. Table 2 reports different types of errors as well as the noise of the measurement per axis. In general, the Ax6 sensors showed greater offset values than the other two sensors, with all three sensors having non-zero axis offset values.

After the offset error correction, the gain values of the different sensors were near 1. The gains of the Opal sensors were the closest to 1 (average of the three measurements were 0.999, 0.999, and 1.000 for the X, Y, and Z axes, respectively), but those of the other sensors were also comparable. The measurement noise was the smallest for the Opal sensors (<0.003 g) but was deemed negligible for all the axes of all the sensors (<0.009 g).

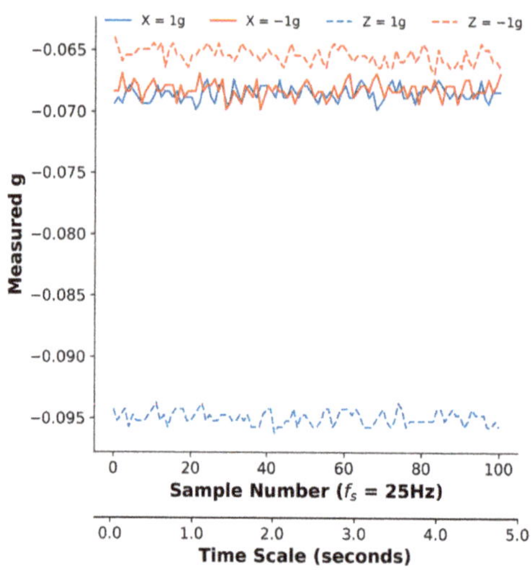

Figure 2. Gravitational acceleration (g) values measured along the Y axis of an Ax6 sensor. The Y axis was a non-gravity axis. Measurement was 5 s long. When the X axis was the gravity axis, the Y axis consistently measured values near $-0.07\,g$, even when the sensor was flipped (solid lines). However, when the Z axis was the gravity axis, the Y values (dashed lines) differed substantially, indicating axis misalignment. The offset of the axis was determined by the mean of the measurements under four different orientations. This sensor's Y axis offset was $-0.074\,g$.

Table 2. Offset, misalignment, gain, and noise of the three measurement axes of sensors.

Error	Axis	Opal v2 *			Ax6 *			Movesense Active HR2 *		
		1	2	3	1	2	3	1	2	3
Offset	X	0.018	0.013	0.013	0.062	0.030	0.065	−0.002	−0.017	0.084
(ideal = 0)	Y	−0.010	−0.023	0.013	−0.074	−0.067	−0.084	−0.029	−0.083	−0.185
	Z	−0.012	−0.004	0.027	0.030	0.018	−0.007	0.114	0.063	0.144
Misalign.	X	0.016	0.008	0.009	0.011	0.009	0.009	0.005	0.008	0.005
(ideal = 0)	Y	0.006	0.006	0.000	0.016	0.005	0.001	0.004	0.021	0.004
	Z	0.006	0.023	0.014	0.011	0.012	0.004	0.004	0.001	0.017
Gain	X	0.999	0.999	0.999	1.002	0.998	0.998	1.009	0.997	1.008
(ideal = 1)	Y	0.999	0.999	0.999	0.998	0.994	0.993	1.001	1.003	1.004
	Z	1.000	1.000	0.999	1.009	1.011	1.013	1.002	1.005	1.009
Noise	X	0.002	0.002	0.002	0.003	0.005	0.009	0.004	0.005	0.004
(ideal = 0)	Y	0.002	0.002	0.003	0.001	0.001	0.002	0.004	0.004	0.008
	Z	0.002	0.001	0.002	0.002	0.002	0.002	0.004	0.004	0.004

* Three different sensors of each type were used (1, 2, and 3). The unit of offset, misalignment, and noise is g.

4.2. Effect of Calibration and Low-Pass Filtering on the Algorithm Output of Data from Different Sensors

The forearm movement data collected with the three sensors were preprocessed and provided to our movement detection algorithm. The efficacy of the two preprocessing steps—calibration and low-pass filtering—were assessed according to their improvements on the reproducibility of movement counts across sensor types. Table 3 shows the movement counts from the three sensors' recordings under different conditions. When the datasets were not processed at all (Raw), the three movement counts differed substantially from one another, ranging from 185 to 269. However, the counts became more similar to each other once the datasets were calibrated and filtered (Opal v2: 195, Ax6: 175, Movesense Active: 213). The Opal sensor's recording generated the most accurate count of movements;

the recordings of the other two sensors either underestimated (Ax6) or overestimated (Active) the number of movements.

Table 3. Movement counts with and without preprocessing steps (reference count: 200).

Sensor	Raw	C	C + F	Noise after C + F (g)		
				X	Y	Z
Opal v2	269	269	195	0.001	0.001	0.001
Ax6	185	196	175	0.001	0.001	0.002
Movesense Active HR2	231	230	213	0.003	0.003	0.003

C: Calibrated gain and offset errors, F: filtered with a first-order low-pass filter (cut-off frequency = 8 Hz).

5. Discussion

The three types of wearable sensors that were tested demonstrated varying levels of measurement errors, including misaligned axes and gain/offset problems. Correcting these errors contributed to generating more comparable results using a threshold-based movement detection algorithm with data from different sensors. This implies that behavioral scientists who consider using commercially available wearable sensors to quantify infant leg movements should inspect the sensors before making use of the raw data. The simple calibration procedure and error-correction methods described herein should correct for the major sources of sensor error.

5.1. Offset Error Is One Main Source of Incorrect Estimation

Sensor offset error appears to be the biggest concern in estimating the number of movements using the norm of linear accelerations coming from IMUs. The gains or sensitivity values of the sensors were all near 1. The measurement noise was also less than 0.01 g, and most misalignment errors were less than 0.02 g. The offset errors, however, were almost always greater than 0.01 g, with some values surpassing 0.1 g.

How does the untreated offset error influence a movement detection algorithm like the one discussed in this article? When calculating the norm, positive or negative offset errors are also squared and added, making the norm at rest greater than 1 g. As a result of axis alignment errors, the measured offsets depend on the sensor orientation, which will change during a recording session, so the detrended norm sits below 0 g at rest at some times and above 0 g at rest at other times. In the former scenario, norm values corresponding to movements are less likely to cross a positive threshold. In the latter, a negative threshold is not crossed (Figure 1A). Offset errors in a given sensor vary greatly in both magnitude and sign. Any attempt to detect movements based on acceleration in a single axis instead of computing the norm count over/underestimates the true acceleration depending on the sign of the offset and the orientation of the sensor axis with respect to both gravity and the infant's limb.

Offset error along measurement axes can explain a recent finding that also supports the need for the calibration of wearable sensors in measuring physical activity [22]. Researchers proposed different sensor measures to estimate physical activity (PA) levels measured with oxygen consumption (VO_2), such as the count data provided through the sensor systems of ActiGraph Llc., Euclidean Norm Minus One (ENMO), or the mean amplitude deviation (MAD) of raw sensor data. Among the measures, the last two are easily calculable for data from any sensor, and distinct values were proposed as thresholds to classify different PA intensities. Weitz et al. [22] investigated the effect of calibration on the estimation of time spent in moderate-to-vigorous physical activity (MVPA) using the two measures. Researchers used low and high thresholds of ENMO [23] and MAD [24] to estimate time spent in MVPA. For a sensor recording of an adult's daily activity, the sum of the time windows where the ENMO or MAD values are between the two thresholds defining MVPA was equal to the daily time spent in MVPA. Researchers reported that the average of the estimated minutes of MVPA per day based on ENMO was significantly overestimated when the sensor data for analyses were not calibrated. Researchers further illustrated that a post hoc auto-calibration [20] decreased ENMO almost by half and thus reduced the estimated MVPA volume. Varying amounts of offset error make the overall Euclidean norm greater than 1 g at rest. Consequently, the chance of ENMO calculated from uncalibrated sensor data crossing the lower threshold of MVPA will be significantly higher than that of the measure coming from calibrated data. Our study and that of Weitz et al. [22] therefore showcase two different examples of how offset error can negatively influence the accuracy and reproducibility of measurements of the frequency of leg movements and intensity of physical activity.

5.2. A Comparison with Alternative Calibration Methods

Calibration methods that estimate errors from the actual recording of movements using wearable sensors have been proposed [9,20,25]. Such methods avoid the requirement for a separate calibration and can be applied to pre-existing records for which the source sensor is unavailable. Preparing datasets for a separate calibration process was described as cumbersome in studies with a high throughput [20]. This is true when thousands of sensors are used in a cohort study. Still, preparing a separate dataset of a sensor would take at most 2 min. The benefit of this method is that the datasets are from static moments at known sensor orientations, making the estimation of misalignment and gain errors reliable, while other more automated methods can speculate at best from the periods of recordings estimated to be the times of non-wear or zero-to-minimal movement such as sleeping. Fortunately, the micro-electromechanical systems (MEMS) technology used in modern IMUs appears to perform stably over time [26]. This suggests that in studies that utilize smaller numbers of sensors, experimenters can collect the calibration data once and keep track of which sensors are used to create which datasets so that their raw data can be corrected before further analysis.

5.3. Additional Aspect to Be Considered

Although not investigated in this report, one other aspect to consider when using multiple wearable sensors simultaneously to quantify movements is clock synchronization between sensors. For example, if researchers want to answer whether an infant's two legs moved simultaneously and in-phase or reciprocally, the samples from the sensors on each leg should have synchronized timestamps. Opal sensors do have a hub device that communicates with sensors to synchronize the internal clocks of the sensors. This is not the case for other sensors. One simple solution to this issue could be to have a noticeable event at the beginning and at the end of a recording (e.g., hitting the sensors together five times) and later interpolate the timestamps of the raw data based on the events.

6. Limitation

While an adult's linear arm movements resemble infant supine kicks in how the relevant joint movements are coordinated, they neither represent the full spectrum of the leg movements infants make during a day, nor are they similar in the actual magnitude values associated with the movements. However, the focus of this article is to demonstrate the problem caused by the measurement errors of commercially available wearable sensors when using an acceleration magnitude threshold-based movement detection algorithm to quantify infant movements and how to address the problem independent of sensor choice. Testing the effect of calibration on each sensor's performance can be carried out by comparing the movement counts derived from uncalibrated and calibrated datasets of the same sensor. The direct comparison of infant movement detection by different sensors is not possible because infant behavior varies between sessions, and it is not feasible to place more than one sensor at a time on an infant's limb.

7. Conclusions

Wearable motion sensors are subject to offset, misalignment, and gain errors and noise. All kinds of errors were present in all three types of sensors tested in this study, with the offset error being the largest in magnitude. The simple calibration and correction method provided here improved the reproducibility of the movement counts obtained by a threshold-based algorithm. Studies that compare data from participants wearing different sensors should obtain calibration data for each sensor and use them to correct the recorded data before quantifying the movements of each participant.

Author Contributions: Conceptualization, J.O., G.E.L. and B.A.S.; methodology, J.O., G.E.L. and B.A.S.; software, J.O.; validation, J.O., G.E.L. and B.A.S.; formal analysis, J.O.; investigation, J.O., G.E.L. and B.A.S.; resources, B.A.S.; data curation, J.O.; writing—original draft preparation, J.O.; writing—review and editing, J.O., G.E.L. and B.A.S.; visualization, J.O. and G.E.L.; supervision, B.A.S.; project administration, B.A.S.; funding acquisition, B.A.S. All authors have read and agreed to the published version of the manuscript.

Funding: This research was funded by the National Institute on Drug Abuse and National Institute On Minority Health And Health Disparities of the National Institutes of Health under Award Number U01DA055362. The content is solely the responsibility of the authors and does not necessarily represent the official views of the National Institutes of Health.

Institutional Review Board Statement: Not applicable.

Informed Consent Statement: Not applicable.

Data Availability Statement: The data presented in this study are available on request from the corresponding author (B.A.S.) as a Data Use Agreement is necessary.

Conflicts of Interest: The authors declare no conflicts of interest.

References

1. Ulrich, B.D.; Ulrich, D.A. Spontaneous leg movements of infants with Down syndrome and nondisabled infants. *Child. Dev.* **1995**, *66*, 1844–1855. [CrossRef] [PubMed]
2. Geerdink, J.J.; Hopkins, B.; Beek, W.J.; Heriza, C.B. The organization of leg movements in preterm and full-term infants after term age. *Dev. Psychobiol.* **1996**, *29*, 335–351. [CrossRef]
3. Jeng, S.-F.; Chen, L.-C.; Yau, K.-I.T. Kinematic analysis of kicking movements in preterm infants with very low birth weight and full-term infants. *Phys. Ther.* **2002**, *82*, 148–159. [CrossRef] [PubMed]
4. McKay, S.M.; Angulo-Barroso, R.M. Longitudinal assessment of leg motor activity and sleep patterns in infants with and without Down syndrome. *Infant Behav. Dev.* **2006**, *29*, 153–168. [CrossRef] [PubMed]
5. Abrishami, M.S.; Nocera, L.; Mert, M.; Trujillo-Priego, I.A.; Purushotham, S.; Shahabi, C.; Smith, B.A. Identification of Developmental Delay in Infants Using Wearable Sensors: Full-Day Leg Movement Statistical Feature Analysis. *IEEE J. Transl. Eng. Health Med.* **2019**, *7*, 2800207. [CrossRef]
6. Gravem, D.; Singh, M.; Chen, C.; Rich, J.; Vaughan, J.; Goldberg, K.; Waffarn, F.; Chou, P.; Cooper, D.; Reinkensmeyer, D.; et al. Assessment of Infant Movement With a Compact Wireless Accelerometer System. *J. Med. Devices* **2012**, *6*, 021013. [CrossRef]
7. Smith, B.A.; Trujillo-Priego, I.A.; Lane, C.J.; Finley, J.M.; Horak, F.B. Daily Quantity of Infant Leg Movement: Wearable Sensor Algorithm and Relationship to Walking Onset. *Sensors* **2015**, *15*, 19006–19020. [CrossRef]
8. Thelen, E.; Fisher, D.M. The organization of spontaneous leg movements in newborn infants. *J. Mot. Behav.* **1983**, *15*, 353–377. [CrossRef]
9. Lötters, J.C.; Schipper, J.; Veltink, P.H.; Olthuis, W.; Bergveld, P. Procedure for in-use calibration of triaxial accelerometers in medical applications. *Sens. Actuators A Phys.* **1998**, *68*, 221–228. [CrossRef]
10. Ru, X.; Gu, N.; Shang, H.; Zhang, H. MEMS Inertial Sensor Calibration Technology: Current Status and Future Trends. *Micromachines* **2022**, *13*, 879. [CrossRef]
11. Tedaldi, D.; Pretto, A.; Menegatti, E. A robust and easy to implement method for IMU calibration without external equipments. In Proceedings of the 2014 IEEE International Conference on Robotics and Automation (ICRA), Hong Kong, China, 31 May–7 June 2014; pp. 3042–3049.
12. Kobsar, D.; Charlton, J.M.; Tse, C.T.F.; Esculier, J.-F.; Graffos, A.; Krowchuk, N.M.; Thatcher, D.; Hunt, M.A. Validity and reliability of wearable inertial sensors in healthy adult walking: A systematic review and meta-analysis. *J. Neuroeng. Rehabil.* **2020**, *17*, 62. [CrossRef] [PubMed]
13. Mancini, M.; Horak, F.B. Potential of APDM mobility lab for the monitoring of the progression of Parkinson's disease. *Expert. Rev. Med. Devices* **2016**, *13*, 455–462. [CrossRef]
14. Storm, F.A.; Heller, B.W.; Mazzà, C. Step detection and activity recognition accuracy of seven physical activity monitors. *PLoS ONE* **2015**, *10*, e0118723. [CrossRef]
15. Mason, R.; Byerley, J.; Baker, A.; Powell, D.; Pearson, L.T.; Barry, G.; Godfrey, A.; Mancini, M.; Stuart, S.; Morris, R. Suitability of a Low-Cost Wearable Sensor to Assess Turning in Healthy Adults. *Sensors* **2022**, *22*, 9322. [CrossRef] [PubMed]

16. Airaksinen, M.; Gallen, A.; Kivi, A.; Vijayakrishnan, P.; Häyrinen, T.; Ilén, E.; Räsänen, O.; Haataja, L.M.; Vanhatalo, S. Intelligent wearable allows out-of-the-lab tracking of developing motor abilities in infants. *Commun. Med.* **2022**, *2*, 69. [CrossRef]
17. Doherty, A.; Jackson, D.; Hammerla, N.; Plötz, T.; Olivier, P.; Granat, M.H.; White, T.; van Hees, V.T.; Trenell, M.I.; Owen, C.G.; et al. Large Scale Population Assessment of Physical Activity Using Wrist Worn Accelerometers: The UK Biobank Study. *PLoS ONE* **2017**, *12*, e0169649. [CrossRef]
18. Skawinski, K.; Montraveta Roca, F.; Findling, R.D.; Sigg, S. Workout Type Recognition and Repetition Counting with CNNs from 3D Acceleration Sensed on the Chest. In *Advances in Computational Intelligence, Proceedings of the 15th International Work-Conference on Artificial Neural Networks, IWANN 2019, Gran Canaria, Spain, 12–14 June 2019*; Rojas, I., Joya, G., Catala, A., Eds.; Springer International Publishing: Cham, Switzerland, 2019; pp. 347–359.
19. Ishii, S.; Luimula, M.; Yokokubo, A.; Lopez, G. VR Dodge-ball: Application of Real-time Gesture Detection from Wearables to ExerGaming. In Proceedings of the 2020 11th IEEE International Conference on Cognitive Infocommunications (CogInfoCom), IEEE, Mariehamn, Finland, 23–25 September 2020; pp. 000081–000082.
20. van Hees, V.T.; Fang, Z.; Langford, J.; Assah, F.; Mohammad, A.; da Silva, I.C.M.; Trenell, M.I.; White, T.; Wareham, N.J.; Brage, S. Autocalibration of accelerometer data for free-living physical activity assessment using local gravity and temperature: An evaluation on four continents. *J. Appl. Physiol.* **2014**, *117*, 738–744. [CrossRef]
21. Thelen, E.; Bradshaw, G.; Ward, J.A. Spontaneous kicking in month-old infants: Manifestation of a human central locomotor program. *Behav. Neural Biol.* **1981**, *32*, 45–53. [CrossRef] [PubMed]
22. Weitz, M.; Morseth, B.; Hopstock, L.A.; Horsch, A. Influence of Accelerometer Calibration on the Estimation of Objectively Measured Physical Activity: The Tromsø Study. *J. Meas. Phys. Behav.* **2024**, *7*. [CrossRef]
23. Hildebrand, M.; VAN Hees, V.T.; Hansen, B.H.; Ekelund, U. Age group comparability of raw accelerometer output from wrist- and hip-worn monitors. *Med. Sci. Sports Exerc.* **2014**, *46*, 1816–1824. [CrossRef]
24. Vähä-Ypyä, H.; Vasankari, T.; Husu, P.; Mänttäri, A.; Vuorimaa, T.; Suni, J.; Sievänen, H. Validation of Cut-Points for Evaluating the Intensity of Physical Activity with Accelerometry-Based Mean Amplitude Deviation (MAD). *PLoS ONE* **2015**, *10*, e0134813. [CrossRef] [PubMed]
25. Bouten, C.V.; Koekkoek, K.T.; Verduin, M.; Kodde, R.; Janssen, J.D. A triaxial accelerometer and portable data processing unit for the assessment of daily physical activity. *IEEE Trans. Biomed. Eng.* **1997**, *44*, 136–147. [CrossRef] [PubMed]
26. Łuczak, S.; Zams, M.; Bagiński, K. Selected Aging Effects in Triaxial MEMS Accelerometers. *J. Sens.* **2019**, *2019*, 5184907. [CrossRef]

sensors

Article

Multilevel Pain Assessment with Functional Near-Infrared Spectroscopy: Evaluating ΔHBO_2 and ΔHHB Measures for Comprehensive Analysis

Muhammad Umar Khan, Maryam Sousani, Niraj Hirachan, Calvin Joseph, Maryam Ghahramani, Girija Chetty, Roland Goecke and Raul Fernandez-Rojas *

Human-Centred Technology Research Centre, Faculty of Science and Technology, University of Canberra, Canberra, ACT 2617, Australia
* Correspondence: raul.fernandezrojas@canberra.edu.au

Abstract: Assessing pain in non-verbal patients is challenging, often depending on clinical judgment which can be unreliable due to fluctuations in vital signs caused by underlying medical conditions. To date, there is a notable absence of objective diagnostic tests to aid healthcare practitioners in pain assessment, especially affecting critically-ill or advanced dementia patients. Neurophysiological information, i.e., functional near-infrared spectroscopy (fNIRS) or electroencephalogram (EEG), unveils the brain's active regions and patterns, revealing the neural mechanisms behind the experience and processing of pain. This study focuses on assessing pain via the analysis of fNIRS signals combined with machine learning, utilising multiple fNIRS measures including oxygenated haemoglobin(ΔHBO_2) and deoxygenated haemoglobin (ΔHHB). Initially, a channel selection process filters out highly contaminated channels with high-frequency and high-amplitude artifacts from the 24-channel fNIRS data. The remaining channels are then preprocessed by applying a low-pass filter and common average referencing to remove cardio-respiratory artifacts and common gain noise, respectively. Subsequently, the preprocessed channels are averaged to create a single time series vector for both ΔHBO_2 and ΔHHB measures. From each measure, ten statistical features are extracted and fusion occurs at the feature level, resulting in a fused feature vector. The most relevant features, selected using the Minimum Redundancy Maximum Relevance method, are passed to a Support Vector Machines classifier. Using leave-one-subject-out cross validation, the system achieved an accuracy of 68.51% $\pm$ 9.02% in a multi-class task (No Pain, Low Pain, and High Pain) using a fusion of ΔHBO_2 and ΔHHB. These two measures collectively demonstrated superior performance compared to when they were used independently. This study contributes to the pursuit of an objective pain assessment and proposes a potential biomarker for human pain using fNIRS.

Keywords: pain assessment; fNIRS; statistical features; SVM; machine learning

1. Introduction

Pain, despite its unpleasantness, acts as an essential biomarker in our bodies, alerting us to potential health issues, injuries, or emotional stress. Pain can be localised to a particular region, like an injury, but it can also be more widespread, as seen in many illnesses [1]. Pain is a significant issue in society as it poses a substantial public health challenge, impacts the quality of life of sufferers, and places a burden on the economy [2,3]. The economic impacts of pain are drastic, imposing a financial burden exceeding AUD 73 billion dollars annually, including AUD 48.3 billion dollars in lost productivity in Australia alone [4,5]. Furthermore, it impacts the day-to-day routines and significantly diminishes the overall quality of life. For instance, low back pain is the leading cause of disability in the world, with over 600 million people living with pain [6]. Therefore, the assessment and management of pain is essential for a wide range of clinical disorders and treatments, and its early diagnosis plays a vital role in mitigating the risk of its progression into chronic conditions or contributing to depression or anxiety [7].

Pain is a subjective experience and its measurement is difficult. In clinical practice, two primary subjective methods are used for pain assessment: self-reports and clinical judgment [8]. The commonly accepted method to assess pain is self-report. Self-reporting techniques aim to gauge a patient's pain using verbal or numerical self-assessment tools, including methods such as visual analogue scales, verbal descriptor scales, numerical rating scales, or the McGill Pain Questionnaire [9,10]. When self-reports are not accessible or may be unreliable, clinical observations can serve as a supplementary or alternative method. Clinical judgment for pain assessment relies on examining and understanding the nature, intensity, and context of the patient's pain experience based on observations [7]. Despite their convenience and utility, subjective reports come with various limitations such as inconsistent measurement scales and variations in how pain is understood by medical professionals and patients. Furthermore, these methods cannot be effectively employed in cases involving children or patients with neurological disorders.

In order to address these limitations, researchers have turned to the analysis of the neurological aspects of pain using objective methods such as neuroimaging [11]. For instance, Wager et al. [12] developed a system that employs machine learning to analyse data obtained from functional magnetic resonance imaging (fMRI). Their work demonstrated the potential to identify a consistent neurological signature of pain at the individual level. While fMRI-based objective assessments of pain have made significant progress in understanding the brain's pain mechanisms, the size and cost of MRI scanners and other conventional neuroimaging tools (such as positron emission tomography) make them impractical for routine clinical use [13]. This limitation has increased the interest in portable neuroimaging devices that offer similar technical advantages to fMRI. One such technology is functional near-infrared spectroscopy (fNIRS), which measures changes in the concentrations of oxygenated hemoglobin (ΔHBO_2) and deoxygenated haemoglobin (ΔHHB)—similar to the blood oxygen level-dependent signal in fMRI. fNIRS is capable of non-invasive measurement of near-infrared light absorption within the range of 700 to 1000 nm through the skull [14]. In contrast to traditional MRI scanners, the portability and compatibility of fNIRS with ferromagnetic and electrical components provide researchers with the option to monitor and study functional brain activity in clinical settings [15,16].

Machine learning has played a pivotal role in neuroimaging-based methods for the study of pain [17,18]. It helps us to better understand the pain by uncovering patterns within clinical and experimental data [19]. Machine learning methods can effectively acquire the ability to map features to known classes, enabling them to predict a pain phenotype class based on a complex set of obtained features. For instance, Brown et al. [20], in an fMRI study, employed the Support Vector Machine (SVM) algorithm to distinguish between painful and non-painful experimental stimuli, achieving an accuracy of 81%. In an EEG study, Gram et al. [21] examined individuals who had received either morphine or a placebo following cold pressor test stimulation. They used the SVM algorithm to classify responders, achieving an accuracy of 71.9%. This classification was based on wavelet coefficients derived from each EEG band. These studies have shown the potential of neuroimaging and machine learning in the identification of pain.

In pain research using fNIRS, machine learning has proven to be effective for the detection and prediction of pain [22]. In a study by Pourshoghi et al. [22], authors used an SVM classifier using B-spline coefficients from functional data analysis. They achieved a classification accuracy of 94% in distinguishing between low-pain and high-pain signals using fNIRS. In Fernandez et al. [23], the results indicate that by using the Gaussian Support Vector Machine (SVM), they achieved an accuracy of 94.17% in classifying the four types of pain within the fNIRS data. Zeng et al. [24] investigated chronic pain's impact on brain function using fNIRS. Machine learning achieved high accuracy in identifying chronic pain patients based on resting-state fNIRS data, suggesting the potential for using functional connectivity features as neural markers for chronic pain diagnosis. Despite the promising results obtained by the mentioned studies, there is still limited research in this field within the literature.

This study employs an approach for pain assessment that leverages the analysis of fNIRS signals in combination with machine learning techniques. This approach utilises fNIRS measurements of ΔHBO_2 and ΔHHB to provide a comprehensive and accurate evaluation of pain levels. While the literature emphasises ΔHBO_2 as a more promising fNIRS measure [25,26], recent studies, as highlighted by Ho et al. [27], indicate that both measures exhibit high accuracy in classification tasks. Therefore, in this study, both ΔHBO_2 and ΔHHB measures have been taken into account. First, the pain information of 30 healthy subjects was collected using quantitative sensory testing (QST). Then, we performed a channel selection process to remove faulty channels from the analysis. Subsequently, ten statistical features from each measure were extracted. Then, we utilised well-known classifiers to identify pain levels using this reduced feature set. This study makes the following contributions: (1) proposing an fNIRS channel selection strategy for rejecting noisy channels based on high-frequency and high-amplitude artifacts; (2) presenting a group of possible features from fNIRS signals for the assessment of pain; (3) identifying that ΔHBO_2 is better at detecting high pain intensity and ΔHHB is good at detecting low pain intensity; and (4) proposing the combination of ΔHBO_2 and ΔHHB as a possible biomarker of human pain. This study contributes to the field of pain assessment and offers new avenues for understanding and quantifying pain in a more precise and objective manner.

2. Materials and Methods

Figure 1 presents the core system block diagram of the proposed fNIRS-based pain assessment system. The system integrates attributes from both ΔHBO_2 and ΔHHB to assess the pain level. Further elaboration on the materials and methodology is provided in the following subsections.

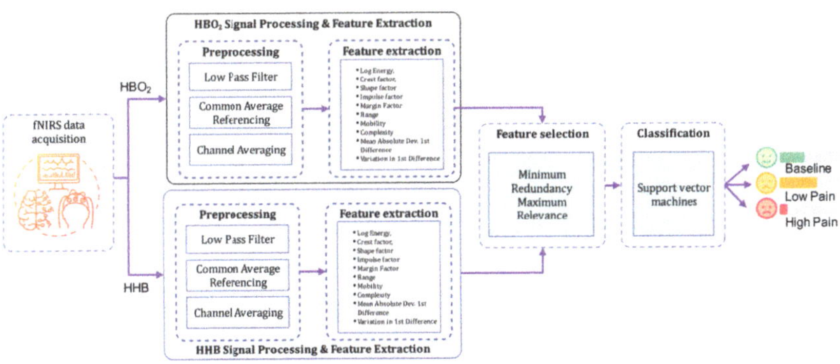

Figure 1. System block diagram of the proposed fNIRS-based pain assessment system.

2.1. Experimental Protocol

In this study, 30 healthy individuals (7 females and 23 males) aged 19 to 52 years (31.7 ± 8.7 yrs) participated. None had unstable medical conditions, chronic pain, or recent medication usage prior to testing. Participants received detailed explanations and provided written informed consent before the start of the experiments. The research, involving human participants, received ethical approval from the University of Canberra's Human Research Ethics Committee (reference number 11837).

The data collection procedure took place at the Human–Machine Interface Laboratory at the University of Canberra, Australia. Participants were seated comfortably with both arms resting on the table. Electrodes from a transcutaneous electrical nerve stimulation (TENS) machine (Medihightec Medical Co., Ltd., Taipei City, Taiwan) were placed on the participants' inner forearm and the back of their right hand. The experimental process consisted of two phases: an initial assessment of individual pain perceptions using the

QST protocol, which determined pain thresholds and tolerances, followed by the pain stimulation phase. We defined the *pain threshold (low pain)* as the lowest stimulus intensity at which stimulation became painful, and *pain tolerance (high pain)* as the highest intensity of pain the participant could endure before reaching a point of intolerable discomfort. In the pain stimulation phase, fNIRS data were acquired and a 60 s baseline recording was obtained before the start of the experiment. A counterbalanced approach was employed, alternating between low and high stimuli intensity and forearm or hand stimulation. Six 10 s stimulus repetitions were recorded for each type of stimulus, followed by 40 s rest intervals. Figure 2 presents a schematic representation of the stimulation and perception of pain.

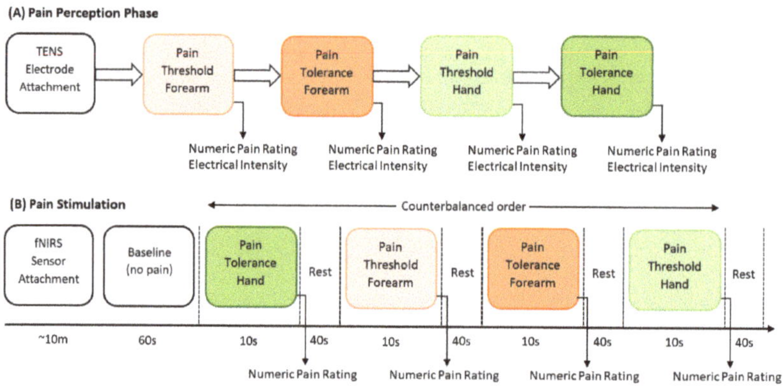

Figure 2. Schematic representation of the experimental procedure.

Changes in ΔHBO_2 and ΔHHB concentration (μmol/L) were measured using a wireless, continuous wave fNIRS device (Artinis Medical Systems, Gelderland, the Netherlands). The fNIRS system includes 24 channels covering the prefrontal cortex (PFC). Optodes (10 sources and 8 detectors) are separated by 35 mm and placed on the frontal lobe (Figure 3). The near-infrared light was emitted by sources with wavelengths of 760 and 840 nm at a sampling rate of 50 Hz. Figure 4 displays the raw fNIRS channels (ΔHBO_2) recorded over a 5 min duration while a subject experienced varying pain intensities.

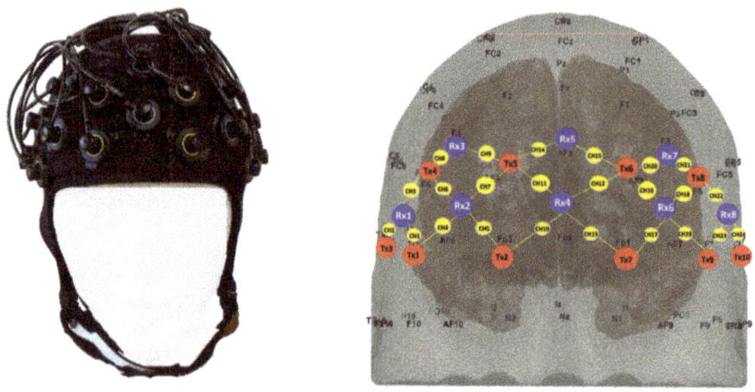

(**a**) (**b**)

Figure 3. fNIRS channel information: (**a**) fNIRS cap. (**b**) Schematic of fNIRS channel locations. Red: Sources; Blue: Detectors; and Yellow: Channels. Specifically, the optodes Tx1, Tx2, Tx7, Tx9, Rx3, and Rx7 were positioned at the following locations on the standard 10–20 EEG system: Tx1: at F8; Tx2: at Fp2; Tx7: at Fp1; Tx9: at F7; Rx3: at F4; and Rx7: at F3.

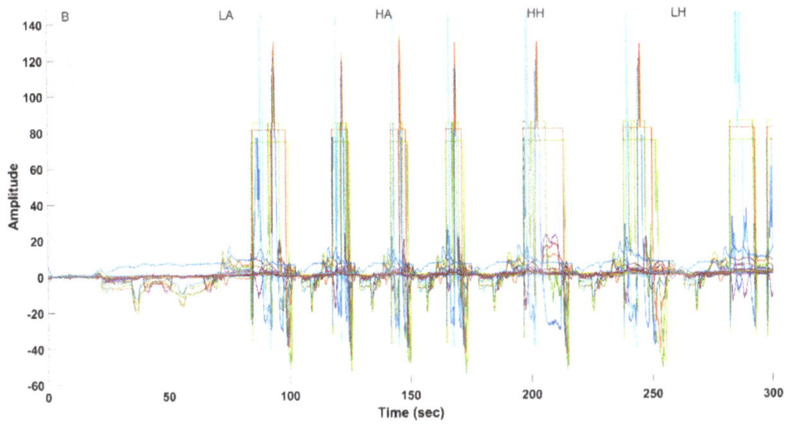

Figure 4. Twenty-two-Channel fNIRS (measuring changes in ΔHBO_2) raw data (excluding two faulty channels) with annotated and highlighted durations for different conditions: B (Baseline), LA (Low Arm Pain), HA (High Arm Pain), HH (High Hand Pain), and LH (Low Hand Pain). The gray background in the figure represents the duration of each experiment phase: Baseline: 60 s, LA, LH, HA, and HH, each lasting 10 s.

2.2. Channel Selection

In the context of processing fNIRS data from 24 channels, as shown in Figure 4, some specific challenges arose. Two of the electrodes related to channels 19 and 23 were found to be malfunctioning, necessitating their exclusion from the analysis. This action was taken to ensure the integrity of the data. Additionally, among the remaining 22 channels, it was observed that certain channels exhibited distinct and undesirable features in the form of high amplitude and sharp peaks resembling square wave artifacts. These peculiar patterns suggest that these channels were significantly contaminated by movement artifacts or other non-neural artifacts. To effectively address this issue and proceed with data analysis, a preliminary step involved the systematic identification of unreliable channels to be excluded from further processing. This selection was accomplished using the relative range (RR) operator threshold. Relative range (Equation (1)) is defined as the ratio of the range of the derivative of an fNIRS channel to the range of the raw channel, as follows:

$$\text{RR} = \frac{\max(x'_{ch}) - \min(x'_{ch})}{\max(x_{ch}) - \min(x_{ch})} \tag{1}$$

where x'_{ch} is the derivative of an fNIRS channel x'_{ch}, which represents the rate of change in a signal. In the context of fNIRS signals, the derivative can highlight regions where the signal changes rapidly, which may correspond to high-amplitude peaks (i.e., spikes) within a channel. As a result, high RR values indicate the presence of these high-amplitude sharp peaks. Experimental findings revealed that channels with an RR exceeding 0.1 (10%) are typically contaminated by these artifacts. With this threshold, the channels contaminated by artifacts were excluded, ensuring that only artifact-free channels were retained for subsequent processing. The raw fNIRS channels ($n = 17$) for ΔHBO_2 measurement selected after the channel selection algorithm are shown in Figure 5. After this step, the data from three subjects were excluded from further processing as the algorithm resulted in the removal of over 70% of their number of channels. For the remaining 27 subjects, the number of retained channels after the selection process ranged from 16 to 22.

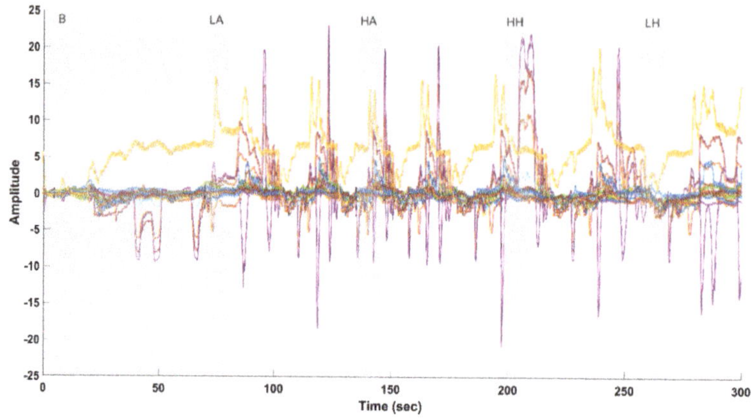

Figure 5. Raw fNIRS channels (measuring changes in ΔHBO_2) selected after the proposed channel selection algorithm featuring the relative range (RR). The intervals for various pain conditions are highlighted and annotated as B (Baseline), LA (Low Arm Pain), HA (High Arm Pain), HH (High Hand Pain), and LH (Low Hand Pain). The gray background in the figure represents the duration of each experiment phase: Baseline: 60 s, LA, LH, HA, and HH, each lasting 10 s.

2.3. Dataset Organisation

After completing the data collection and channel selection process, all recorded data were segmented into 10-second intervals for each class. This resulted in six observations for the baseline class per subject, 12 observations for the low pain class per subject, and 12 observations for the high pain class per subject. In order to address the class observation imbalance, six additional observations from the rest periods of each subject, prior to the pain stimulation, were included in the baseline class. Consequently, the dataset consisted of a total of 972 observations. Each subject contributed 12 observations for each class, resulting in a cumulative total of 36 observations per subject. The dataset included 324 observations for each of the Baseline (B), Low Pain (LP), and High Pain (HP) classes.

2.4. Signal Processing: Filtration and Averaging

To suppress the noise and pulsation in fNIRS data (ΔHBO_2 and ΔHHB), as shown in Figure 6, each available fNIRS channel was passed through a 4th order Butterworth infinite impulse response low-pass filter with a cut-off frequency of 0.16 Hz [23].

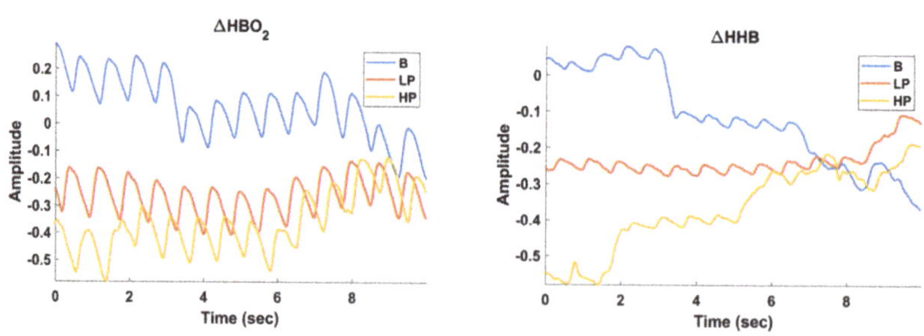

Figure 6. Raw 10-Second Data Segments for Baseline (B), Low Pain (LP), and High Pain (HP) Classes, displayed for Channel 1 of ΔHBO_2 (**Left**) and ΔHHB (**Right**).

During fNIRS data acquisition, there can be various common noise sources that affect the measurements. These noise sources can include changes in blood flow unrelated to neural activity, motion artifacts, and systemic physiological changes such as heart rate and respiration [23]. These sources can introduce noise into the fNIRS data. Common Average

Referencing (CAR) [28] involves calculating the averages from all available channels across the scalp for each wavelength (ΔHBO_2 and ΔHHB). This average is then subtracted, for each wavelength from the signal of each individual channel. This effectively subtracts out the common noise components shared by all channels. Equation (2) shows the channel-averaging scheme:

$$h_{\text{avg}}(k) = \frac{1}{M} \sum_{j=1}^{M} H(k,j) \tag{2}$$

where h is the average of fNIRS measure H (ΔHBO_2 or ΔHHB), M is the total number of channels for each participant, k is the discrete time for which the signal is recorded, and j is the channel number. The preprocessed version of both fNIRS measures, i.e., ΔHBO_2 and ΔHHB for various experimental conditions, is displayed in Figure 7.

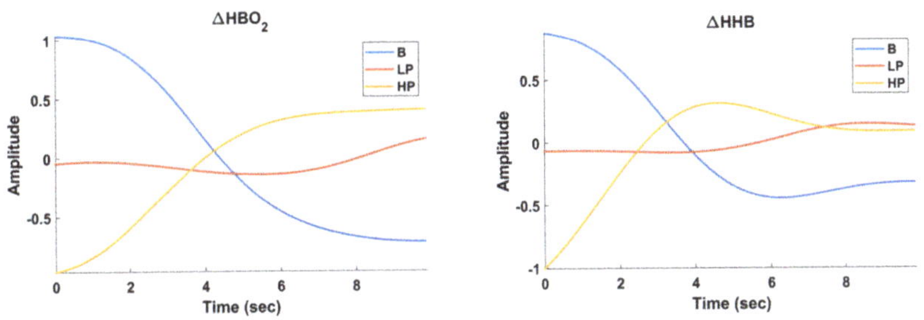

Figure 7. Preprocessed 10-Second Data Segments for Baseline (B), Low Pain (LP), and High Pain (HP) Classes, displayed for ΔHBO_2 (**Left**) and ΔHHB (**Right**). The processing pipeline encompasses low-pass filtering, Common Average Referencing (CAR) for each filtered channel, and the final step of averaging across all channels, culminating in a consolidated vector representation.

2.5. Feature Extraction

The ΔHBO_2 and ΔHHB signals display distinct characteristics associated with the pain intensities. Amplitude, as an indicator of pain intensity, increases with more painful stimuli, signifying higher neural activity and oxygen demand. Variation in these signals highlights the dynamic nature of pain experiences, showcasing rapid and substantial fluctuations over time. Complexity in ΔHBO_2 and ΔHHB responses uncovers the intricate interactions between brain regions and physiological systems involved in pain processing [29]. The dynamics of ΔHBO_2 and ΔHHB responses reveal the timing of pain intensity, from pain onset to apex, and then, to recovery. Moreover, the stability of these signals distinguishes sustained pain from transient changes, providing insights into the persistence of pain perception. To extract the fNIRS signal information related to intensity, dynamics, stability, complexity, and variation-like characteristics [30], we have carefully chosen features [31,32] such as Log Energy, Crest Factor, Shape Factor, Impulse Factor, Margin Factor, Mobility, Complexity, Mean Absolute Deviation of First Difference, Range, and Variation in First Difference as defined in Table 1. These features are extracted from both ΔHBO_2 and ΔHHB signals and fused at the feature level to create a fused feature vector.

2.6. Feature Selection

Feature selection is crucial for improving model efficiency by focusing on important features, reducing dimensionality, and ultimately improving the overall performance in machine learning tasks. In this work, the *Minimum Redundancy Maximum Relevance (MRMR)* algorithm [33] is utilised. MRMR identifies the most informative features for a given task by considering both their relevance to the target variable and their redundancy with respect to each other. It evaluates the mutual information between features and the target, ranking them by relevance while also measuring the redundancy between features. The algorithm then selects features that achieve the right balance between relevance and redundancy, resulting in a subset of features that can improve model performance with reduced features.

Table 1. Details of statistical features used in this study. The feature vector F comprises all ten features, with h as the preprocessed signal (ΔHBO_2 or ΔHHB), h' as the derivative of h, and $\bar{h'}$ as the mean of h'. h_{peak}, h_{rms}, and h_{am} denote the peak, root mean square, and absolute mean of the input signal h, respectively, while $var(.)$ represents the variance.

Features	Definitions
Log Energy	$F_1 = \sum_{i=1}^{n} log(h_i^2)$
Crest Factor	$F_2 = \frac{h_{peak}}{h_{rms}}$
Shape Factor	$F_3 = \frac{h_{rms}}{h_{am}}$
Impulse Factor	$F_4 = \frac{h_{peak}}{h_{am}}$
Margin Factor	$F_5 = \frac{h_{peak}}{{h_{am}}^2}$
Mobility	$F_6 = \sqrt{\frac{var(h')}{var(h)}}$
Complexity	$F_7 = \frac{F_6(h')}{F_6(h)}$
Mean Absolute Deviation of First Derivative	$F_8 = \frac{1}{n}\sum_{i=1}^{n} \lvert h'_i - \bar{h'} \rvert$
Range	$F_9 = \max(h) - \min(h)$
Variation in First Derivative	$F_{10} = \sqrt{\frac{1}{n}\sum_{i=1}^{n} (h'_i - \bar{h'})^2}$

2.7. Classification

In the context of pain level assessment, the classification focus was on distinguishing between various pain classes: Baseline (B), Low Pain (LP), and High Pain (HP). To achieve this, we employed a reduced feature set consisting of statistical features extracted from both ΔHBO_2 and ΔHHB signals. We utilised well-known classifiers such as Discriminant (Disc) [34], K-Nearest Neighbour (KNN) [35], and Support Vector Machine (SVM) [36] to identify pain levels using the feature set. We employed parameter optimisation, carefully tuning the classifiers using a Bayesian approach [37]. This data-driven decision-making process is supported by an acquisition function known as 'expected improvement per second plus', which underwent 50 iterations. We identified the hyperparameters for each classification algorithm that minimised the 10-fold cross-validation loss across the entire dataset [38].

The classification performance was evaluated using a leave-one-subject-out cross-validation (LOSOCV) approach [39]. In LOSOCV, the model's effectiveness is assessed by withholding one individual's data from the dataset for testing, while the data from the remaining participants undergoes 10-fold cross validation. This process is repeated iteratively for each subject in the dataset, ensuring that each subject serves as the test set exactly once. The performance metrics consisting of accuracy (Acc), sensitivity (Sen), specificity (Spec), and F1 score (F1) and obtained in each iteration were averaged to provide a comprehensive assessment of the model's overall performance. Additionally, we systematically tested the identification of the best-performing model with varying numbers of features based on their MRMR rank. Thus, the combination of feature engineering, hyperparameter optimisation, and classification algorithms proves to be a powerful toolkit for decoding pain levels based on fNIRS signals.

2.8. Statistical Analysis

The obtained features were also analysed using statistical analysis to identify significant differences in the obtained features across the different experimental conditions for both ΔHBO_2 and ΔHHB independently. This information will help validate our hypothesis, indicating that the obtained features encompass pain-related data from the experimental conditions. First, the data were examined for normality and homogeneity using the Kolmogorov–Smirnov tests. Focusing on the ten extracted features from ΔHBO_2 and

ΔHHB measurements for the classification of the pain level, differences were analysed using Analysis of Variance (ANOVA). A post hoc Bonferroni test was carried out for multiple comparisons. The significant level was set to $p < 0.05$. All statistical analyses were performed using SPSS version 29.

3. Results

In this section, the outcomes of the proposed multi-class fNIRS-based pain assessment system are presented. The results of the system are demonstrated via the independent utilisation of ΔHBO_2 and ΔHHB signals, along with employing combined haemoglobin measures. Ten features are extracted from each measure and are passed to the three classifiers (Disc, KNN, and SVM). In the case of $\Delta HBO_2 + \Delta HHB$, the features from each measure are fused before the classification stage, resulting in a total of 20 features in this case. The selection of classifiers for each experiment was made following extensive hyperparameter tuning, as detailed in Table 2.

Activation levels of fNIRS using both ΔHBO_2 and ΔHHB measurements for different experimental conditions are presented in Figure 8. As shown, the highest activation in the prefrontal cortex for ΔHBO_2 (first row) is recorded for HA (High Arm pain), while LH (Low Arm pain) exhibits the lowest concentration level compared to other conditions and with a very similar activation level to the baseline. Similar to ΔHBO_2, the most elevated activation in ΔHHB measures is observed in the HA condition. However, other conditions do not exhibit a significant increase.

Table 2. Optimised hyperparameters for different classification algorithms via Bayesian Optimisation in the context of distinguishing between Baseline (B), Low Pain (LP), and High Pain (HP).

Model	Parameters	ΔHBO_2	ΔHHB	$\Delta HBO_2 + \Delta HHB$
Disc	Discriminant Type	Pseudo Linear	Linear	Diagonal Linear
	Gamma	7.55×10^{-4}	0.0025	0.006
	Delta	3.51×10^{-5}	2.96×10^{-5}	2.12×10^{-5}
KNN	Number of Neighbours	211	1	25
	Distance	Chebychev	Cosine	City Block
	Distance Weight	Inverse	Inverse	Equal
	Exponent	–	–	–
	Neighbour Search	KD-Tree	Exhaustive	Exhaustive
	Standardisation	Yes	Yes	Yes
SVM	Coding	One vs. All	One vs. All	One vs. One
	Box Constraint	2.1888	10.3923	980.4894
	Kernel Scale	–	–	13.2018
	Kernel Function	Polynomial	Polynomial	Gaussian
	Polynomial Order	3	3	–
	Standardise	Yes	Yes	Yes

3.1. Classification Results

The results in terms of performance metrics for each measure are presented in Table 3. For the ΔHBO_2 measure, the SVM classifier performs remarkably well as compared to that of Disc and KNN, achieving the highest accuracy of 64.67%. It exhibits outstanding sensitivity (92.85%) and specificity (97.22%), underlining its ability to effectively identify pain instances while maintaining high precision. For the ΔHHB measure, the SVM classifier again excels with the highest accuracy of 62.28%. It maintains remarkable sensitivity (92.87%) and specificity (97.07%), showcasing its effectiveness in pain assessment. The KNN classifier exhibits an accuracy of 41.83%, whereas the Disc classifier displays an accuracy of 50.94%. The combined $\Delta HBO_2 + \Delta HHB$ measure, when paired with the SVM classifier, outperforms other classification algorithms with an accuracy of 66.55%. Sensitivity (93.8%) and specificity (96.14%) remain high, highlighting the SVM's effectiveness in pain assessment. The F1 Score of 96.98% emphasises the balanced performance. On the other hand, the KNN classifier, with an accuracy of 40.19%, shows lower performance, and the Disc classifier, with an accuracy of 56.23%, exhibits moderate performance. The SVM classifier consistently achieves high accuracy, sensitivity, specificity, and F1 Score, with the $\Delta HBO_2 + \Delta HHB$ measure performing the best among all measures.

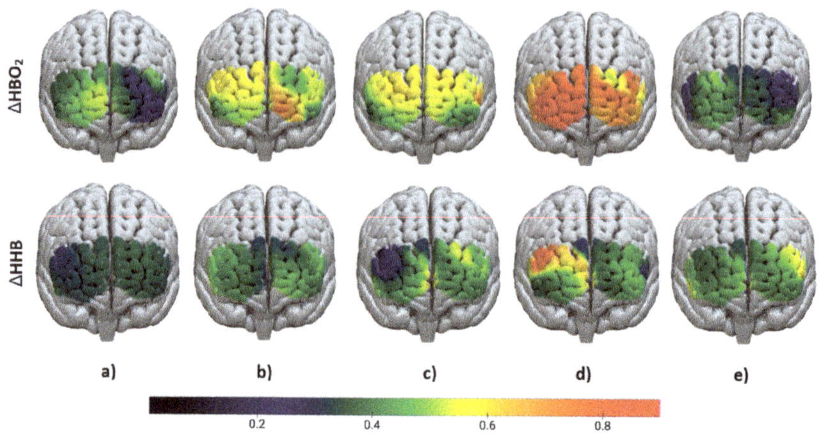

Figure 8. Haemodynamic changes shown using fNIRS for ΔHBO_2 (first row) and ΔHHB (second row) measures: (**a**) Baseline, (**b**) HH (High Hand Pain), (**c**) LH (Low Hand Pain), (**d**) HA (High Arm Pain), and (**e**) LA (Low Arm Pain). The color bar signifies the change in concentration of ΔHBO_2 and ΔHHB (Δµmol). These calculations are derived from the averages across all subjects for each respective channel.

Table 3. System performance metrics (Acc: Accuracy, Sen: Sensitivity, Spec: Specificity, and F1 Score) for different classification algorithms (Disc, KNN, and SVM) across various measures, with each measure having a different feature vector length denoted by #.

Measure	Model	#	Acc	Sen	Spec	F1 Score
ΔHBO_2	Disc	10	51.78 ± 9.94	74.78 ± 19.43	73.30 ± 11.91	85.98 ± 9.60
	KNN		41.74 ± 7.73	56.96 ± 17.03	64.81 ± 13.29	75.53 ± 7.85
	SVM		64.67 ± 5.99	92.85 ± 8.25	97.22 ± 3.65	96.67 ± 3.76
ΔHHB	Disc	10	50.94 ± 7.60	73.57 ± 12.12	75.77 ± 9.53	85.35 ± 6.42
	KNN		41.83 ± 8.34	44.36 ± 14.30	74.23 ± 9.10	73.14 ± 5.75
	SVM		62.28 ± 5.86	92.87 ± 8.24	97.07 ± 3.62	96.63 ± 3.83
$\Delta HBO_2 + \Delta HHB$	Disc	20	56.23 ± 6.84	76.32 ± 11.62	79.32 ± 10.81	87.24 ± 5.73
	KNN		40.19 ± 8.09	43.63 ± 13.77	68.06 ± 10.01	70.99 ± 6.07
	SVM		66.55 ± 7.36	93.8 ± 6.38	96.14 ± 3.04	96.98 ± 3.08

Following the acquisition of reference values using the full feature set (see Table 3), the feature set underwent a feature selection process using MRMR to minimise redundancy and enhance the discriminative power. Table 4 presents the performance metrics for each measure after applying MRMR. The results provide insights into how feature selection impacts the performance of pain assessment models. In the ΔHBO_2 measure, the feature selection process has notably influenced the performance of different classifiers. The SVM classifier, with nine selected features, achieves the highest accuracy of 65.71% with improved sensitivity (93.18%) and specificity (95.99%). The KNN classifier, with seven selected features, exhibits enhanced accuracy at 44.22%, although it still falls behind SVM.

In the ΔHHB measure, feature selection has similarly enhanced the performance of the classifiers. The SVM classifier, with nine selected features, maintains its position as the top-performing classifier with an accuracy of 63.42% along with improved sensitivity (94.44%) and specificity (97.22%). The combined $\Delta HBO_2 + \Delta HHB$ measure benefits from feature selection, particularly in the SVM classifier with 15 selected features. It achieves the highest accuracy at 68.51%, emphasising the significance of choosing both hemoglobin measures. Sensitivity (94.7%), specificity (94.29%), and the F1 Score also reflect notable improvements. The KNN classifier, with 18 selected features, shows an accuracy of 40.8%. These findings emphasise the crucial role of both measures, particularly in the combined ($\Delta HBO_2 + \Delta HHB$) measure, where the SVM classifier emerges as the optimal choice for precise and well-balanced pain assessment.

Table 4. System performance metrics (Acc: Accuracy, Sen: Sensitivity, Spec: Specificity, and F1 Score) with MRMR-based selected features for different classification algorithms (Disc, KNN, and SVM) applied to each measure, with the feature vector length denoted by #.

Measure	Model	#	Acc	Sen	Spec	F1 Score
ΔHBO_2	Disc	10	51.78 ± 9.94	74.78 ± 19.43	73.30 ± 11.91	85.98 ± 9.60
	KNN	7	44.22 ± 8.16	55.36 ± 15.38	70.22 ± 13.16	76.30 ± 7.20
	SVM	9	65.71 ± 5.97	93.18 ± 8.03	95.99 ± 4.24	96.77 ± 3.67
ΔHHB	Disc	10	50.94 ± 7.6	73.57 ± 12.12	75.77 ± 9.53	85.35 ± 6.42
	KNN	10	41.83 ± 8.34	44.36 ± 14.3	74.23 ± 9.10	73.14 ± 5.75
	SVM	9	63.42 ± 6.85	94.44 ± 8.33	97.22 ± 3.27	97.40 ± 3.84
$\Delta HBO_2 + \Delta HHB$	Disc	20	56.23 ± 6.84	76.32 ± 11.62	79.32 ± 10.81	87.24 ± 5.73
	KNN	18	40.8 ± 7.26	44.58 ± 15.27	68.83 ± 9.34	71.72 ± 6.34
	SVM	15	68.51 ± 9.02	94.70 ± 5.77	94.29 ± 4.92	97.33 ± 2.92

Table 5 lists the features corresponding to the optimal results for each measure. In the approach using a fusion of haemoglobin measures ($\Delta HBO_2 + \Delta HHB$), among the fifteen selected features, nine belong to ΔHBO_2, highlighting its greater contribution compared to the six features from ΔHHB.

Table 5. List of selected features for each measure, with # indicating the number of features.

Measure	#	Selected Features
ΔHBO_2	9	Mobility, Complexity, Range, Shape Factor, Variation in First Derivative, Impulse Factor, Mean Absolute Deviation of First Derivative, Log Energy, Crest Factor.
ΔHHB	9	Crest Factor, Complexity, Shape Factor, Mobility, Range, Variation in First Derivative, Log Energy, Mean Absolute Deviation of First Derivative, Margin Factor.
$\Delta HBO_2 + \Delta HHB$	15	ΔHBO_2: Mobility, Complexity, Range, Shape Factor, Variation in First Derivative, Impulse Factor, Mean Absolute Deviation of First Derivative, Log Energy, Crest Factor. ΔHHB: Crest Factor, Complexity, Shape Factor, Mobility, Range, Variation in First Derivative.

In pain assessment, class-wise performance is also crucial because it enables the accurate identification of different pain levels, helping clinicians in making treatments based on individual pain experiences and needs. Analysing the class-wise performance of each measure, as depicted in Figure 9, highlights the superior effectiveness of the SVM classifier, particularly in accurately classifying instances of Baseline (B), Low Pain (LP), and High Pain (HP) compared to other classification methods such as Disc and KNN. Notably, the ΔHBO_2 measure demonstrates its strength in achieving higher classification accuracy for High Pain (HP) instances, while the ΔHHB measure excels in classifying Low Pain (LP) cases. However, it is important to emphasise the significance of identifying the absence of pain (B) in pain assessment, and here, the ΔHHB measure proves better at predicting pain-free observations compared to the ΔHBO_2 measure. The fusion of both ΔHBO_2 and ΔHHB effectively integrates this information, yielding improved results for both LP and HP classes. In summary, the fusion of both fNIRS measures enhances class-wise accuracies in pain assessment, contributing to a more comprehensive and precise pain perception evaluation.

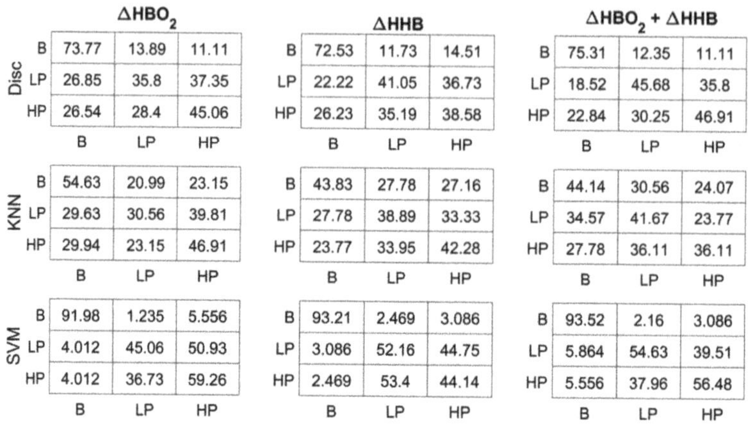

Figure 9. Class-wise accuracy (%) assessment of different measures using Disc; KNN; and SVM classifiers using confusion charts.

3.2. Statistical Analyses

The results regarding the comparison of the statistically significant ΔHBO_2 feature in the different experimental conditions are provided in Table 6. Among ten different features for ΔHBO_2 measurement, Log Energy, Crest Factor, Shape Factor, and Range exhibit significant differences compared to other features in distinguishing between experiment conditions, as indicated by their respective p-values ($F_{(2,972)} = 3.078$, $p = 0.046$, $F_{(2,972)} = 3.264$, $p = 0.039$, $F_{(2,972)} = 3.466$, $p = 0.032$, $F_{(2,972)} = 10.179$, $p < 0.001$, respectively). For ΔHHB measures, three features, Log Energy, Margin Factor, and Range, showed significant differences in identifying pain levels as compared to other features ($F_{(2,972)} = 3.127$, $p = 0.044$, $F_{(2,972)} = 4.134$, $p = 0.016$, $F_{(2,972)} = 4.558$, $p = 0.011$, respectively). The results of the post hoc test for the comparison of pain levels for statistically significant features of the ΔHHB measure have been provided in Table 7.

Table 6. Post Hoc Test Results for Different Levels of Pain in Various Features of ΔHBO_2 (Only comparisons with significant ($p \leq 0.05$) values are reported.)

Feature	Group One	Group Two	Mean Diff.	Std. Error	Sig.	Lower Bound	Upper Bound
Log Energy	No Pain	Low Pain	73.67	41.110	0.073	−7.00	154.33
		High Pain	96.99 *	41.110	0.018	16.33	177.66
	Low Pain	No Pain	−73.67	41.110	0.073	−154.33	7.00
		High Pain	23.33	40.966	0.569	−57.06	103.71
	High Pain	No Pain	−96.99 *	41.110	0.018	−177.66	−16.33
		Low Pain	−23.33	40.966	0.569	−103.71	57.06
Crest factor	No Pain	Low Pain	−0.02	0.039	0.685	−0.09	0.06
		High Pain	0.078 *	0.039	0.044	0.00	0.15
	Low Pain	No Pain	0.02	0.039	0.685	−0.06	0.09
		High Pain	0.094 *	0.039	0.015	0.02	0.17
	High Pain	No Pain	−0.078 *	0.039	0.044	−0.15	0.00
		Low Pain	−0.094 *	0.039	0.015	−0.17	-0.02
Shape factor	No Pain	Low Pain	−0.017 *	0.006	0.008	−0.03	0.00
		High Pain	−0.01	0.006	0.067	−0.02	0.00
	Low Pain	No Pain	0.017 *	0.006	0.008	0.00	0.03
		High Pain	0.01	0.006	0.402	−0.01	0.02
	High Pain	No Pain	0.01	0.006	0.067	0.00	0.02
		Low Pain	−0.01	0.006	0.402	−0.02	0.01

Table 6. *Cont.*

Feature	Group One	Group Two	Mean Diff.	Std. Error	Sig.	Lower Bound	Upper Bound
Impulse factor	No Pain	Low Pain	−0.05	0.055	0.386	−0.16	0.06
		High Pain	0.08	0.055	0.167	−0.03	0.19
	Low Pain	No Pain	0.05	0.055	0.386	−0.06	0.16
		High Pain	0.125 *	0.055	0.024	0.02	0.23
	High Pain	No Pain	−0.08	0.055	0.167	−0.19	0.03
		Low Pain	−0.125 *	0.055	0.024	−0.23	−0.02
Range	No Pain	Low Pain	−0.165 *	0.037	$p \leq 0.001$	−0.24	−0.09
		High Pain	−0.129 *	0.037	0.001	−0.20	−0.06
	Low Pain	No Pain	0.165 *	0.037	$p \leq 0.001$	0.09	0.24
		High Pain	0.04	0.037	0.33	−0.04	0.11
	High Pain	No Pain	0.129 *	0.037	0.001	0.06	0.20
		Low Pain	−0.04	0.037	0.33	−0.11	0.04

*: the mean difference is significant at 0.05 level.

Table 7. Post Hoc Test Results for Different Levels of Pain in Various Features of ΔHHB (Only comparisons with significant p-values are reported).

Feature	Group One	Group Two	Mean diff.	Std. Error	Sig.	Lower Bound	Upper Bound
Log Energy	No Pain	Low Pain	104.153 *	49.629	0.036	6.770	201.535
		High Pain	110.774 *	49.629	0.026	13.391	208.156
	Low Pain	No Pain	−104.153 *	49.629	0.036	−201.535	−6.770
		High Pain	6.62	49.4554	0.894	−90.420	103.662
	High Pain	No Pain	−110.774 *	49.629	0.026	−208.156	−13.391
		Low Pain	−6.62	49.455	0.894	−103.662	90.420
Margin Factor	No Pain	Low Pain	1.629 *	0.572	0.004	0.506	2.752
		High Pain	0.621	0.572	0.277	−0.501	1.744
	Low Pain	No Pain	−1.629 *	0.572	0.004	−2.752	−0.506
		High Pain	−1.007	0.570	0.078	−2.126	0.111
	High Pain	No Pain	−0.621	0.572	0.277	−1.744	0.501
		Low Pain	1.007	0.570	0.078	−0.111	2.126
Range	No Pain	Low Pain	−0.106 *	0.04	0.00	−0.18	−0.04
		High Pain	−0.072 *	0.04	0.04	−0.14	0.00
	Low Pain	No Pain	0.106 *	0.04	0.00	0.04	0.18
		High Pain	0.03	0.04	0.34	−0.04	0.10
	High Pain	No Pain	0.072 *	0.04	0.04	0.00	0.14
		Low Pain	−0.03	0.04	0.34	−0.10	0.04

*: the mean difference is significant at 0.05 level.

4. Discussions

To the best of the authors' knowledge, this is the first study that deals with the objective assessment of pain via fNIRS within a comprehensive exploration of ΔHBO_2 and ΔHHB measures. The findings reveal an association between pain intensities and distinct statistical patterns in haemoglobin concentrations. Considering the overall system accuracy, the ΔHBO_2 measure demonstrated better performance than the ΔHHB measure in the multiclass scenario used in this study. However, when examining accuracies for specific classes, ΔHBO_2 excels in identifying High Pain signals, while ΔHHB demonstrates better accuracy for Low Pain observations. Upon a comparison of both fNIRS measures, it can be concluded that the fusion of ΔHBO_2 and ΔHHB measures at the feature level emerges as an effective method for the categorisation of the three pain intensities in our experimental conditions.

Based on the classification results, it can be deduced that the SVM classification algorithm is most effective when used with the selected statistical features across all the measures in pain assessment. Both of the fNIRS measures are considered to be reliable in evaluating pain, with ΔHBO_2 demonstrating slightly higher accuracy than the ΔHHB measure when used independently. However, the most optimal results are obtained when combining both ΔHBO_2 and ΔHHB, suggesting that a combination of these two measures offers the best performance for pain assessment in our experimental conditions. While ΔHBO_2 provides insights into the oxygenated haemoglobin concentration, which can indicate changes in blood flow and tissue activity, ΔHHB reveals deoxygenated haemoglobin levels reflecting variations in tissue oxygen consumption. By integrating these two measures, a more holistic understanding of the physiological responses to pain is achieved. This combined approach allows for a more robust assessment as it captures both the supply and demand aspects of oxygen delivery, thus enhancing the ability to detect and interpret changes in pain perception.

Existing studies on pain assessment using neuroimaging methods have primarily focused on binary classifications, mainly distinguishing between pain and no pain. However, the development of approaches capable of distinguishing various signatures of pain has been neglected so far. This limitation is significant given the diverse origins (e.g., peripheral, emotional, and phantom pain), varying intensities, and durations of pain experienced in the human body. Different types of pain are carried to the central nervous system by different sensory receptors, responding to various stimuli associated with pain, such as temperature, chemical, or pressure [15]. Hence, there is a need for machine learning models that can effectively differentiate between multiple pain signatures at varying intensities, offering greater relevance for real-world scenarios. In contrast, our study addresses this gap by focusing on multilevel pain classification, considering pain originating from different locations of the body, specifically the hand and arm. This is particularly important for patients who are unable to communicate verbally, such as elderly people recovering from a stroke or with advanced dementia, and when the source of pain is not readily apparent.

In examining activation levels across different pain conditions, our focus on ΔHBO_2 and ΔHHB measures provides valuable insights into the neural responses associated with pain perception. As depicted in Figure 8, the most pronounced increase in ΔHBO_2 levels occurs in response to High Arm (HA) pain, emphasising the sensitivity of this measure to high pain intensities. Similarly, increased activation is evident in High Hand (HH) pain compared to Low Hand (LH) and Low Arm (LA) pains. These outcomes are consistent with prior research, reinforcing the notion that elevated pain levels are associated with a more pronounced neural response in fNIRS studies investigating ΔHBO_2 [22,40]. The significant increase in activation during various conditions compared to the baseline, except for Low Arm (LA) pain, is noteworthy. This could, in part, account for the lower accuracy observed in identifying the low pain (LP) class using ΔHBO_2 across all three classifiers. Activation levels in ΔHHB data exhibited minimal fluctuations across diverse conditions, with the exception of the HA condition, where the highest activation, akin to ΔHBO_2, was observed. These findings highlight the superior efficacy of ΔHBO_2 as a more reliable measure for pain assessment compared to ΔHHB, when used independently. However, when combined in a feature fusion scheme, they collectively obtained better accuracy than when used independently.

While our proposed system demonstrated better performance in identifying different pain levels, it presents some limitations. First, the channel selection algorithm employed in our study served the purpose of rejecting channels saturated with artifacts and noise. However, it may automatically discard channels containing valuable pain-related information. To address this, a more advanced preprocessing algorithm should be considered, capable of mitigating noise in unreliable channels without outright rejection. This would ensure that potentially relevant information is retained in the dataset for more comprehensive pain assessment. Second, it is evident in our preprocessing stage, where we opted to average out all channels to generate a single time series vector. This approach, while simplifying the data, has the drawback of suppressing information inherent in individual channels. In our future work, we will conduct analysis by defining specific regions of interest based on functional areas of the brain, which can provide insights into the localised functions and responses to pain associated with different brain regions. Finally, our investigation into fNIRS data primarily focused on the time domain, emphasising the extraction and assessment of simple statistical features. However, by exclusively focusing on the time domain, we may have overlooked valuable information present in other domains. To broaden the scope of the analysis, we should consider additional domains, such as frequency or cepstral domains, throughout the stages of preprocessing, feature extraction, and evaluation.

5. Conclusions

In this study, we introduced a multilevel pain intensity assessment using fNIRS data, compiling a novel dataset from healthy individuals experiencing varying induced pain levels in distinct body locations. Analysing ΔHBO_2 and ΔHHB measures, we found that ΔHBO_2 outperformed ΔHHB overall but excelled in predicting high and low pain classes, respectively. Combining both measures significantly improved the performance, demonstrating the potential of fNIRS for multilevel pain assessment. The system achieved 68.51% $\pm$ 9.02% accuracy, 94.7% $\pm$ 5.77% sensitivity, and 94.29% $\pm$ 4.92% specificity in diagnosing no pain, low pain, and high pain observations, respectively. Future research aims to explore integrating fNIRS with other sensor modalities, analysing pain-related information in different fNIRS domains, and effectively pinpointing the site of pain.

Author Contributions: Conceptualisation, R.F.-R.; methodology, M.U.K., M.S. and R.F.-R.; formal analysis, M.U.K. and M.S.; investigation, M.U.K., M.S. and R.F.-R.; data interpretation: M.U.K., M.S. and R.F.-R.; resources, R.F.-R. and R.G.; data collection, N.H., C.J. and R.F.-R.; writing—original draft preparation, M.U.K. and M.S.; writing—review and editing, M.G., G.C., R.G. and R.F.-R.; supervision, R.F.-R. and R.G.; project administration, R.F.-R.; funding acquisition, R.F.-R. All authors have read and agreed to the published version of the manuscript.

Funding: This research received no external funding.

Institutional Review Board Statement: The experimental procedures involving human subjects described in this paper were approved by the University of Canberra's Human Ethics Committee (Number: 11837).

Informed Consent Statement: Informed consent was obtained from all subjects involved in the study.

Data Availability Statement: The data that support the findings of this study are available from R.F.R., upon reasonable request.

Conflicts of Interest: The authors declare that they have no known competing financial interests or personal relationships that could have appeared to influence the work reported in this paper.

References

1. Rajesh, M.; Muthu, J.S.; Suseela, G. iPainRelief-A pain assessment and management app for a smart phone implementing sensors and soft computing tools. In Proceedings of the 2013 International Conference on Information Communication and Embedded Systems (ICICES), Chennai, India, 21–22 February 2013; pp. 434–441.
2. Dueñas, M.; Ojeda, B.; Salazar, A.; Mico, J.A.; Failde, I. A review of chronic pain impact on patients, their social environment and the health care system. *J. Pain Res.* **2016**, *9*, 457–467. [CrossRef]
3. Blyth, F.M.; March, L.M.; Brnabic, A.J.; Jorm, L.R.; Williamson, M.; Cousins, M.J. Chronic pain in Australia: A prevalence study. *Pain* **2001**, *89*, 127–134. [CrossRef] [PubMed]
4. Painaustralia. *The Cost of Pain in Australia*; Painaustralia: Deakin, Australia, 2019.
5. van Leeuwen, M.T.; Blyth, F.M.; March, L.M.; Nicholas, M.K.; Cousins, M.J. Chronic pain and reduced work effectiveness: The hidden cost to Australian employers. *Eur. J. Pain* **2006**, *10*, 161–166. [CrossRef] [PubMed]
6. Ferreira, M.L.; de Luca, K.; Haile, L.M.; Steinmetz, J.D.; Culbreth, G.T.; Cross, M.; Kopec, J.A.; Ferreira, P.H.; Blyth, F.M.; Buchbinder, R.; et al. Global, regional, and national burden of low back pain, 1990–2020. its attributable risk factors, and projections to 2050: A systematic analysis of the Global Burden of Disease Study 2021. *Lancet Rheumatol.* **2023**, *5*, e316–e329. [CrossRef] [PubMed]
7. Naranjo-Hernández, D.; Reina-Tosina, J.; Roa, L.M. Sensor technologies to manage the physiological traits of chronic pain: A review. *Sensors* **2020**, *20*, 365. [CrossRef] [PubMed]
8. Cowen, R.; Stasiowska, M.K.; Laycock, H.; Bantel, C. Assessing pain objectively: The use of physiological markers. *Anaesthesia* **2015**, *70*, 828–847. [CrossRef] [PubMed]
9. Breivik, H.; Borchgrevink, P.C.; Allen, S.M.; Rosseland, L.A.; Romundstad, L.; Breivik Hals, E.; Kvarstein, G.; Stubhaug, A. Assessment of pain. *Br. J. Anaesth.* **2008**, *101*, 17–24. [CrossRef]
10. Turk, D.C.; Melzack, R. *Handbook of Pain Assessment*; Guilford Press: New York, NY, USA, 2011.
11. Luo, J.; Zhu, H.Q.; Gou, B.; Wang, X.Q. Neuroimaging assessment of pain. *Neurotherapeutics* **2022**, *19*, 1467–1488. [CrossRef] [PubMed]
12. Wager, T.D.; Atlas, L.Y.; Lindquist, M.A.; Roy, M.; Woo, C.W.; Kross, E. An fMRI-based neurologic signature of physical pain. *N. Engl. J. Med.* **2013**, *368*, 1388–1397. [CrossRef]

13. Hu, X.S.; Nascimento, T.D.; Bender, M.C.; Hall, T.; Petty, S.; O'Malley, S.; Ellwood, R.P.; Kaciroti, N.; Maslowski, E.; DaSilva, A.F. Feasibility of a real-time clinical augmented reality and artificial intelligence framework for pain detection and localization from the brain. *J. Med. Internet Res.* **2019**, *21*, e13594. [CrossRef]
14. Fernandez Rojas, R.; Huang, X.; Ou, K.L. Toward a functional near-infrared spectroscopy-based monitoring of pain assessment for nonverbal patients. *J. Biomed. Opt.* **2017**, *22*, 1–12. [PubMed]
15. Fernandez Rojas, R.; Brown, N.; Waddington, G.; Goecke, R. A systematic review of neurophysiological sensing for the assessment of acute pain. *NPJ Digit. Med.* **2023**, *6*, 76. [CrossRef]
16. Yücel, M.A.; Aasted, C.M.; Petkov, M.P.; Borsook, D.; Boas, D.A.; Becerra, L. Specificity of hemodynamic brain responses to painful stimuli: A functional near-infrared spectroscopy study. *Sci. Rep.* **2015**, *5*, 9469. [CrossRef]
17. Lemm, S.; Blankertz, B.; Dickhaus, T.; Müller, K.R. Introduction to machine learning for brain imaging. *Neuroimage* **2011**, *56*, 387–399. [CrossRef]
18. Joseph, G.; Mcculloch, C.; Nevitt, M.; Link, T.; Sohn, J. Machine learning for predicting knee osteoarthritis progression over 8 years using combined MR imaging features, demographics, and clinical factors: Data from the osteoarthritis initiative. *Osteoarthr. Cartil.* **2021**, *29*, S45–S46. [CrossRef]
19. Lötsch, J.; Ultsch, A. Machine learning in pain research. *Pain* **2018**, *159*, 623. [CrossRef]
20. Brown, J.E.; Chatterjee, N.; Younger, J.; Mackey, S. Towards a physiology-based measure of pain: Patterns of human brain activity distinguish painful from non-painful thermal stimulation. *PloS ONE* **2011**, *6*, e24124. [CrossRef] [PubMed]
21. Gram, M.; Graversen, C.; Olesen, A.E.; Drewes, A. Machine learning on encephalographic activity may predict opioid analgesia. *Eur. J. Pain* **2015**, *19*, 1552–1561. [CrossRef]
22. Pourshoghi, A.; Zakeri, I.; Pourrezaei, K. Application of functional data analysis in classification and clustering of functional near-infrared spectroscopy signal in response to noxious stimuli. *J. Biomed. Opt.* **2016**, *21*, 101411. [CrossRef] [PubMed]
23. Fernandez Rojas, R.; Huang, X.; Ou, K.L. A machine learning approach for the identification of a biomarker of human pain using fNIRS. *Sci. Rep.* **2019**, *9*, 5645. [CrossRef]
24. Zeng, X.; Tang, W.; Yang, J.; Lin, X.; Du, M.; Chen, X.; Yuan, Z.; Zhang, Z.; Chen, Z. Diagnosis of Chronic Musculoskeletal Pain by Using Functional Near-Infrared Spectroscopy and Machine Learning. *Bioengineering* **2023**, *10*, 669. [CrossRef] [PubMed]
25. Sousani, M.; Rojas, R.F.; Preston, E.; Ghahramani, M. Towards a Multi-Modal Brain-Body Assessment in Parkinson's Disease: A Systematic Review in fNIRS (February 2023). *IEEE J. Biomed. Health Inform.* **2023**, *27*, 4840–4853. [CrossRef]
26. Udina, C.; Avtzi, S.; Durduran, T.; Holtzer, R.; Rosso, A.L.; Castellano-Tejedor, C.; Perez, L.M.; Soto-Bagaria, L.; Inzitari, M. Functional near-infrared spectroscopy to study cerebral hemodynamics in older adults during cognitive and motor tasks: A review. *Front. Aging Neurosci.* **2020**, *11*, 367. [CrossRef] [PubMed]
27. Ho, T.K.K.; Kim, M.; Jeon, Y.; Kim, B.C.; Kim, J.G.; Lee, K.H.; Song, J.I.; Gwak, J. Deep learning-based multilevel classification of Alzheimer's disease using non-invasive functional near-infrared spectroscopy. *Front. Aging Neurosci.* **2022**, *14*, 810125. [CrossRef] [PubMed]
28. Pfurtscheller, G.; Bauernfeind, G.; Wriessnegger, S.C.; Neuper, C. Focal frontal (de) oxyhemoglobin responses during simple arithmetic. *Int. J. Psychophysiol.* **2010**, *76*, 186–192. [CrossRef] [PubMed]
29. Karunakaran, K.D.; Peng, K.; Berry, D.; Green, S.; Labadie, R.; Kussman, B.; Borsook, D. NIRS measures in pain and analgesia: Fundamentals, features, and function. *Neurosci. Biobehav. Rev.* **2021**, *120*, 335–353. [CrossRef]
30. Naseer, N.; Hong, K.S. fNIRS-based brain-computer interfaces: A review. *Front. Hum. Neurosci.* **2015**, *9*, 3. [CrossRef]
31. Khan, M.U.; Aziz, S.; Hirachan, N.; Joseph, C.; Li, J.; Fernandez-Rojas, R. Experimental Exploration of Multilevel Human Pain Assessment Using Blood Volume Pulse (BVP) Signals. *Sensors* **2023**, *23*, 3980. [CrossRef]
32. Usman, A.; Choudhry, M.A. A precision detection technique for power disturbance in electrical system. *Electr. Eng.* **2022**, *104*, 781–796. [CrossRef]
33. Attivissimo, F.; D'Alessandro, V.I.; De Palma, L.; Lanzolla, A.M.L.; Di Nisio, A. Non-Invasive Blood Pressure Sensing via Machine Learning. *Sensors* **2023**, *23*, 8342. [CrossRef]
34. Li, T.; Zhu, S.; Ogihara, M. Using discriminant analysis for multi-class classification: An experimental investigation. *Knowl. Inf. Syst.* **2006**, *10*, 453–472. [CrossRef]
35. Feng, C.; Zhao, B.; Zhou, X.; Ding, X.; Shan, Z. An Enhanced Quantum K-Nearest Neighbor Classification Algorithm Based on Polar Distance. *Entropy* **2023**, *25*, 127. [CrossRef] [PubMed]
36. Hearst, M.A.; Dumais, S.T.; Osuna, E.; Platt, J.; Scholkopf, B. Support vector machines. *IEEE Intell. Syst. Their Appl.* **1998**, *13*, 18–28. [CrossRef]
37. Zhang, C.; Yang, H.; Fan, C.C.; Chen, S.; Fan, C.; Hou, Z.G.; Chen, J.; Peng, L.; Xiang, K.; Wu, Y.; et al. Comparing Multi-Dimensional fNIRS Features Using Bayesian Optimization-Based Neural Networks for Mild Cognitive Impairment (MCI) Detection. *IEEE Trans. Neural Syst. Rehabil. Eng.* **2023**, *31*, 1019–1029. [CrossRef]
38. Elsayad, A.M.; Nassef, A.M.; Al-Dhaifallah, M. Bayesian optimization of multiclass SVM for efficient diagnosis of erythemato-squamous diseases. *Biomed. Signal Process. Control.* **2022**, *71*, 103223. [CrossRef]

39. Pauli, M.P.; Pohl, C.; Golz, M. Balanced leave-one-subject-out cross-validation for microsleep classification. *Curr. Dir. Biomed. Eng.* **2021**, *7*, 147–150. [CrossRef]
40. Fernandez Rojas, R.; Liao, M.; Romero, J.; Huang, X.; Ou, K.L. Cortical network response to acupuncture and the effect of the hegu point: An fNIRS study. *Sensors* **2019**, *19*, 394. [CrossRef]

sensors

MDPI

Article

Design Decisions for Wearable EEG to Detect Motor Imagery Movements

Ana Carretero * and Alvaro Araujo

B105 Electronic Systems Lab, ETSI de Telecomunicación, Universidad Politécnica de Madrid, 28040 Madrid, Spain; araujo@b105.upm.es

* Correspondence: anacp@b105.upm.es; Tel.: +34-910-67-22-42 (ext. 72242)

Abstract: The objective of this study was to make informed decisions regarding the design of wearable electroencephalography (wearable EEG) for the detection of motor imagery movements based on testing the critical features for the development of wearable EEG. Three datasets were utilized to determine the optimal acquisition frequency. The brain zones implicated in motor imagery movement were analyzed, with the aim of improving wearable-EEG comfort and portability. Two detection algorithms with different configurations were implemented. The detection output was classified using a tool with various classifiers. The results were categorized into three groups to discern differences between general hand movements and no movement; specific movements and no movement; and specific movements and other specific movements (between five different finger movements and no movement). Testing was conducted on the sampling frequencies, trials, number of electrodes, algorithms, and their parameters. The preferred algorithm was determined to be the FastICACorr algorithm with 20 components. The optimal sampling frequency is 1 kHz to avoid adding excessive noise and to ensure efficient handling. Twenty trials are deemed sufficient for training, and the number of electrodes will range from one to three, depending on the wearable EEG's ability to handle the algorithm parameters with good performance.

Keywords: design decisions; detection; EEG; imagery motion patterns; wearable

1. Introduction

Electroencephalography (EEG) is a method that records brain signals, and it is used in many configurations and applications, such as epilepsy, sleep disorders, brain–computer interfaces (BCIs), and mental states.

The way to obtain these signals is through an electrode cap that covers the head and allows the attachment of the electrodes to the scalp. There are currently many electrode caps that are available on the market [1], the main limitations of which are their size and weight, as these characteristics make them uncomfortable for users and, in some cases, unportable.

These devices do not allow for complete monitoring throughout the day; thus, clinical treatments are restricted, and the integration with actuators in a closed-loop system is not feasible, which is why they are regularly used in hospitals and research labs for specific tasks with limited amounts of time.

In order to avoid these limitations, wearable EEGs (wEEGs) have emerged as a solution to lead to new treatments, diagnostics, and applications, and they are expected to improve the patient's quality of life and comfort with their treatment, as well as to make it possible to register real behaviors over the course of a day. If a small device only covers the brain zones that a doctor needs to monitor, then it could be worn for 24 h, seven days a week. In the future, it is expected that this small device will become like a wearable band-aid for recording signals that are useful for the diagnosis and treatment of certain diseases. Moreover, wearable band-aids can be used not only in the hospital but also at home, during work, or even during exercise.

In recent years, in addition to electrode caps, different wearable EEGs have appeared on the market [1–4] for various applications. The BitBrain Diadem has been used to test the user experience [5]. The B-Alert X-10 has been used to test the motivation of users when they perform specific exercises [6]. The Emotiv Insight has been used to control some external smart home devices [7] and the cursor of an application [8].

There are others, such as the emotive EPOC X, Muse S Headband, and Neurosity Crown, that have been tested for emotion recognition [9] and that use emotions (in the case of emotive EPOC X) to improve the virtual reality experience [10]. Ganglion and Cyton + Daisy are the devices that are the most similar to the conception of wearable EEG. They are used to study, among other things, motor imagery movements [11,12]. The issue with these devices is that they function as EEG interfaces, meaning that they do not process data to make decisions about the movement. It may be possible for the Ganglion to make some decisions because it has an integrated microcontroller, but not complex decisions. Data portability is limited to having a connection with a Bluetooth dongle.

Other wearable-EEG initiatives using minimal electrodes are focused on stress monitoring while driving to obtain real-execution-time responses without the need for calibration [13,14]; additionally, emotional recognition has demonstrated good performance [15]. These studies offer valuable contexts for applied processing techniques.

In line with these EEG interfaces, the most recently developed wearable EEG is the Galea, which can measure the physiological response of the user when they are experiencing virtual content to obtain data from the parasympathetic and sympathetic nervous systems using more electrodes on the face around the eyes [16].

To summarize all of the technology presented, there are two recognized sides: the open-source option, led by the OpenBCI company, and the private sphere, led by BitBrain. Both reduce the number of device channels and select the ones related to the final configuration desired by the user.

The four key aspects that wearable EEG must comply with are comfort, autonomy, portability, and wireless operation, which are determined, first, according to the wearable-EEG size, with the number of electrodes being important for the wearer. Considering that this device is positioned on the head, if it is large, heavy, or has wires, it is useless because of the user's discomfort, which is why, in this work, we did not consider caps, as they are uncomfortable and cover the head the whole time. The user could feel awkward wearing the device outside the hospital. Moreover, they are tedious and difficult to mount on and unmount from the head, not only because of their size but also because of the gels needed to obtain good contact with the scalp. The amount of gel used is directly related to the number of electrodes; thus, reducing the number of electrodes reduces the amount of gel in contact with the head and, obviously, the size of the wearable EEG. Consequently, working with wearable EEGs as small as can be created will reduce these inconveniences.

In addition, there are computational and memory limitations because of power consumption, which means that the sampling frequency and processing algorithm need to be carefully selected. These last two aspects involve a trade-off between the size of the wearable EEG device and its computational capacity. The balance in this trade-off is possible because many electrodes are not needed to obtain a good performance; however, the computational load has to be strong enough to support complex algorithms [17].

In this work, we focused on the BCIs for motor imagery movements because this type of signal can benefit not only individuals who are mobile but also those who are not. This is useful for people who do not have a continuous neural connection between the brain and one or more limbs (due to spinal cord injury). After detection, the device knows the movement that the person wants to execute; then, this information can be delivered to the specific muscles and nerves required to initiate the intended movement as it is thought in the brain.

Consequently, the objective of this work is to provide clarity regarding the decisions about the number of electrodes required for the wearable EEG, considering the developed algorithms and their complexities to address the computation problems. The ultimate goal is to design an optimal motor imagery acquisition device that strikes a balance between the obtained information and usability.

Because this information is sensitive, in the future, ethics and data protection will be carefully considered to ensure the security of users [18].

2. Materials and Methods

2.1. Datasets and Sampling Frequencies

We looked for datasets according to the following criteria: more than one record per subject; more than one motor imagery movement; and data recorded with different sampling frequencies.

Each dataset works with its own frequency. For this reason, in this work, three datasets were selected. We tested different user-record sampling frequencies because this limited the signal bandwidth and the number of samples to be processed, with higher frequencies allowing for more samples per subject record. Hence, more information could be extracted, and the results are more reliable. However, conversely, higher frequencies require more data-processing resources, and, furthermore, having more information does not necessarily produce better results.

These datasets will be explained from low to high sampling frequencies, which were the main reason we selected them.

2.1.1. PhysioNet Dataset (PDS)

The PhysioNet database [19] contains recorded images of 109 subjects opening, closing, and relaxing their left and right fists while wearing a 64-electrode cap. There are three records of the same task for each subject.

The sampling frequency is 160 Hz, which allows for the study of the influence of the use of a low frequency to record the data, which implies a bandwidth of 80 Hz. The duration is approximately two minutes per record, which means that this dataset has 160 samples per second.

2.1.2. Nature Dataset (NDS)

The Nature database contains the recorded movements of the five right-hand fingers of 13 subjects with a sampling frequency of 1 kHz, which results in more data than in the previous database, and 22 input channels (electrodes) [20].

In this case, the bandwidth is not 500 Hz but 100 Hz because a 0.53–100 Hz bandpass filter was applied to these recordings. The data between the frequencies of the filter are still sampled at 1 kHz, which implies more data than in the previous database. Specifically, there are 1000 samples per second.

The record duration is 55 min, and each trial has an average duration of three seconds: 1 s of movement signal and 1.5–2.5 s of relaxation. Although there is only one record, it can be split up because of its long duration.

2.1.3. Our Own Dataset (OODS)

To test higher frequencies, our own dataset was recorded using an ad hoc PCB that is able to connect two electrodes (one in a selectable position and the reference behind the ear) [21]. The recording time is two seconds, and the sampling frequency is 9.524 kHz; thus, there are 9524 samples per second.

There are four movements that were recorded four times: the opening/closing of the left/right fists. Relaxation was also recorded ten times.

2.1.4. Common Decisions in Dataset-Preprocessing Stage

Based on the assumption that every subject is different and must be treated independently [12], for the PDS and NDS, three subjects were considered, while for OODS, just one subject was considered. Moreover, 50/60 Hz artifacts were removed, and the movement signals were separated from relaxation for flexible combinations in further analyses.

2.2. Motor Imagery Brain Zone

The brain zones, according to the electrode positions, were selected using Brodmann areas [22]. These areas are a way of mapping the cortex and its distinguished functions. Because we worked under the motor imagery paradigm in this study, the brain zones implicated in the imagery movements were as follows: the primary motor cortex, which is the main source of motor activation; the supplementary motor area (SMA), which is involved in motor learning and planning, as well as the motor activation of the hand; and the pre-motor cortex, which also mainly relates to motor action planning. There are other sections of the brain that are involved with motor imagery movement, but their main functions are not related to it: the dorsal anterior cingulate area, which is also involved in motor action planning, and the opercular area, which is involved in the motor action of the fingers and toes. In addition, there are the angular and occipitotemporal areas, which are involved in visual processing and object recognition, which are important because most of our tests need object recognition to subsequently execute the movement.

The work in [23] was used to link the Brodmann areas related to motor imagery movement with the electrodes involved in the 10–5 electrode configuration.

For the main zones implicated in motor imagery (MIZ), the implicated electrodes are the FC3, FC2, FC4, C5, C3, C4, and C6 electrodes. For the secondary sections, the more relevant electrodes are the F7, F8, P7, and P8 electrodes. The dorsal anterior cingulate area is a depth zone and does not match with any electrode. Handedness is important in the main zone because the use of the left side of the body activates the right side of the brain and vice versa. In contrast, the use of the right side of the body activates the left side of the brain.

2.3. Preprocessing

The dataset signals were ready for use, and no additional filtering or preprocessing was needed.

2.4. Detection Algorithms

The idea is to run the algorithms inside the device within the execution time, which produces the thought-of imagery movement result in a moment (or as fast as possible). This means that there is no server to save the data, process them, and return them with the result. Consequently, the algorithm selection has to be conducted cautiously in terms of its performance.

In terms of wearable EEGs, there are some limitations in the chips that must be considered. These chips, which control the logic and process of the detection, have limited functions, a limited size, limited complexity, and a limited economic cost to satisfy wearable-EEG demands.

The selection of the algorithms was based on these premises, which assume that they have to be as simple and fast as possible, although, in fact, nowadays, many chips perform well in terms of these requirements.

Two detection algorithms were selected for implementation: the Fast Independent Component Analysis (fastICA) algorithm and the Common Spatial Pattern (CSP) algorithm. Both of them have been used in similar experiments [24].

Other techniques were used in combination with these algorithms. For the fastICA algorithm, a correlation was applied for the resultant components. For the CSP algorithm, a previous filter bank (FB) was used.

All of the developed code was implemented in MATLAB 2023.

2.4.1. FastICACorr Algorithm

Figure 1 presents the FastICACorr algorithm process, which we use to explain the two parts of the FastICACorr algorithm.

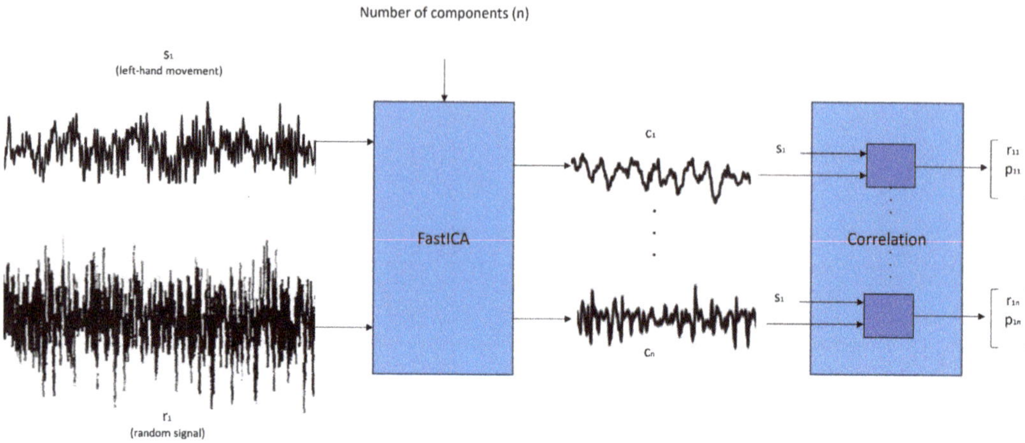

Figure 1. FastICACorr algorithm block representation.

FastICA Algorithm

The reason we selected the FastICA algorithm is that its use can be extended to obtain characteristics from a signal. The FastICA algorithm can provide information about the differences between different signals for a subsequent comparison. Moreover, its output can be generated in a short lapse of time, and this time depends on the complexity selected when it is implemented.

In this work, we used this algorithm to separate the brain signals into all of their possible components because EEG signals are formed by many mixed signals.

The Independent Component Analysis (ICA) algorithm is based on the separation of the original source signals that conform to the treated signal. In terms of EEGs, the observed signals are the potentials presumably generated by mixing some of the underlying brain activity components.

The "fast" property is focused on finding the maximum of the non-Gaussianity, which, for this case, is measured via negentropy approximation.

All of the formal expressions for implementing the FastICA algorithm were based on those in [25]. In addition, to obtain good negentropy approximation values, the three non-quadratic functions of the reference work were implemented.

In this work, we wanted to study the performance of just one electrode as well. The input of this algorithm is a vector formed by one-person single-imagery-movement trials (s_1 of the left-hand movement in Figure 1) with all of the electrodes to be studied in every case. For example, for 2000 samples per trial, if there are 4 trials for one person per image of open-hand movement for one electrode, then the vector will be 4 (trials) $\times$ 2000 (samples). If there are two electrodes, then the vector will be 8 $\times$ 2000. This vector was compared with a random signal (r_1 in Figure 1) to avoid the influence of any other predefined signals.

In this way, the information of each electrode is added to the general information; thus, if the electrode addition decreases the performance, then the added electrodes are not contributing to improving the results.

The outputs of this algorithm are the "components" (from c_1 to c_n in Figure 1, with n as the number of components), which can be considered unique keys decoded for every input signal that is introduced.

Correlation

Correlation is a technique that shows a mutual connection between two variables, and, to measure it, we used the corrcoef command in MATLAB, which outputs two matrices: the "r" matrix, which shows the correlation coefficients, and the "p" matrix, which can be used to assess the hypothesis on the relation between the observed phenomena, as seen in Figure 1.

In this work, we compared the fICA input signal with every output component obtained from the fICA. For example, if we had 20 output components (n = 20 in Figure 1), then we calculated the correlation between the fICA input signal that generated these components (s_1 of Figure 1) and the 20 output components obtained (from c_1 to c_{20}). Thus, we had 20 pairs of r–p matrices.

These matrices are symmetrical because the order for calculating the correlation was not significant in these cases. The values on the diagonal are one because a variable always has a perfect correlation with itself. In conclusion, one paired r–p value was used per calculated correlation.

As the output and input are signals with many samples, this technique allows for a reduction from a number of samples to a pair of them.

2.4.2. FB-CSSP Algorithm

Figure 2 shows the scheme of this algorithm, which is the image that we will use for its explanation.

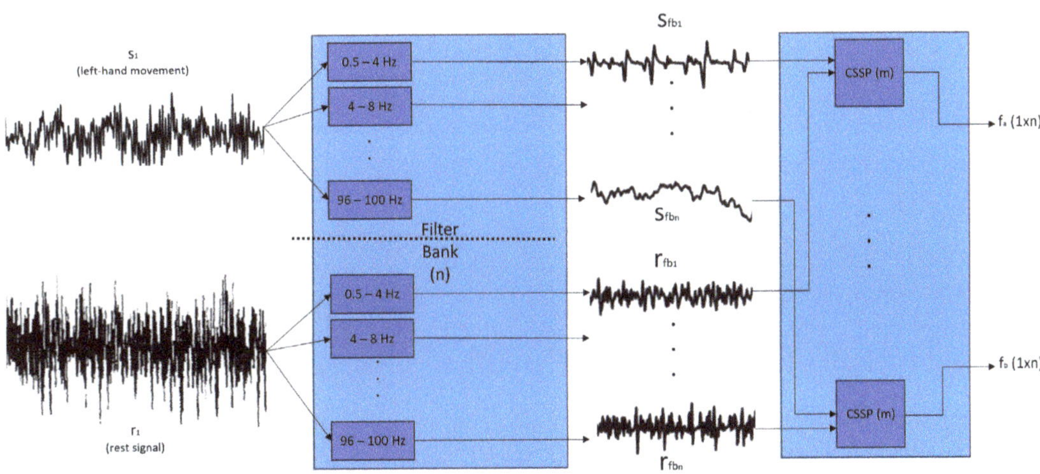

Figure 2. FB-CSSP algorithm block representation.

Filter Bank (FB)

A filter bank is a group of parallel filters with different frequencies and a narrow bandwidth that allows for the separation of an input signal into many different signals related to the input. The separation of some different signals is needed because, in this way, the CSSPs will have more data to build a reliable output. In addition, because the filters cover the whole frequency spectrum of each database, the algorithm generates the output using all possible frequencies, from the lowest to the highest. In the case of Figure 2, as an example to understand the process of this filter bank, the frequency limit is 100 Hz.

To obtain more samples to include in the CSP algorithm, a filter bank is proposed. This filter is formed by a group of bandpass filters that start at 0.5 Hz and increase until the database frequency limit is reached.

When selecting the bandpass filter bandwidth, we separated the frequencies into two groups: known bands and unknown bands. The known bands are the delta (0.5–4 Hz), theta (4–8 Hz), alpha (8–12 Hz), beta (12–30 Hz), and gamma (30–45 Hz) bands. The unknown bands start at 55 Hz or 65 Hz (depending on the precedence of the data) and finish at the frequency limit.

Following the assumptions of the first uses of this algorithm [26], for the known bands, sub-bands are generated with a width of 4 Hz. The sub-beta bands are as follows: 12–16 Hz; 16–20 Hz; 20–24 Hz; 24–28 Hz; 28–32 Hz. The sub-gamma bands are as follows: 30–34 Hz; 34–38 Hz; 38–42 Hz; 42–46 Hz. The delta, theta, and alpha bands are their own bands.

Among the unknown bands, as these frequencies are not classified, the width of the filters is 5 Hz for compatibility with the frequency limit.

Common Spatial Signal Patterns (CSSPs)

The Common Spatial Pattern algorithm was the basis for the CSSP algorithm of this study because it focuses on decreasing the dimension of the data using linear transformations. In this manner, a low-subspace projection matrix is created, the rows of which conform to the weights for the included channels. Additionally, this data reduction is controllable via the "m" parameter, which is the reduced-dimension parameter. The limitation of this algorithm is that only two different types of input signals can be tested at once.

In our work, execution with just one electrode was required, which is why this algorithm needed to be adapted. Following the basic CSP formulas in [27], in this "Signal-Patterns" algorithm, the inputs are the output signals of the filter bank that belong to some of the trials of one person performing one movement for every electrode included in the test. That is, the vector has as many input signals as the number of filters generated from the filter bank (from S_{fb1} to S_{fbn} in Figure 2). As this algorithm also needs to compare two signals, for testing, all of the electrode configurations were selected instead of only one electrode. The rationale for mixing the post-filtered electrode signals in the same execution is the same as for the FastICA algorithm: to determine the influence of the included electrodes on the result.

In the input, the comparison is between the imagery movement signals and the rest signals.

The outputs, in this case, are two vectors with comparative features between the two input signals. One vector (f_a from Figure 2) comes from a comparison of the S_{fb1} and r_{fb1} inputs, and the other vector (f_b from Figure 2) comes from a comparison of the rfb1 and S_{fb1} inputs (inverted comparison). As the outputs are concrete vectors (f_a $(1 \times n)$ and f_b $(1 \times n)$ in Figure 2, where n is the number of filters), a post-correlation technique to reduce the data output is not needed in this case.

2.5. Classification Algorithms

The Classification Learner tool in MATLAB was employed to classify the results from the detection algorithms [28], and 12 classification algorithms were tested to determine the one with the best fit to the included data. A k-fold cross-validation of ten was used.

Moreover, a Support Vector Machine algorithm was implemented for a simple assessment of the results.

The outputs generated by the classifier are the success percentage for two-class classification and the confusion matrix for more-than-two-class classification.

The performances of the classification algorithms used were not deeply explored, as they were primarily utilized to verify the functionality of the detection algorithms, and, moreover, they exhibited low latency. The only precondition here was to use enough data for classification using all of the classes.

3. Results

The results of the various tests are presented below. Due to the numerous variables involved in the experiments (types of movements, algorithm parameters, number of electrodes, etc.), the objective was to narrow them down according to the obtained results, always seeking to optimize the information/low-resource ratio. All of the experiments were performed on three subjects. Because the results were very similar, it was not necessary to increase the number of subjects.

In the preliminary tests, the FB-CSSP algorithm showed low reliability in the specificity of the movement results because every brain zone yielded the same result when comparing movement with no movement. Consequently, the FastICACorr algorithm was used to analyze the sampling frequencies for the different databases.

The results were divided into three subsections to explore the different movement concepts: general movement (all finger movements considered as a single category) was compared with no movement/rest; specific movements, involving the separation of each finger movement, were compared with no movement; and each finger movement and no movement were compared with each other. Six categories were used: thumb, index, middle, ring, and pinkie (movement), as well as rest (no movement).

3.1. General Movement vs. Rest

These tests were aimed at determining whether a general movement can be detected and differentiated from no movement, involving the finger and fist movements discussed in the Datasets and Sampling Frequencies Section.

3.1.1. FastICACorr Algorithm

The FastICACorr algorithm allows for the obtainment of varying numbers of components, representing the output signals explained in the algorithm definition. Tests were executed for 10 and 20 components to observe the differences in the algorithm's performance. The results obtained from one electrode (C4) and one subject for the three databases are shown in Table 1, which helped to determine the database for the subsequent tests.

Table 1. Success rates for one electrode for different databases (Fast Independent Component Analysis CORR algorithm).

DDBB (fs)	10 Components	20 Components
OODS (9.5 kHz)	84.80%	93.60%
NDS (1 kHz)	90.60%	95.10%
PDS (160 Hz)	92.00%	95.63%

The NDS was selected for all of the tests because the difference from the PDS is assumed to be marginal, and the sampling frequency of 1 kHz offers increased flexibility and more data to analyze. The results for both algorithm complexities are coherent: more complexity implies a better success rate.

The initial tests were for one electrode; the C3 and C4 electrodes were used because they are optimally positioned for motor imagery detection. Tests were conducted for 20, 50, 100, and 150 trials, each with ten components (Figure 3 and Table A1 of Appendix A). The aim was to assess the influence of the trials on the success rate.

In Figure 3, it is evident that the success rate decreased with 100 and 150 trials. Consequently, these trial quantities were excluded from the subsequent tests.

Figure 4 shows the test results for one, two, three, and four electrodes for 10 components ("10c") and 20 components ("20c") with 20 and 50 trials. The selected electrodes were the C3, C4, P3, and P4 electrodes, maintaining proximity to the closest main zone of motor imagery detection. The aim of these tests was to assess the impact of the FastICACorr algorithm components on the success rate and the influence of increasing the number of electrodes.

When comparing the number of components in Figure 4, 20 components are considered optimal in all cases because of their higher success rates. Increasing the trial number does not consistently enhance the success rate. While there are cases in which the success rate slightly improves with 50 trials, in other cases, it worsens. Moreover, more trials imply higher performance demands.

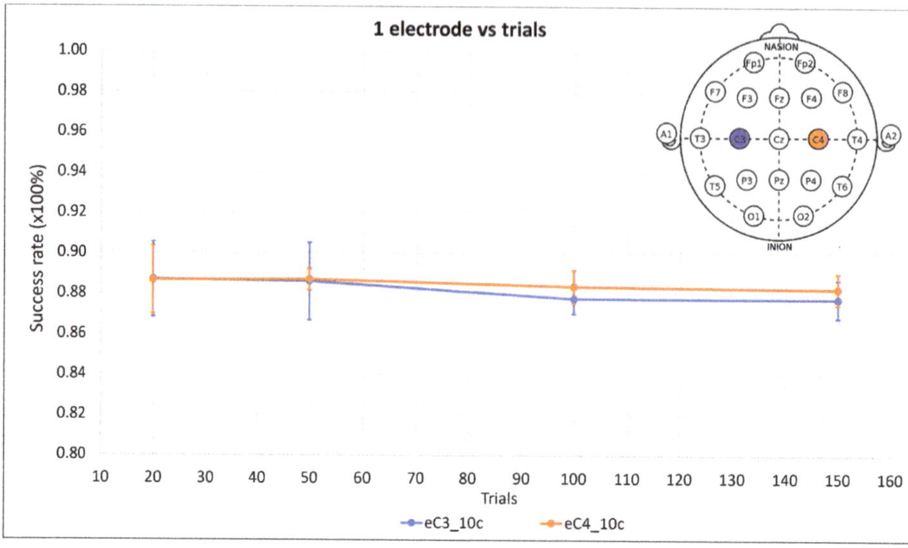

Figure 3. Success rate means of three subjects with their standard deviations as error bars for 20, 50, 100, and 150 trials for ten components of the FastICACorr algorithm for C3 and C4 electrodes treated independently. The positions of the electrodes on the scalp are shown.

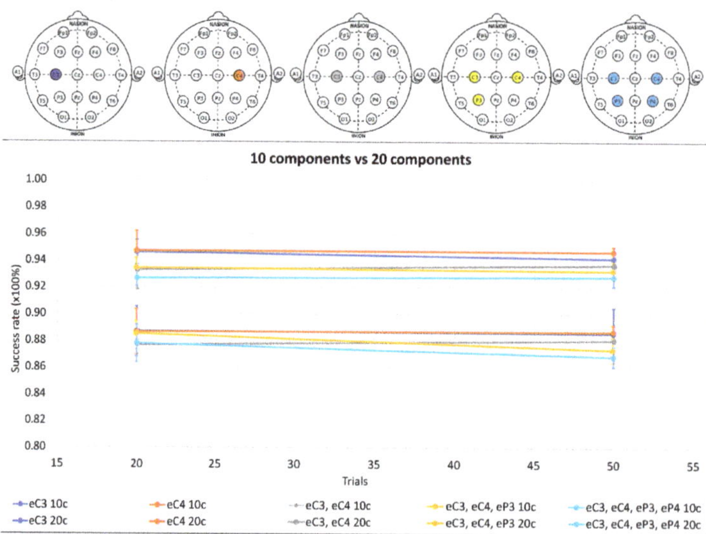

Figure 4. Success rate means of three subjects with error bars as standard deviations for 20 and 50 trials for one, two, three, and four electrodes. Results for ten components of the FastICACorr algorithm are denoted by "10c". Results for 20 components of the FastICACorr algorithm are denoted by "10c". Results for 20 components of the FastICACorr algorithm are denoted by "20c".

Consequently, 20 trials are deemed the most favorable. Therefore, 20 components and 20 trials were established as constant values for the subsequent test cases. All of the relevant data are available in Tables A1 and A2 of Appendix A.

To identify the minimum number of electrodes needed to provide sufficient information for achieving a good performance, nine electrodes were tested: the C3,

Cz, C4, P3, Pz, P4, F3, Fz, and F4 electrodes. The results are illustrated in Figure 5, alongside the outcomes for one, two, three, and four electrodes (Tables A1–A3 of Appendix A). The first electrode was C3 (Figure 5a), as movement is imagined for the right hand, but the influence of the movement on the right zone of the brain was also assessed (Figure 5b).

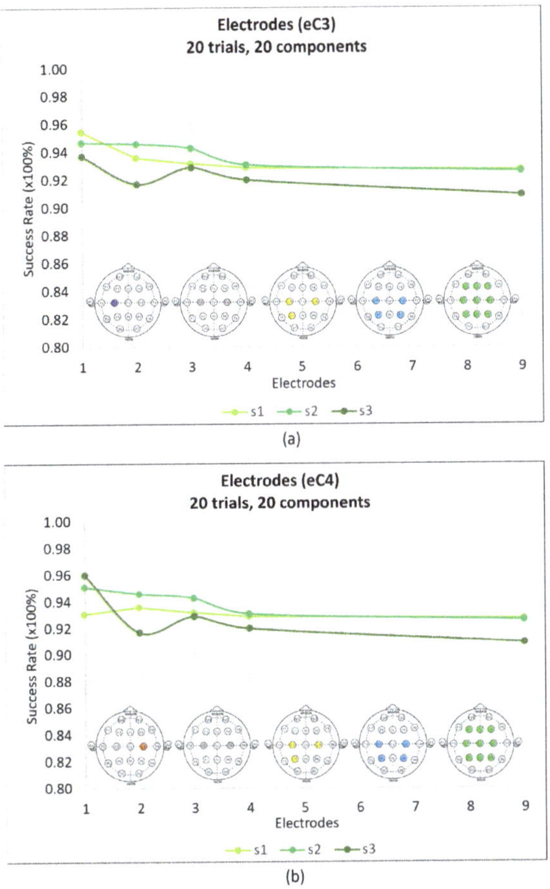

Figure 5. Success rates for different electrode configurations. (**a**) Electrode comparison for 20 trials, 20 components, and three subjects. Points reflect one (C3), two (C3, C4), three (C3, C4, P3), four (C3, C4, P3, P4), and nine (C3, Cz, C4, P3, Pz, P4, F3, Fz, and F4) electrodes. The headings below the subject lines contain the corresponding electrodes. (**b**) The same as (**a**); however, in this case, the first electrode is C4.

Algorithm Complexity

As this algorithm incorporates a parameter that controls the number of generated components, the time taken to obtain each component was examined. Figure 6 illustrates the time taken to obtain each component from 1 to 20, being irrelevant the relation between the color-component. The consistent results across the different trial groups, electrodes, and movement scenarios are indicative of the generic behavior of the FastICACorr components.

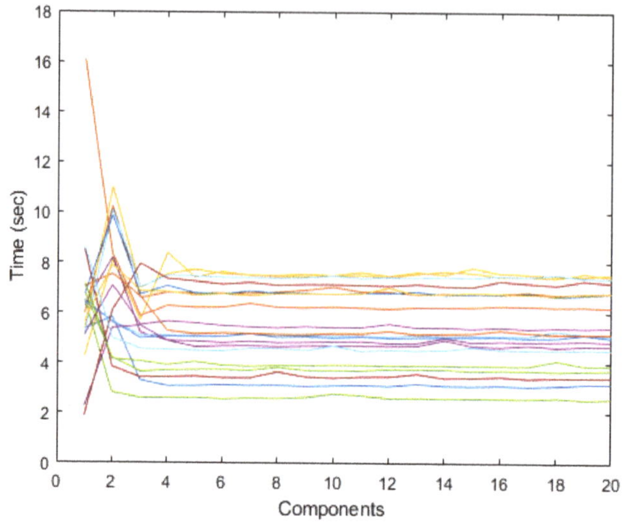

Figure 6. Time in seconds for each of the 20 components for 20 trials. Every color represents one component.

Brain Zone Tests

This subsection describes the tests aimed at determining the significance of the correct zone selection for detecting motor imagery movement. The zone has to be as small and consolidated as possible to meet the wearable-EEG requirements.

The motor imagery zones were tested with two and three electrodes, incorporating not only electrodes belonging to the MIZ but also those randomly selected from other scalp zones (the far-away zones (FAZs)). Testing the FAZs reinforces the importance of the zones and enhances the algorithm's reliability.

Following the motor imagery brain zone, because the NDS electrode configuration is a 10–20-electrode system, the object recognition zones (P7 and P8) were covered by the T5 and P3 electrodes in the left hemisphere and by the T6 and P4 electrodes in the right hemisphere. The F7 and F8 electrodes were also tested for the influence of the finger motor execution zone in the imagery. In conclusion, the C3 and C4 electrodes were tested in cooperation with the P3 and P4, T5 and T6, and F7 and F8 electrodes for 20 trials and 20 components (Figure 7). The graph also presents the results for the C3 and C4 electrodes to allow for a comparison with the electrode addition.

For the FAZs, we selected the Cz and Pz electrodes. For comparison, the overall best results across the three subjects in Figure 5 were obtained with the following groups: C3, C4, P3; C3, C4, F7; and C3, C4, T6. The C3 and C4 electrodes were selected to compare the electrode addition. All of the groups are shown in Figure 8 for 20 components and 20 trials.

Following the electrode analysis, tests for the other algorithms were executed.

3.1.2. FB-CSSP Algorithm

To determine the optimal algorithm for our study, tests were executed for the FB-CSSP algorithm with one, two, and three electrodes, using 20 and 150 trials, as shown in Table 2. These parameters facilitated a comparison with the results of similar tests using the FastICACorr algorithm.

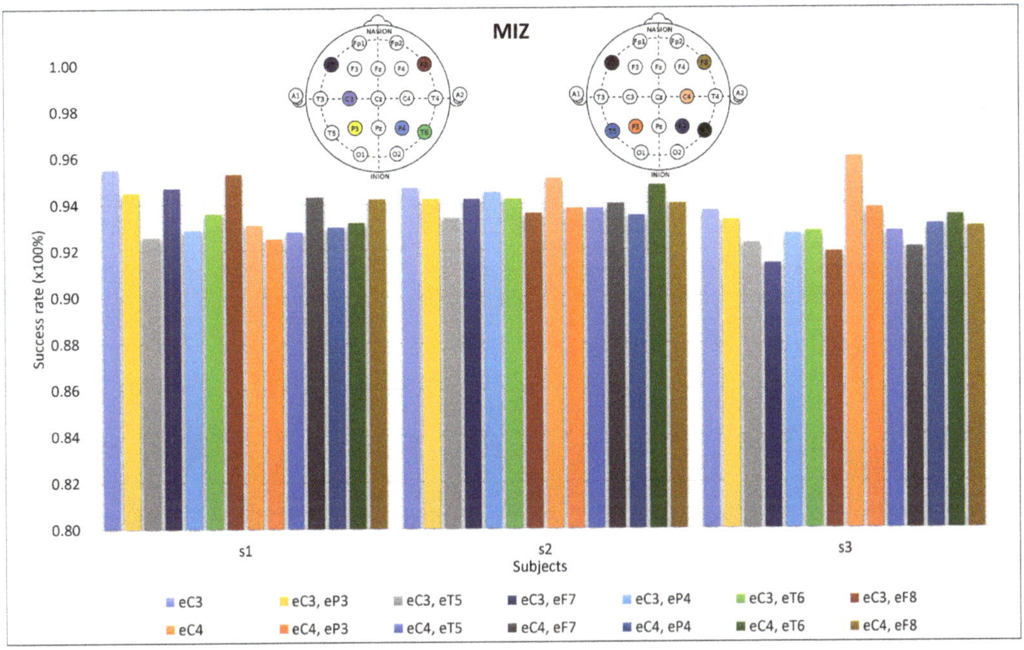

Figure 7. Success rates for the angular area and occipitotemporal area in conjunction with the main motor imagery zone. The first eight are related to the C3 electrode (positions on the electrode-filled head on the upper left), and the other eight are related to the C4 electrode (positions on the electrode-filled head on the upper right). The C3 and C4 electrodes are shown for comparison with the other lines.

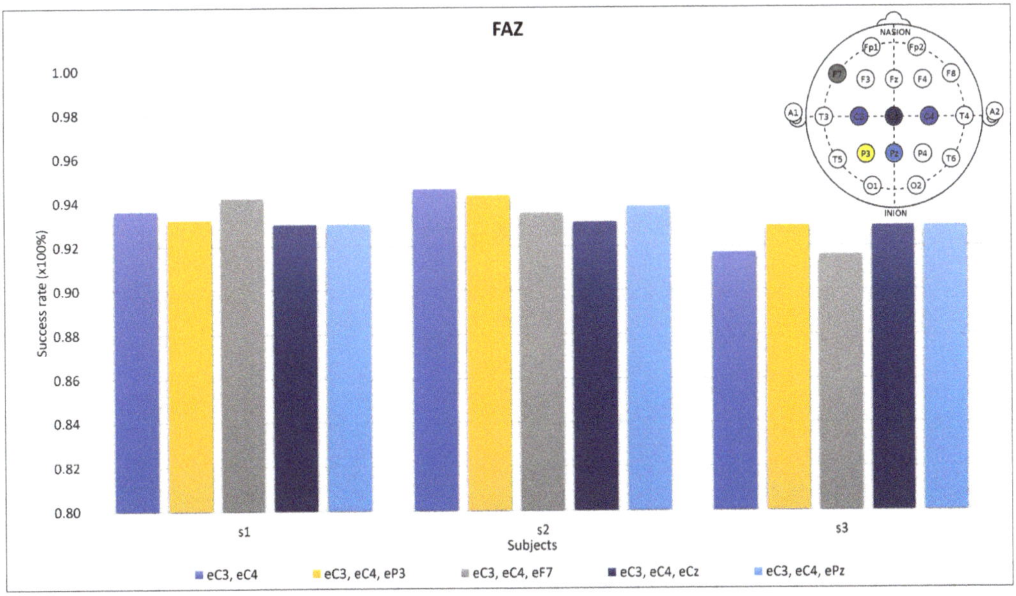

Figure 8. Success rates for far-away zones (FAZs) (last two blue bars) and the main motor imagery zone (first blue, yellow, and gray bars). The electrode-filled head on the upper right reflects the positions.

Table 2. Success rates for one (C3 and C4), two (C3 and C4 together), and three (C3, C4, and P3) electrodes for 20 and 150 trials and three subjects for FB-CSSP algorithm.

Electrodes	C3		C4		C3, C4		C3, C4, P3	
Trials	20	150	20	150	20	150	20	150
Subject 1 (%)	100	100	100	100	100	100	100	100
Subject 2 (%)	100	100	100	100	100	100	100	100
Subject 3 (%)	100	100	100	100	100	100	100	100

The configurable "m" parameter (reduced-dimension parameter) was set to one for all of the tests, as setting it to two and five resulted in decreased success rates.

Brain Zone Tests

The motor imagery zones were examined by combining the MIZ and FAZs in the tests featuring 20 trials and two and four electrodes (Table 3).

Table 3. Success rates for two and four electrodes for far-away electrodes for FB-CSSP algorithm.

Electrodes	C3, Cz, C4, F7	C4, F7
Trials	20	20
Subject 1 (%)	100	100
Subject 2(%)	99.80	100
Subject 3 (%)	100	100

3.2. *Specific Movement vs. No Movement*

We used specific-movement-vs.-no-movement tests to differentiate the finger movements from no movement.

3.2.1. FastICACorr Algorithm

The five right-handed NDS finger movements (thumb, index, middle, ring, and pinkie) were assessed and compared to no movement.

Success rates were obtained for one, two, three, and four electrodes with 10 and 20 components and 20 and 50 trials, as reflected in Figure 9.

3.2.2. FB-CSSP Algorithm

Tests were performed for the C3 and C4 electrodes individually for 20 and 150 trials, with the "m" parameter set to one. The results are presented in Tables 4 and 5. The idea was to test the algorithm's influence in the trials and the capability of detecting the five finger movements.

Table 4. Success rates for C3 electrode for different movements for 20 and 150 trials for FB-CSSPs algorithm.

Movement vs. No Movement	Thumb	Index	Middle	Ring	Pinkie
Trials			**20**		
Subject 1 (%)	60.7	52.4	65.5	60.7	72.6
Subject 2 (%)	61.9	67.9	61.9	59.5	56
Subject 3 (%)	59.5	58.3	57.1	73.8	52.4
Trials			**150**		
Subject 1 (%)	73.8	57.1	61.9	50	56
Subject 2 (%)	71.4	59.5	66.7	65.5	66.7
Subject 3 (%)	63.1	60.7	69	66.7	77.4

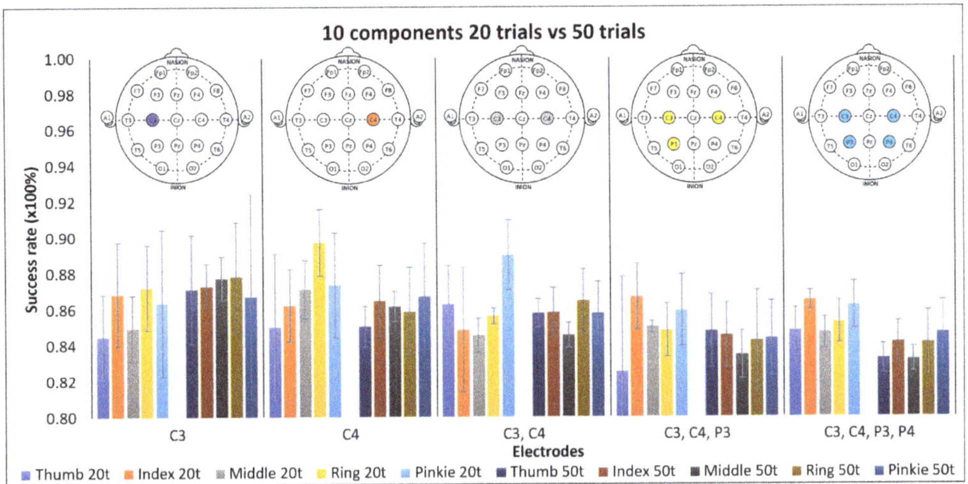

(a)

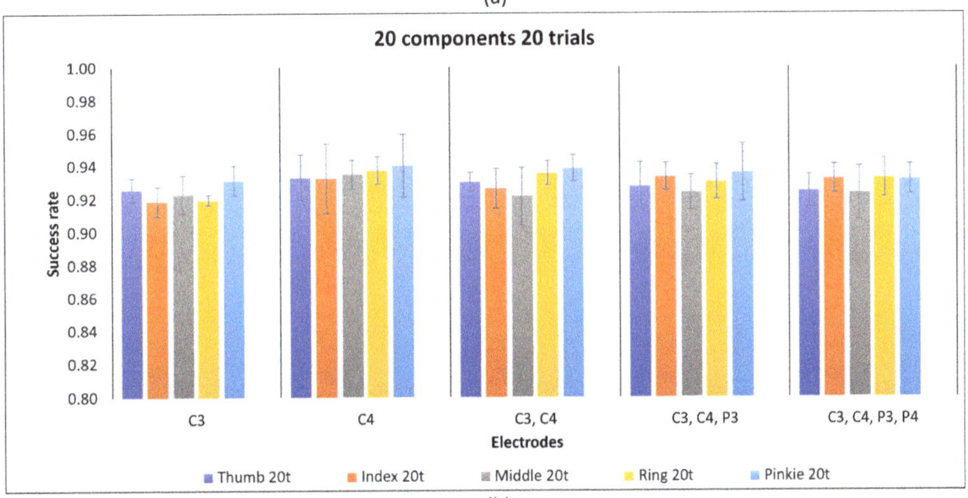

(b)

Figure 9. Five finger movements vs. no movement. Mean success rates of three subjects with their standard deviations as error bars are shown. (**a**) Ten components for 20 and 50 trials for the five groups of electrodes selected. (**b**) Twenty components for 20 and 50 trials for the five groups of electrodes selected.

Table 5. Success rates for C4 electrode for different movements for 20 and 150 trials for FB-CSSPs algorithm.

Movement vs. No Movement	Thumb	Index	Middle	Ring	Pinkie
Trials			20		
Subject 1 (%)	70.2	64.3	71.4	50	63.1
Subject 2 (%)	57.1	61.9	50	50	56
Subject 3 (%)	53.6	64.3	61.9	54.8	58.3
Trials			150		
Subject 1 (%)	64.3	51.2	54.8	50	53.60
Subject 2 (%)	75	71.4	63.1	65.5	56
Subject 3 (%)	58.3	58.3	69	75	77.4

3.3. Specific Movement vs. Specific Movement

These tests were aimed at distinguishing a movement when all possible movements were grouped together, including no movement.

FastICACorr Algorithm

The finger movements were compared with each other, resulting in 15 distinct comparisons: thumb–index; thumb–middle; thumb–ring; thumb–pinkie; index–middle; index–ring; index–pinkie; middle–ring; middle–pinkie; ring–pinkie; rest–thumb; rest–index; rest–middle; rest–ring; and rest–pinkie. Confusion matrices are used to present this information, as they offer a clear and organized representation of the movement success rates, along with the percentages indicating instances when another movement was incorrectly recognized.

The executed tests were consistent with those in the previous section. The results for 20 components and 20 trials are shown for one, two, three, and four electrodes. For simplicity, the confusion matrices for one subject are presented in Figure 10.

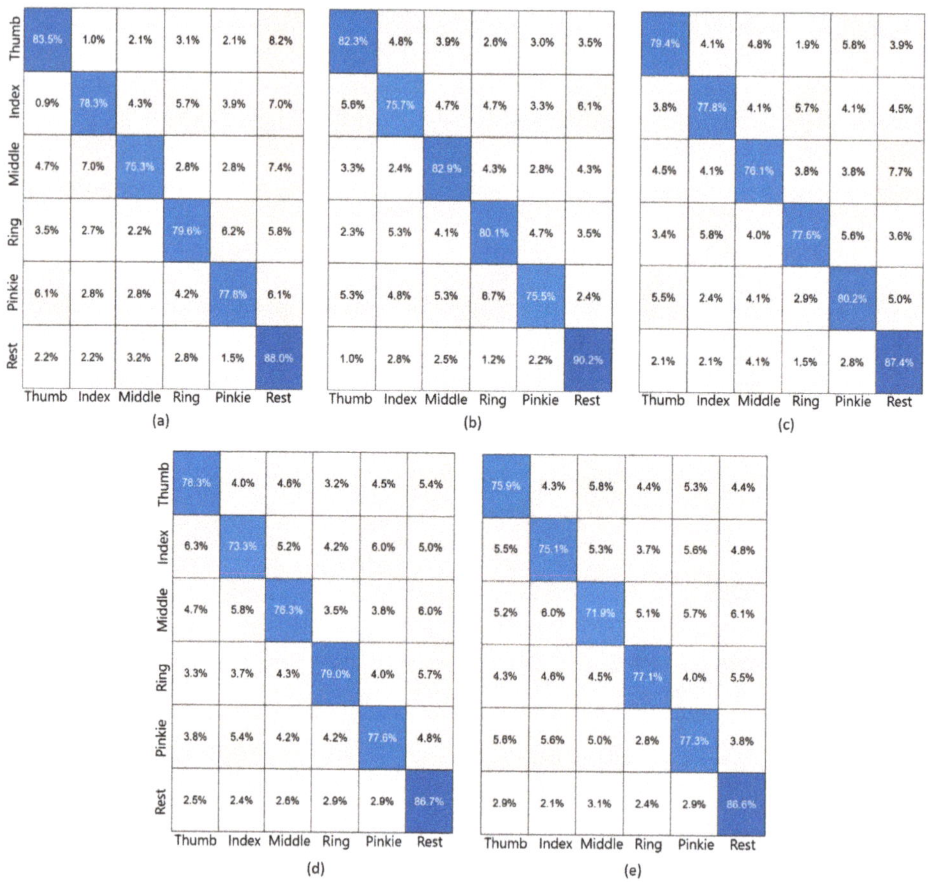

Figure 10. Confusion matrices for specific movement vs. specific movement for 20 trials and 20 components. The movements range from thumb movement to rest. Every box shows the success rate (in percentage) when compared one by one. Blue diagonal boxes show the proportion of times when the movement was correctly detected. Nude pink boxes show the proportion of times that the main movement was wrongly detected as the movement in the nude pink box. (**a**) C3 electrode; (**b**) C4 electrode; (**c**) C3 and C4 electrodes; (**d**) C3, C4, and P3 electrodes; (**e**) C3, C4, P3, and P4 electrodes.

4. Discussion

The idea of developing wearable EEG comes from the need for a small device with a band-aid size that can be used for 24 h, 7 days a week. This system will be very useful to doctors to improve the diagnosis and treatment of some mental health conditions. Moreover, it will collect some data that are not currently collected because of the new cases in which these devices can be used. Furthermore, these decisions are expected to happen within a shorter lapse of time, which is the reason for studying the algorithm performance to fix the wearable-EEG specifications.

To explain the decisions for these wearable-EEG specifications, this section is divided into sub-stages for clarity and organization.

4.1. Sampling Frequency

The sampling frequency is an important consideration in a system because higher frequencies imply higher system performance requirements, such as computational costs and battery consumption. The minimum operating frequency value allows for improvement and the extension of the wearable-EEG life.

Observing Table 1, the PDS with 160 Hz is the dataset with the best results. The second-best dataset, very close to the PDS, is the NDS with 1 kHz, and the third-best dataset is OODS with approximately 9.5 kHz. This implies that lower frequencies improve the results due to the lower number of samples. An excess of data could lead to noise and misunderstandings for the algorithms, consequently reducing the success rate.

While one sample per 6.25 ms is sufficient for good results, having more samples provides the flexibility to work on, implement, and ensure reliable algorithm results.

The difference between the best two database success rates is less than 2%, and, in line with the points discussed above, the selected frequency was 1 kHz for the NDS.

4.2. Detection Algorithms

The FastICACorr algorithm was capable of extracting differentiable features from a signal in all of the cases presented, making it suitable for real-world wearable-EEG scenarios.

The success rates improve when the number of obtained components increases. However, the main drawback of this algorithm is its high computational cost in terms of the time required to obtain these components. When the computational resources are reduced, the time required to obtain the components increases.

Observing the processing times in Figure 6, obtaining a component takes two–eight seconds, stabilizing after the fourth component. These first four components are highly variable, depending on the signal. For higher numbers of components, the time is constant and quantifiable.

The numbers of trials and algorithm components were also studied, as these two parameters directly influence the computational cost of this algorithm. In Figure 3, the success rate evolution shows no improvement when the number of trials is more than 50. This can be explained by the saturation of the data and the inclusion of noise when the upper success rate limit is reached. Thus, 20 and 50 trials are preferable.

Delving into the selection of 20 or 50 trials, Figure 4 shows that although 50 trials may seem better, the increase from 20 to 50 was not observed consistently in all subjects and all electrodes, and in some cases, it was the opposite.

Regarding the number of components, according to Figure 4, using twenty components is always better than using ten, and the increase in the success rate justifies the extra performance cost that this imposes on the system.

As the performance is an important consideration for this study, and it must be balanced with the extra cost of the algorithm components, it is preferable to have fewer trials and more algorithm components, which is why, for the results in Figure 4, all of the tests were executed in advance for 20 trials and 20 components.

Longer processing times are acceptable if they are spent in the training phase of the system. After that, the device will have been trained on every movement and non-movement; thus, it will not take long to obtain the result for a new incoming signal.

The FB-CSSP algorithm is a low-computational-cost algorithm in terms of the time that it takes to finish processing. There are two reasons why this algorithm is not useful for this work. First, it is an algorithm based on a comparison of two types of signals, and the features obtained are based on this comparison. For example, movement one and movement two will have characteristics based on this comparison; however, if the detection of another movement from a random signal is wanted, comparisons with all of the different saved signals have to be performed. The second reason is that this algorithm shows the same percentage, 100%, with every electrode on the scalp because the difference between movement and no movement at the output is still high at every point. Moreover, when a specific movement is trying to be recognized, the success rate is low in comparison with that of the FastICACorr algorithm. These characteristics make the algorithm less specific for movement differentiation. Additionally, the tree structure that it generates with a random signal input with many possible movements produces an increase in the time–computational costs. Nevertheless, this algorithm is useful for other applications that fit with its characteristics.

All of these points led to the selection of the FastICACorr algorithm as the preferable algorithm. However, online tests in real time need to be conducted to corroborate its use for wearable devices.

4.3. Motor Imagery Brain Zone

Primary-motor-cortex and supplementary-motor-area data from (20 trials, 20 components) the C3 and C4 electrodes independently achieved mean success rates of 94.63% and 94.73% for general movement vs. no movement, respectively, as shown in Figure 5 and Table A1. When the C3 and C4 electrodes were combined with the P3 electrode, as shown by the points for two and three electrodes in Figure 5, the success rates experienced varied outcomes, with decreases in some cases and stability and increases in others. However, for four electrodes, the success rates diminished, and this trend persisted for nine electrodes. This suggests that using a single electrode is preferable, as additional electrodes do not yield improvements. Furthermore, fewer electrodes reduce the time–computational costs of the algorithms and allow for higher algorithmic complexity. Based on these considerations, the decision was made to exclude four and nine electrodes from this work, opting for a single electrode, either C3 or C4, depending on the user, and considering the different motor imagery zones to test the performances of two and three electrodes.

The different motor imagery zones discussed in the Motor Imagery Brain Zone (MIZ) Section of Materials and Methods were tested using the P3, P4, T5, and T6 electrodes. Figure 7 shows that the P3 electrode had a better performance than the T5 electrode for all the subjects when compared with the C3 electrode. However, when compared with the C4 electrode, not all of the subjects had better P3 performance than T5 performance. The P4 and T6 electrodes are contraries, and the T6 electrode had a better performance than the T4 electrode with the C4 electrode in all subjects, but not with the C3 electrode.

According to the data, it seems that subject three has a left prominence while subject one has a right prominence, and subject two has the same prominence for the two sides, which is why these two electrodes are the main ones implicated in the motor imagery movements. Continuing with the focus on the subjects, the same zones do not have the same implications in the movements for the different subjects. Subject one is predominant in F8, F7, and P3 electrodes; subject two is predominant in T6, F7, and P3 electrodes; and subject three is predominant in P3, T6, and P4 electrodes.

The differentiation of the zones was tested using the far-away zones (FAZs) for the Cz and Pz electrodes in Figure 6. The different prominences remain out of scope when the C3 and C4 electrodes are involved because they unify both sides. The success rates when using these zones have similar results to those of some MIZ combinations (Figure 7). In fact, some electrodes selected for the MIZ have smaller success rates than those of some groups of FAZs. In most cases, across all tested users, the combination of two electrodes outperformed the combination of three electrodes, resulting in the decision to discard the use of three electrodes. When comparing one and two electrodes, the C3 and C4 electrodes consistently exhibited the highest success rates for all three subjects. Therefore, for this work, the use of two electrodes in combination was dismissed.

A preliminary user study is essential to determine individual preferences, and if a user requires more than one electrode, then the optimal zones must be selected based on specific tests.

4.4. Electrode-vs.-FastICACorr-Component Trade-Off

The success rate increases when the number of algorithm components increases. The best performance is for one electrode and 20 components. Adding electrodes generally decreases the success rate; consequently, the highest supported complexity for the algorithm is justified.

The best balance in the trade-off between the electrodes and algorithm components is 20 components for the C3 or C4 electrode.

4.5. Movement Differentiation

Finger motor imagery detection is feasible with one, two, three, and four electrodes with the C3 or C4 electrode and 20 components, yielding optimal results.

For 20 trials and ten components, the C4 electrode outperformed the C3 electrode, while, for 50 trials, the C3 electrode surpassed the C4 electrode, on average. However, this difference between 20 and 50 trials in the C3 electrode was not high, and for the other electrode groups, 20 trials was best; thus, the use of 50 trials was discarded for this movement differentiation.

Given that the C3 electrode is typically associated with the right hand, additional training is needed for an accurate focus on specific movements. Moreover, the success rate decreases with the inclusion of more electrodes.

For 20 trials and 20 components, the success rate was between 92% and 94% for all electrode groups. The C4 electrode was slightly better in terms of the mean, supporting the decision that one electrode was enough to obtain the best performance for this study.

As previously mentioned, 20 components consistently outperformed 10 components for each electrode group, which supports the idea that the weight of the components is more significant than the number of electrodes for the different movements.

These results also suggest that a specific movement corresponds to a specific zone within the MIZ because specific fingers, such as the thumb and pinkie, had better results with the C3–C4-electrode combination. The involvement of more than two electrodes that were within the common MIZ but far from their specific zones resulted in decreased success rates for all fingers.

4.6. Subject Independence and Requirements

Many hypotheses from [12] were reaffirmed in these experiments.

First, different subjects have different minds. Although the tasks and conditions were the same, it is not assumable that their brains worked in the same manner, which is called brain plasticity.

Subjects have differences in their success rates, starting with the dominant hand, necessitating adjustments for optimal success rates. These changes are not of significant concern because the wearable EEG will have a calibration phase to adapt to specific users.

Continuing with this hypothesis, distinct EEG signals arise from different subjects, emphasizing the importance of considering brain plasticity in the development of future devices.

A configurable wearable device that accommodates different subjects is crucial, ensuring scalability and mobility around the brain for diverse needs and reinforcing the idea of a comfortable and portable device for users.

5. Conclusions

This study comprehensively analyzed two algorithms, emphasizing the importance of making optimal design decisions for wearable EEG, considering the trade-off between performance and resource efficiency. The design decisions were obtained from the grouping characteristics discussed.

Regarding the sampling frequency, a system capable of obtaining data between 1 and 6.25 ms suffices for good success rates. Working within this margin depends on the device requirements imposed, such as the computational cost and battery life.

For general movement, 20 trials are adequate for good performance rates, while specific movements benefit from 50 trials over 20.

The FastICACorr algorithm was selected for continued testing on online cases, with 20 components selected due to the computational costs. This factor is inversely proportional to the number of electrodes: more components require fewer electrodes, and the weight of the components is greater than the weight of the electrodes in wearable EEG.

Depending on the requirements of the device and the user's brain, the system will implement one electrode, either the C3 or C4 electrode. If a specific subject requires more electrodes, the C3 and C4 electrodes can be used together, and if additional electrodes are needed, the P3, P4, or T6 electrodes can be added.

A wearable device with these design decisions can be implemented to improve comfort and portability for users at any place and at any moment.

The general conclusion is that a few-electrode wearable EEG is feasible with algorithms that generate outputs in real execution time. Consequently, a portable and comfortable device for patients with specific requirements can be developed.

Author Contributions: Conceptualization, A.C. and A.A.; methodology, A.C. and A.A.; software, A.C.; validation, A.C.; formal analysis, A.C. and A.A.; investigation, A.C.; resources, A.C. and A.A.; data curation, A.C. and A.A.; writing—original draft preparation, A.C.; writing—review and editing, A.C. and A.A.; visualization, A.C. and A.A.; supervision, A.A.; project administration, A.A.; funding acquisition, A.A. All authors have read and agreed to the published version of the manuscript.

Funding: This research was partially funded by Comunidad de Madrid under the grant P022/BMD-7236 MINA-CM project.

Institutional Review Board Statement: Not applicable.

Informed Consent Statement: Not applicable.

Data Availability Statement: Carretero Pérez, Ana (2024), "EEG Motor imagery open/close hands with C4 electrode dataset", Mendeley Data, V1, https://doi.org/10.17632/bfn2pksz45.1 (accessed on 22 July 2024) [21].

Conflicts of Interest: The authors declare no conflicts of interest.

Appendix A

Table A1. Success rates for one electrode for different trials and components for FastICACorr algorithm.

Electrodes	C3					
Components	10				20	
Trials	20	50	100	150	20	50
Subject 1 (%)	86.8	90.8	88.6	88.8	95.5	93.9
Subject 2 (%)	90.5	87.3	87.5	87.5	94.7	94.8
Subject 3 (%)	88.8	87.7	87.2	87	93.7	93.9

Table A1. *Cont.*

Electrodes	C4					
Components	10				20	
Trials	20	50	100	150	20	50
Subject 1 (%)	87.8	89	89.3	89.1	93.1	94.9
Subject 2 (%)	90.6	88.1	88	88.1	95.1	95.1
Subject 3 (%)	88	89	87.8	87.6	94.5	94.8

Table A2. Success rates for two, three, and four electrodes for different trials and components for FastICACorr algorithm.

Electrodes	C3, C4			
Components	10		20	
Trials	20	50	20	50
Subject 1 (%)	87.7%	88.7%	93.6%	93.4%
Subject 2 (%)	88%	87.3%	94.6%	94.2%
Subject 3 (%)	87.3%	88.2%	91.7%	93.5%
Electrodes	C3, C4, P3			
Components	10		20	
Trials	20	50	20	50
Subject 1 (%)	88.8	88	93.2	93
Subject 2 (%)	89.3	86.3	94.3	94
Subject 3 (%)	87.5	87.8	92.9	92.8
Electrodes	C3, C4, P3, P4			
Components	10		20	
Trials	20	50	20	50
Subject 1 (%)	87.2	87.6	92.9	92.8
Subject 2 (%)	89.4	86.1	93.1	93.5
Subject 3 (%)	86.8	86.9	92	92.1

Table A3. Success rates for two electrodes for different trials and components for FastICACorr algorithm.

Electrodes	**C3, C4, P3, P4, F3, F4, Cz, Pz, Fz**
Components	20
Trials	20
Subject 1 (%)	92.7
Subject 2 (%)	92.6
Subject 3 (%)	90.9

References

1. Niso, G.; Romero, E.; Moreau, J.T.; Araujo, A.; Krol, L.R. Wireless EEG: A survey of systems and studies. *NeuroImage* **2023**, *269*, 119774. [CrossRef] [PubMed]
2. Sourov, I.H.; Ahmed, F.A.; Opu, T.I.; Mutasim, A.K.; Bashar, M.R.; Tipu, R.S.; Amin, A.; Islam, K. EEG-Based Preference Classification for Neuromarketing Application. *Comput. Intell. Neurosci.* **2023**, *2023*, 4994751. [CrossRef]
3. He, C.; Chen, Y.-Y.; Phang, C.-R.; Stevenson, C.; Chen, I.-P.; Jung, T.-P.; Ko, L.-W. Diversity and Suitability of the State-of-the-Art Wearable and Wireless EEG Systems Review. *IEEE J. Biomed. Health Inform.* **2023**, *27*, 3830–3843. [CrossRef] [PubMed]
4. Emish, M.; Young, S.D. Remote Wearable Neuroimaging Devices for Health Monitoring and Neurophenotyping: A Scoping Review. *Biomimetics* **2024**, *9*, 237. [CrossRef] [PubMed]
5. Gaspar-Figueiredo, D.; Abrahão, S.; Insfrán, E.; Vanderdonckt, J. Measuring User Experience of Adaptive User Interfaces using EEG: A Replication Study. *ACM Int. Conf. Proc. Ser.* **2023**, *10*, 52–61. [CrossRef]

6. Lee, S.; Schoenwether, B.; Garner, C.; Nachtegall, A.; Kim, D.; Roth, E. EFFECTS OF MOVEMENT SONIFICATION AUDITORY FEEDBACK ON REPETITIONS AND BRAIN ACTIVITY DURING THE BENCH PRESS. *ISBS Proc. Arch.* **2023**, *41*, 72. Available online: https://commons.nmu.edu/isbs/vol41/iss1/72 (accessed on 2 November 2023).
7. Arachchige, M.D.J.; Nafea, M.; Nugroho, H. A hybrid EEG and head motion system for smart home control for disabled people. *J. Ambient. Intell. Humaniz. Comput.* **2023**, *14*, 4023–4038. [CrossRef]
8. Siow, E.K.S.; Chew, W.J.; Mun, H.K. Human Computer Interface (HCI) using EEG signals. *J. Phys. Conf. Ser.* **2023**, *2523*, 012012. [CrossRef]
9. Moontaha, S.; Schumann, F.E.F.; Arnrich, B. Online Learning for Wearable EEG-Based Emotion Classification. *Sensors* **2023**, *23*, 2387. [CrossRef] [PubMed]
10. Farokhah, L.; Sarno, R.; Fatichah, C. Comparative study of emotion elicitation between 4K UHD screen and virtual reality using EEG. *AIP Conf. Proc.* **2023**, *2609*, 40004. [CrossRef]
11. ASaibene; Caglioni, M.; Corchs, S.; Gasparini, F. EEG-Based BCIs on Motor Imagery Paradigm Using Wearable Technologies: A Systematic Review. *Sensors* **2023**, *23*, 2798. [CrossRef] [PubMed]
12. Arora, A.; Malkin, R. HANDLINK: A Dexterous Robotic Hand Exoskeleton controlled by Motor Imagery (MI). *J. Adv. Med. Med. Res.* **2022**, *34*, 427–436. [CrossRef]
13. Ronca, V.; Di Flumeri, G.; Vozzi, A.; Giorgi, A.; Arico, P.; Sciaraffa, N.; Babiloni, F.; Borghini, G. Validation of an EEG-based Neurometric for online monitoring and detection of mental drowsiness while driving. In Proceedings of the Annual International Conference of the IEEE Engineering in Medicine and Biology Society, EMBS, Glasgow, UK, 11–15 July 2022; pp. 3714–3717. [CrossRef]
14. Sciaraffa, N.; Di Flumeri, G.; Germano, D.; Giorgi, A.; Di Florio, A.; Borghini, G.; Vozzi, A.; Ronca, V.; Varga, R.; van Gasteren, M.; et al. Validation of a Light EEG-Based Measure for Real-Time Stress Monitoring during Realistic Driving. *Brain Sci.* **2022**, *12*, 304. [CrossRef] [PubMed]
15. Abdel-Hamid, L. An Efficient Machine Learning-Based Emotional Valence Recognition Approach Towards Wearable EEG. *Sensors* **2023**, *23*, 1255. [CrossRef] [PubMed]
16. Bernal, G.; Hidalgo, N.; Russomanno, C.; Maes, P. Galea: A physiological sensing system for behavioral research in Virtual Environments. In Proceedings of the 2022 IEEE Conference on Virtual Reality and 3D User Interfaces, VR 2022, Christchurch, New Zealand, 12–16 March 2022; pp. 66–76. [CrossRef]
17. Carretero, A.; Araujo, A. Analysis of Simple Algorithms for Motion Detection in Wearable Devices. In Proceedings of the Annual International Conference of the IEEE Engineering in Medicine and Biology Society, EMBS, Glasgow, UK, 11–15 July 2022. [CrossRef]
18. Drew, L. The rise of brain-reading technology: What you need to know. *Nature* **2023**, *623*, 241–243. [CrossRef] [PubMed]
19. EEG Motor Movement/Imagery Dataset. Available online: https://archive.physionet.org/pn4/eegmmidb/ (accessed on 11 November 2020).
20. Kaya, M.; Binli, M.K.; Ozbay, E.; Yanar, H.; Mishchenko, Y. Data descriptor: A large electroencephalographic motor imagery dataset for electroencephalographic brain computer interfaces. *Sci. Data* **2018**, *5*, 180211. [CrossRef] [PubMed]
21. Carretero, A. EEG Motor Imagery Open/Close Hands with C4 Electrode Dataset. Mendeley Data, (V1). Available online: https://data.mendeley.com/drafts/bfn2pksz45 (accessed on 18 July 2024).
22. Strotzer, M. One Century of Brain Mapping Using Brodmann Areas. *Clin. Neuroradiol.* **2009**, *19*, 179–186. [CrossRef] [PubMed]
23. Guttmann-Flury, E.; Sheng, X.; Zhu, X. Channel selection from source localization: A review of four EEG-based brain-computer interfaces paradigms. *Behav. Res. Methods* **2023**, *55*, 1980–2003. [CrossRef] [PubMed]
24. Arpaia, P.; Esposito, A.; Natalizio, A.; Parvis, M. How to successfully classify EEG in motor imagery BCI: A metrological analysis of the state of the art. *J. Neural Eng.* **2022**, *19*, 031002. [CrossRef] [PubMed]
25. Hyvärinen, A.; Oja, E. Independent component analysis: Algorithms and applications. *Neural Netw.* **2000**, *13*, 411–430. [CrossRef] [PubMed]
26. Novi, Q.; Guan, C.; Dat, T.H.; Xue, P. Sub-band common spatial pattern (SBCSP) for brain-computer interface. In Proceedings of the 3rd International IEEE EMBS Conference on Neural Engineering, Kohala Coast, HI, USA, 2–5 May 2007; pp. 204–207. [CrossRef]
27. Wang, Y.; Gao, S.; Gao, X. Common spatial pattern method for channel selection in motor imagery based brain-computer interface. In Proceedings of the Annual International Conference of the IEEE Engineering in Medicine and Biology—Proceedings, Shanghai, China, 17–18 January 2006; pp. 5392–5395. [CrossRef]
28. The MathWorks Inc. *MATLAB Version: 9.14.0.2254940 (R2023a)*; The MathWorks Inc.: Natick, MA, USA, 2023; Available online: https://www.mathworks.com (accessed on 18 July 2024).

sensors

MDPI

Article

Designing a Hybrid Energy-Efficient Harvesting System for Head- or Wrist-Worn Healthcare Wearable Devices

Zahra Tohidinejad [1], Saeed Danyali [1,*], Majid Valizadeh [1], Ralf Seepold [2], Nima TaheriNejad [3] and Mostafa Haghi [3,*]

[1] Department of Electrical Engineering, Ilam University, Ilam 69315-516, Iran; z.tohidinejad@ilam.ac.ir (Z.T.); m.valizadeh@ilam.ac.ir (M.V.)
[2] Ubiquitous Computing Laboratory, HTWG Konstanz—University of Applied Sciences, 78462 Konstanz, Germany; ralf.seepold@htwg-konstanz.de
[3] Institute of Computer Engineering, Heidelberg University, 69120 Heidelberg, Germany; nima.taherinejad@ziti.uni-heidelberg.de
* Correspondence: s.danyali@ilam.ac.ir (S.D.); mostafa.haghi@ziti.uni-heidelberg.de (M.H.)

Abstract: Battery power is crucial for wearable devices as it ensures continuous operation, which is critical for real-time health monitoring and emergency alerts. One solution for long-lasting monitoring is energy harvesting systems. Ensuring a consistent energy supply from variable sources for reliable device performance is a major challenge. Additionally, integrating energy harvesting components without compromising the wearability, comfort, and esthetic design of healthcare devices presents a significant bottleneck. Here, we show that with a meticulous design using small and highly efficient photovoltaic (PV) panels, compact thermoelectric (TEG) modules, and two ultra-low-power BQ25504 DC-DC boost converters, the battery life can increase from 9.31 h to over 18 h. The parallel connection of boost converters at two points of the output allows both energy sources to individually achieve maximum power point tracking (MPPT) during battery charging. We found that under specific conditions such as facing the sun for more than two hours, the device became self-powered. Our results demonstrate the long-term and stable performance of the sensor node with an efficiency of 96%. Given the high-power density of solar cells outdoors, a combination of PV and TEG energy can harvest energy quickly and sufficiently from sunlight and body heat. The small form factor of the harvesting system and the environmental conditions of particular occupations such as the oil and gas industry make it suitable for health monitoring wearables worn on the head, face, or wrist region, targeting outdoor workers.

Keywords: energy harvesting; hybrid; medical wearable sensor nodes; health monitoring; power management

1. Introduction

With the global spread of the coronavirus (COVID-19) disease, the importance of the remote monitoring of human health has been further highlighted. In this context, wearable devices have gained additional attention in healthcare [1,2]. These devices can seamlessly integrate into the daily lives of individuals, and through the continuous monitoring of vital signs, serve as an effective solution for early disease detection [3]. This early diagnosis can result in preventive measures and prompt therapeutic solutions, preventing the progression of the disease and reducing some of the costs associated with emergency and hospital care [4,5]. Additionally, the use of wearable devices in remote areas can enhance the efficiency of healthcare services, quality of life (QoL), and overall well-being [6]. With surveillance, assessment, and continuous data provision, these devices assist in the early detection of the users' health issues, transforming into a promising approach for preventive healthcare [7].

Wearable technologies encompass electronic devices like smartwatches, wristbands, and augmented reality glasses, often including various physiological/non-physiological sensors, a data processing unit, and communication components. They can collect a diverse range of data including heart rate (HR), blood pressure (BP), respiratory rate, blood oxygen saturation (SpO_2), body movement and physical activity, and more, by using sensors such as photoplethysmograms (PPGs) and accelerometers; these sensors serve as valuable tools in providing real-time health tracking data, especially for individuals dealing with chronic illnesses such as asthma, chronic obstructive pulmonary disease (COPD), diabetes, mobility impairments, and cardiovascular diseases, contribute to identifying movement-related disorders such as Parkinson's disease, and are particularly useful for individuals dealing with Alzheimer's disease, respectively [8,9]. In the realm of wearable technology, the proliferation of unobtrusive, non-intrusive, and non-invasive devices signifies a key advancement in user-centered design [10–12]. These features are particularly important in continuous health monitoring [13,14]. There is a hindrance in the practical use of wearable devices, stemming from issues related to battery recharging and battery life, among other aspects such as accuracy, reliability, and clinical use [6,15–21].

With the reduction in size and power consumption of electronic circuits, integrating data processing units, and communication components, real-time monitoring and analysis have been enhanced, which allow healthcare professionals to remotely monitor their patients' progress and make informed decisions regarding their medical care [17].

Despite the above-mentioned advantages of using wearable devices and some of the most popular and contributing sensors with the applications, these electronic devices rely on batteries to power themselves [22]. Therefore, a primary concern in these scenarios revolves around the necessity of a continuous power source. This dilemma has led to the emergence of innovative solutions, particularly in the field of energy harvesting approaches and power management systems. Energy harvesting techniques involve capturing and utilizing environmental energy sources such as solar energy [23], thermal energy [24], vibrations [25], kinetic energy [26], and radio frequency (RF) [27] energy to supply power to wearable devices. Additionally, efficient power management systems optimize the use of available energy, ensure a longer battery life, reduce the need for frequent recharging, and store the harvested energy from the environment or the individual's body in an energy storage unit, which is ultimately used to power the consumer load [28].

The power density of various energy sources is provided in Table 1.

Table 1. Power densities of different energy sources [6].

Energy Sources	Power Density
Ambient light	100 mW/cm^2 (direct sun) 100 µW/cm^2 (indoor illumination)
Thermoelectric	60 µW/cm^2
Radio frequency	1 µW/cm^2 (ambient) 15 µW (external)
Human	1000 µW/cm^2 (biochemical) 4 µW/cm^3 (biomechanical—microgenerator) 200 µW/cm^3 (biomechanical—piezoelectric)

Accordingly, solar energy in outdoor environments has the highest power. However, in many cases, it may not fully meet the needs due to the instability of these sources and their temporal-spatial unavailability. To address this issue, the use of hybrid energy harvesting systems has been introduced as an effective solution, capable of harvesting energy from multiple sources [29]. Hybrid energy harvesting systems have the advantage of providing a more reliable and stable power source by harnessing energy from various sources. By integrating various energy harvesters such as photovoltaic (PV) panels, thermoelectric generators (TEG), kinetic harvesters, RF harvesters, and more, these systems can ensure a continuous flow of energy, even when a single source is unavailable or not producing sufficient power. This not only increases the overall power density, but also enhances the system's stability and efficiency [29].

Our contributions in this work are as follows:

- Designing a double-source hybrid PV/TEG energy harvesting system to achieve the maximum power point tracking (MPPT) for both input sources, during battery charging and loading, addressing 96% efficiency in energy conversion.
- Validating the harvesting system with two DC/DC boost converters, which operate with MPPT and can charge or supply a common battery/load simultaneously or individually, showcasing the capability to turn into a self-powered wearable device.
- Implementing a low-cost, compact form factor, and universal harvesting system compatible with a wide range of wearable healthcare devices in different mode of wearability such as wrist- or head-worn systems.

The rest of this paper is organized as follows. Section 2 reviews the related work. In Section 3, we describe the materials, methods, and study design. Section 4 presents the experimental results. This is followed by the discussion in Section 5, and we present out conclusions in Section 6.

2. Related Work

Several single-source and hybrid energy harvesting systems for healthcare monitoring applications have been introduced. Yaoguang et al. [30] proposed a wearable TEG that harvests heat from the human body. The structural architecture of the TEG device consisted of 12 TEG modules electrically linked in series and connected in parallel via copper strips. This wrist-worn TEG device captured body heat and generated more power while walking compared to stationary. A requirements analysis and performance evaluation of wearable sensors in medical applications was presented in [26], addressing the fundamental issue of piezoelectric kinetic energy harvesting devices. In [31], a kinetic energy harvesting device was used instead of an accelerometer to assess calorie consumption, as kinetic energy is generated when the user expends calories through bodily movements. A wearable sensor system for long-term health monitoring was described in [28], where the device measured the temperature, HR, SpO_2, and human body acceleration in real-time. In [32], a rotating piezoelectric energy harvesting device was tested and developed, capable of generating a maximum power of 7 μW when worn on the arm during activities. Additionally, ref. [33] described the development of a flexible piezoelectric generator designed to harvest energy from the dynamic movement of the ear canal. In [34], a scalable triboelectric energy harvesting system for electronic textiles was proposed to extract energy from daily human movement. The suggested energy harvester is scalable, stretchable, and wearable. The system's output power is enhanced by the capacitor capacity and the mechanical input frequency, providing guidance for practical applications. However, a drawback of this system is its ability to power wearable electronic devices only during human movement, with significant power losses due to high rectifier losses. In [35], a wearable sensor system in the form of glasses was proposed. This system employs algorithms to detect and identify chewing cycles using a piezoelectric pressure sensor placed on the temporalis muscle. The study suggests the potential for the further miniaturization of electronic devices to improve user comfort. Additionally, further research is needed to explore new methods of integrating sensors such as embedding them in the handles of glasses. In [36], a novel hybrid energy harvesting technology was presented to power wearable electronic devices. The study developed a flexible and wearable energy harvesting device that combined solar and RF energy. This work represents the first flexible and wearable hybrid system of solar and RF energy harvesting that was experimentally tested on the human body. Furthermore, to increase the reading range of active radio-frequency identification (RFID) tags and provide a compact multifunctional structure, a hybrid solar and RF energy harvesting system was introduced in [37]. This system includes components such as monocrystalline solar panels, a charging circuit, a rectifier, the EM4325 chip as the receiver antenna, and an RFID tag. The authors in [38] proposed a flexible TEG module that is appropriate for biomedical and wearable devices due to its high-power density on a small scale and flexibility due to its flexible form factor. In [39], various methods were examined to utilize the heat and mechanical energy of the human body for wearable energy harvesting. The focus was on harvesters such as TEG, PV, piezoelectric, electromagnetic, and electrostatic harvesters. This work included hybrid energy harvesters that hybrid the conversion of two or more energy sources to achieve the maximum power density.

Typically, in hybrid energy harvesting systems, Schottky diodes such as the 1N5817 are used to combine input sources, but they suffer from high power losses and voltage drops. This technique, known as "OR-ing", can be applied either before or after the voltage conversion stage. Applying this technique before the voltage conversion stage allows for the use of a single voltage converter for both sources. However, it creates a parallel structure of energy sources and requires the sources to have the same internal impedance. As a result, the load is supplied by the energy source with the higher voltage until the other source surpasses it. Therefore, at any given time, only one energy source can be utilized [28,40,41].

To overcome this drawback, we present a hybrid energy harvesting system designed to power a wide range of wearable medical healthcare sensors including those worn on the head and wrist. The primary objective of this system is to design and implement a compact and efficient double-source energy harvester that simultaneously utilizes solar and body heat sources, thereby extending the battery and system lifespan. This energy harvester relies on compact PV panel and TEG modules. The collected energy is stored in a 3.7 V, 300 mAh lithium battery, facilitating system charging. Additionally, two ultra-low-power DC/DC boost converters are employed to efficiently manage and convert the generated power from the PV and TEG sources. These converters ensure that the harvested energy is effectively converted to the voltage and current levels required for the system load, optimizing the use of available energy and increasing the overall battery life. Furthermore, the use of diminutive PV panel and TEG modules contributes to a streamlined and lightweight design, making it well-suited for various wearable applications.

3. Materials and Methods

The design and implementation of the hardware harvesting system and health-related sensors and components include a PV panel, TEG module, lithium battery, and two BQ25504 (Texas Instruments, Dallas, TX, USA) ultra-low-power DC/DC boost converters, MAX30102 (Analog Devices, Norwood, MA, USA), and MPU6050 (TDK InvenSense, San Jose, CA, USA).

We considered two ultra-low-power DC/DC boost converters for the PV and TEG sources, with identical configurations for battery settings. The ultra-low-power DC/DC boost converter controls the output voltage of the PV panel and TEG module. The low-power IC BQ25504 is the basis for this converter and the proposed structure, accepting a maximum absolute input and output voltage range from −0.3 V to 5.5 V.

3.1. Health-Related Sensors

PPG is a non-invasive technique used to estimate vital signs such as HR, SPO_2, and BP. The MAX30102 sensor provides a convenient and efficient solution for real-time monitoring. It comes in a compact form factor, has low power consumption initialization, a wide range of voltage support, and supporting inter-integrated circuit (I2C) protocol [42–44]. This sensor consumes 6 mA in measurement mode and 2.7 mA in its sleep mode. The MAX30102 chip has compact dimensions of 5.6 mm × 3.3 mm × 1.55 mm.

Another common example of sensors used in wearable technologies for health monitoring is the accelerometer. This can measure physical activity by detecting changes in acceleration. By using the gyroscope and accelerometer present in MPU6050, rotation along all three axes, static acceleration due to gravity, and dynamic acceleration due to motion can be measured. The current consumption is 4.8 mA when active.

3.2. *Proposed Hybrid Energy Harvesting System, and Hardware Specification*

The proposed block diagrams of the hybrid energy harvesting system are shown in Figure 1a,b. The system includes a PV panel for solar energy harvesting, a TEG module for body thermal energy capture, and two DC/DC boost converters. These converters are independent at their input terminals to connect and boost one of the input DC voltages of the PV panel and TEG module. In the later stage, they become parallel at their two output terminals to connect a battery storage and microcontroller unit (NodeMCU Board (Espressif Systems, Shanghai, China)) as a load.

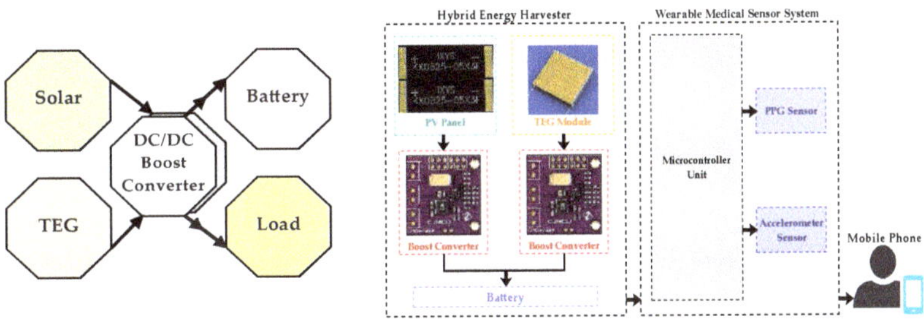

(**a**) (**b**)

Figure 1. (**a**) The block diagram of the hybrid energy harvesting system; (**b**) proposed development of the hybrid energy harvesting system.

3.2.1. Solar Panel

We chose solar cells, model KXOB25-05X3F (×10) from IXYS (Milpitas, CA, USA), each with dimensions of 23 mm × 8 mm × 1.8 mm and a maximum power of 30.7 mW under standard conditions: temperature of 25 °C and 1000 W/m^2 irradiance. This PV panel consisted of a series and parallel connection of these ten cells with an area of 1840 mm^2, every two cells were connected in series, and ultimately, five strings were connected in parallel. This model has a 25% power conversion efficiency. The electrical characteristics of this cell are given in Table 2. All values were measured at the standard condition: 1 sun (=100 mW/cm^2), Air Mass 1.5, 25 °C.

Table 2. KXOB25-05X3F electrical characteristics.

Symbol	Cell Parameter	Typical Ratings
V_{OC}	Open circuit voltage	2.07 V
I_{SC}	Short circuit current	19.5 mA
V_{mpp}	Voltage at MPP	1.67 V
I_{mpp}	Current at MPP	18.4 mA
P_{mpp}	Maximum peak power	30.7 mW
H	Solar cell efficiency	25%

3.2.2. TEG Module

We used the IMC06-126-03 TEG (RMT Ltd, Moscow, Russia) module with dimensions of 16 mm × 16 mm and a thickness of 1.4 mm. The energy harvested from TEG depends on the performance parameters of the TEG module, the TEG module's cross-sectional area, and the material's Seebeck coefficient. In the temperature range of the human body, which is the hot side, the cross-section of the TEG modules must be increased to increase the generated power. Increasing the number of TEG modules effectively improves the cross-sectional area that contributes to power generation, allowing for larger power outputs. Connecting TEG modules in series is a useful solution. Four TEG modules chained in series formed the total area of 1024 mm^2. The polarity of the TEG depends on the direction of the cold and hot sides. These TEGs were used to harvest energy from the body heat. The performance parameters are provided in Table 3 for the TEG cold side in dry air at 27 °C. During the experiments, measurements, and estimations in this work, we only considered the first column information as validation.

Table 3. 1MC06-126-03TEG performance data.

Symbol	Parameter	Values at Hot Side Temperature		
		35 °C	55 °C	85 °C
T_{cold}	Cold side temperature, (°C)	27	27	27
Opt_{η}	Optimum efficiency, (%)	0.40	1.36	2.71
P_{OPT}	Optimum power, (mW)	20	233	964
V_{OPT}	Optimum voltage, (V)	0.244	0.868	1.825
V_{OC}	Open circuit voltage, (V)	0.43	1.51	3.18
I_{SC}	Short circuit current, (A)	0.19	0.63	1.24

3.2.3. DC/DC Converter and Power Management Unit

The DC voltage produced by PV/TEG energy harvesters is typically low. DC/DC power converters such as boost converters were used in the power management circuit to increase the voltage to the required level. We identified and compared the influencing characteristics of several ICs from different companies such as Texas Instruments (TI), STMicroelectronics, Analog Devices, and E-peas on their relevance to the PV and TEG input sources (see Figure 2) [42–49].

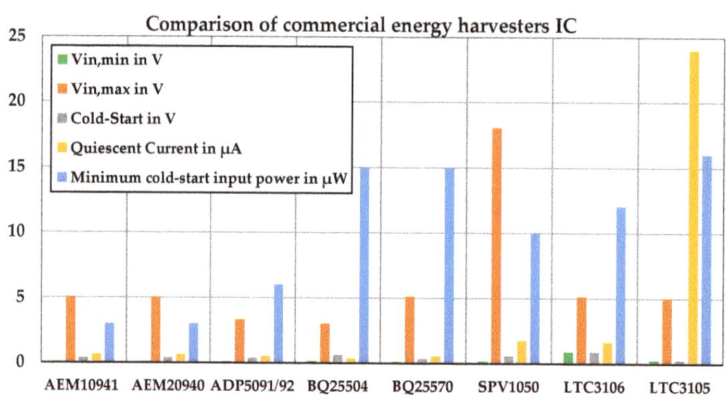

Figure 2. Comparison of several examples of commercial energy harvesting ICs used for PV/TEG energy sources.

A low quiescent current is crucial for maintaining efficiency in low-energy harvesting systems. According to Figure 2, BQ25504 and BQ25570 (Texas Instruments, Dallas, TX, USA) as well as AEM10941 and AEM20940 (E-peas, Mont-Saint-Guibert, Belgium) exhibited low quiescent current consumption, with BQ25504 being the most efficient. In comparison, ADP5091/92, LTC3105/06 (Analog Devices, Norwood, MA, USA), and SPV1050 (STMicroelectronics, Geneva, Switzerland) had a medium to high quiescent current. Taking into account other factors such as low prices and availability, we selected BQ25504.

3.2.4. Energy Storage Unit

The harvested energy was stored in a rechargeable lithium battery with dimensions of 40 mm × 11 mm × 4 mm, capacity of 300 mAh, and a nominal voltage of 3.7 V, which reaches 4.2 V when fully charged.

3.3. *Proposed Multi-Port Energy Harvesting Circuit*

The details of the proposed multi-port energy harvesting power circuit are shown in Figure 3.

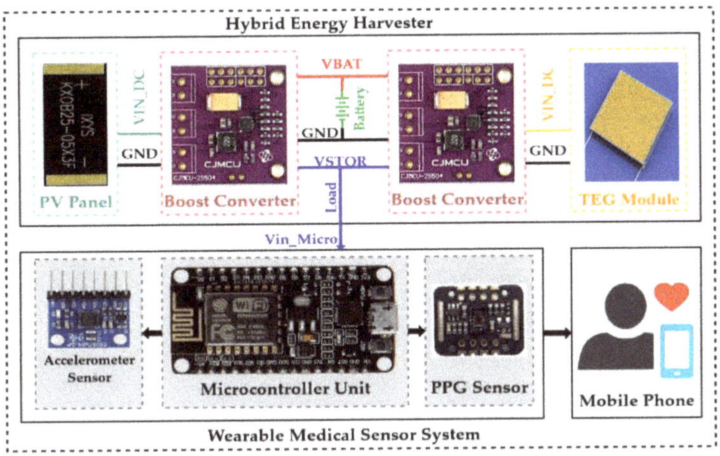

Figure 3. The proposed multi-port energy harvesting system.

Both the V_{BAT} and V_{STOR} pins of both converters were connected to each other (output parallel). As a result, a single battery terminal and a single load terminal were achieved by these common pins to integrate the system input battery and output load. The battery overvoltage (OV) and undervoltage (UV) configurations of both boost converters were designed to be the same (see Figure 4). Therefore, each power source operates with the MPPT and can charge/supply a common battery/load simultaneously or individually.

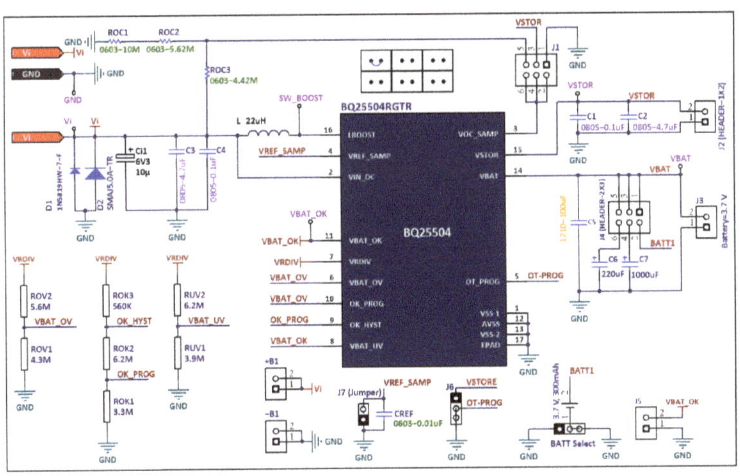

Figure 4. Circuit schematic of the BQ25504 ultra-low-power DC/DC boost converter.

4. Experimental Results

The proposed sensor node included one MAX30102 sensor and one MPU6050 with two operating modes: active mode, in which MAX30102 was activated for 10 S at a sampling rate of 200 Hz, and sleep mode for 180 S. The MPU6050 module worked continuously with a sampling rate of 50 Hz. We tested and measured the battery lifetime of the sensor node under the condition that the energy harvesting system was disconnected. The total operation time of the node was (T). Therefore, in the active mode, the current consumption of the node was measured as: I_{ON} = 36 mA for a period of T_{ON} = 10 S. Furthermore, in sleep mode, we recorded the current consumption of I_{Sleep} = 32 mA during T_{Sleep} = 180 S. Thus, the average current consumption was I_{ave} = 32.21 mA.

With the node's operating voltage of 3.3 V, the power and energy were 106.29 mW and 382.6 Joules, respectively. As a result, with the battery capacity of 300 mAh, the total battery lifetime (T_{BAT}) was calculated as 300 mAh/32.21 mA = 9.31 h.

To calculate the energy produced by the PV panel (E_{PV}), we considered three environmental conditions in which a subject/user wore the sensor node and worked comfortably.

- Sunny day facing the sun;
- Sunny day back to the sun;
- Shady or cloudy conditions.

We performed the experiments for 10, 60, and 120 min (see Table 4).

Table 4. PV panel testing conditions on a sunny day and measured energy.

Test Conditions	Sunny Day: Facing the Sun			Sunny Day: Back to the Sun			Shadow		
	10 min	1 h	2 h	10 min	1 h	2 h	10 min	1 h	2 h
V_{PV} in V	2.92	2.91	2.94	2.8	2.7	2.8	2.09	2.38	2.18
I_{PV} in mA	71.2	63.4	80.1	51.7	48.5	50	5.3	9	4.7
P_{PV} in mW	207.9	184.5	235.5	144.76	131	140	11.07	21.42	10.246
V_{BATT} in V	3.75	3.96	3.96	3.83	3.81	4.04	3.93	3.87	3.81
I_{avrage} in mA [1]	32.21	33.21	33.21	33.26	34.15	32.21	28.21	32.21	28.21

[1] $[(I_{STOR\text{-}ON} \times T_{active\text{-}Sensor}) + (I_{STOR\text{-}OFF} \times T_{sleep\text{-}Sensor})]/T$.

Assume that a worker in the oil, gas, or petrochemical industry is engaged in outdoor activities for 4 h in direct sunlight: 2 h with their back to the sun, and 2 h in shaded conditions. The average power of the PV panel in direct sunlight is 235.5 mW; when this person is engaged in activities with their back to the sun, it is 140 mW; and in shaded conditions, the power is 10.25 mW. Therefore, the E_{PV} can be calculated using Equation (1), which equals 4472.88 joules.

$$\begin{gathered} E_{PV} = P_{PV} \times t \\ E_{PV} = [(2.94\ V \times 80.1\ mA \times 4\ h) + (2.8\ V \times 50\ mA \times 2\ h) \\ +(2.18\ V \times 4.7\ mA \times 2\ h)] \times 3600\ S = 4472.88\ Joules \end{gathered} \tag{1}$$

The battery has a capacity of 300 mAh, and the maximum voltage is 4.2 V. Therefore, using Equation (2), the battery has a stored energy of 1260 joules, where C_{BAT} is the battery capacity and V_{BAT} is the battery voltage.

$$\begin{gathered} E_{BAT} = C_{BAT} \times V_{BAT} \\ E_{BAT} = 300\ mAh \times 4.2\ V = 1260\ mWh \end{gathered} \tag{2}$$

The estimated charging time (T_{CH}) of the battery by the PV energy harvester is measured as:

$$\begin{gathered} T_{CH} = E_{BAT}/E_{PV} \\ T_{CH} = 1260/4472.88 = 0.28\ day = 6.72\ h \end{gathered} \tag{3}$$

In the same manner, the output energy of the TEG module is measured as follows:

$$E_{TEG} = P_{TEG} \times t_{TEG} \tag{4}$$

In a scenario where the output power of the TEG module (P_{TEG}) is considered as 82.2 mW for a period of 8 h, the generated TEG energy at a temperature difference (ΔT) of 8 °C ($\Delta T = T_{hot} - T_{cold}$ = 35 °C − 27 °C = 8 °C) can be calculated as E_{TEG} = 2367 joules. Table 5 shows the details of the TEG module test conditions.

Table 5. Details of the TEG module testing conditions in temperature difference $\Delta T = T_{hot} - T_{cold} = 35\ °C - 27\ °C = 8\ °C$.

Test Conditions	Indoor	
	10 min	**1 h**
V_{TEG} in V	0.96	0.96
I_{TEG} in mA	82.2	82.2
P_{TEG} in mW	78.912	78.912
V_{BATT} in V	3.93	3.92
I_{avrage} in mA [1]	32.21	33.21
Skin temperature	35	35
Environment temperature	27	27

[1] $[(I_{STOR\text{-}ON} \times T_{active\text{-}Sensor}) + (I_{STOR\text{-}OFF} \times T_{sleep\text{-}Sensor})]/T$.

4.1. The Results of the Hybrid Energy Harvesting System

The conversion efficiency of the PV panel harvester was calculated at the maximum measured input power of 1840 mW. This amount was 1000 W/m^2 of sunlight intensity, and according to the measured PV panel area of 0.01840 m^2, the maximum peak power output at the standard conditions was 307 mW, and the conversion efficiency was 16.68%. The system energy harvesting prototype is shown in Figure 5, when it is worn on the human body.

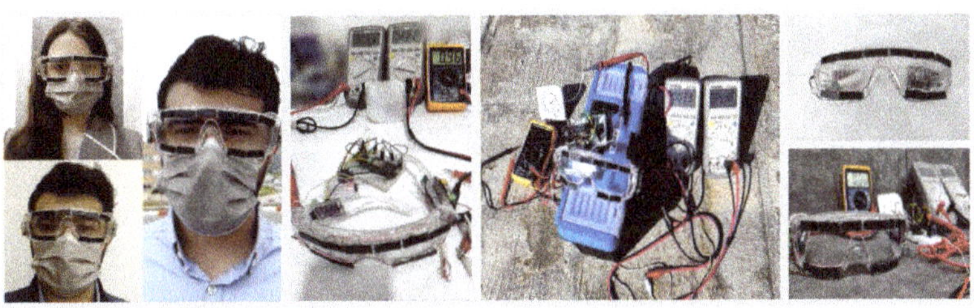

Figure 5. Hardware implementation of the prototype energy harvesting system on glasses as a wearable device, which are worn on the human body.

We also evaluated the PV and TEG energy harvesting system under various resistive loads. Each step is explained below.

4.1.1. First Experimental Stage: Wearable Sensor Node

Figure 6 shows the maximum measured power of the PV energy harvester at different hours on 27 and 28 August 2023, respectively (for further details, see Table 4). The measurements were taken under various conditions: a sunny day facing the sun, a sunny day with the back to the sun, and a shadow day where $P_{TEG} = 0$. According to this comparison, negative battery power indicates that the battery is charging, while positive battery power indicates that the battery is discharging, meaning that power consumption is being supplied from the battery.

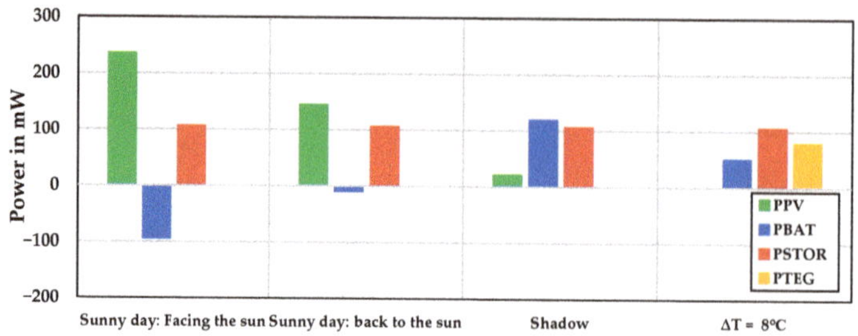

Figure 6. Exchange of PV/TEG power, battery, and wearable sensor node power in different weather conditions.

We estimated the losses of the BQ25504 sensor according to the input and output power and then calculated the efficiency as follows:

$$P_{IN} = P_{loss} + P_{OUT} \tag{5}$$

$$\eta = \frac{P_{OUT}}{P_{IN}} = \begin{cases} if\ P_{BAT} > 0 \rightarrow \begin{cases} P_{OUT} = P_{STOR} \\ P_{IN} = P_{PV} + P_{TEG} + |P_{BAT}| \end{cases} \\ if\ P_{BAT} < 0 \rightarrow \begin{cases} P_{OUT} = P_{STOR} + |P_{BAT}| \\ P_{IN} = P_{PV} + P_{TEG} \end{cases} \end{cases} \tag{6}$$

The maximum conversion efficiency on a sunny day facing the sun (η) was 85%, where η is the ratio of output power to input power. The PV energy harvester charges the battery and provides a current for the sensor node. On a sunny day with the back to the sun, the PV panel alone supports the total power required for the sensor node, effectively turning it into a self-powered system without the battery (see Table 4). The maximum conversion efficiency in this scenario was 80%.

According to Figure 6, the PV panel could not fully meet the sensor node's power consumption in shadow conditions. Therefore, the battery acted as a backup, supplying the additional power required by the sensor node.

Figure 6 also shows the TEG module test conditions at a temperature difference of 8 °C, with $P_{PV} = 0$. In this scenario, the battery supplemented the system load alongside the TEG module. The maximum conversion efficiency of the TEG module (η) was 82%.

4.1.2. Second Experimental Stage: PV Energy Harvesting System under the Various Resistive Loads

Figure 7 illustrates the values of the PV input power, battery power, and output power under various resistive loads on a sunny day. It also shows the person entering into the shade for a few minutes under specific conditions and then returning to the sunlight.

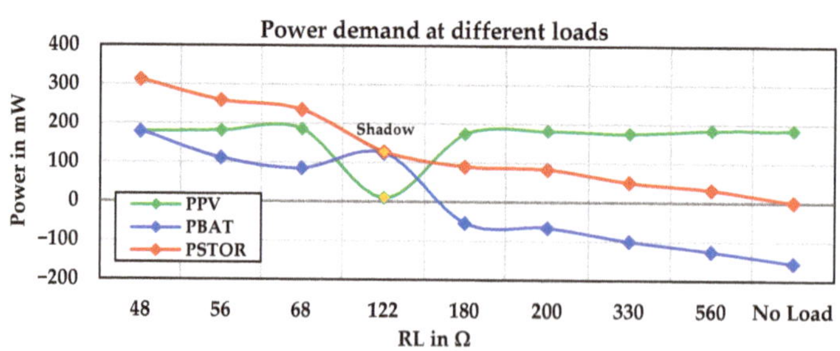

Figure 7. The power demand of the PV energy harvesting system under the various resistive loads.

According to the diagram, in the no-load state, the energy harvested from the PV panel is stored in the battery. As the resistive load values are varied from 31 mW to 312 mW, the battery is charged at five points, achieving an efficiency of 83%. At other points, due to the high load demand, the battery discharges, resulting in an efficiency of 87%, as the system's power consumption is supplemented by the battery. Figure 8 shows the efficiency and power losses of the PV energy harvesting system under various resistive loads on sunny days and specifically in shadow conditions.

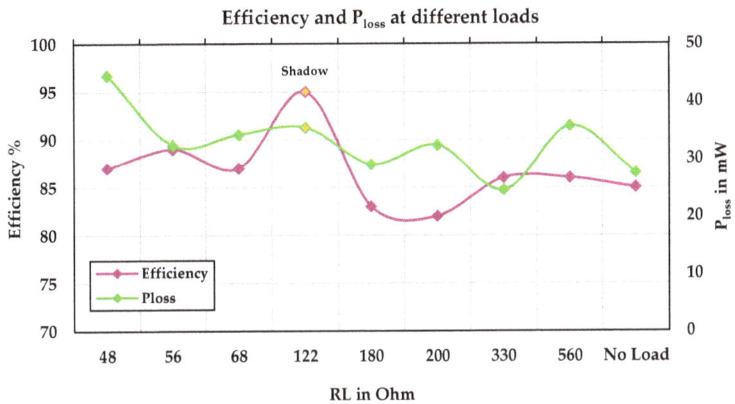

Figure 8. Efficiency and power losses of the PV energy harvesting system under various resistive loads.

4.1.3. Third Experimental Stage: TEG Energy Harvesting System under the Various Resistive Loads

Figure 9 illustrates the TEG input power, battery power, and output power under various resistive loads at a temperature difference of 8 °C. At three points, the battery is also being charged in addition to the load being supplied by the TEG modules. Figure 10 shows the efficiency and power losses of the TEG energy harvesting system under various resistive loads at a temperature difference of 8 °C.

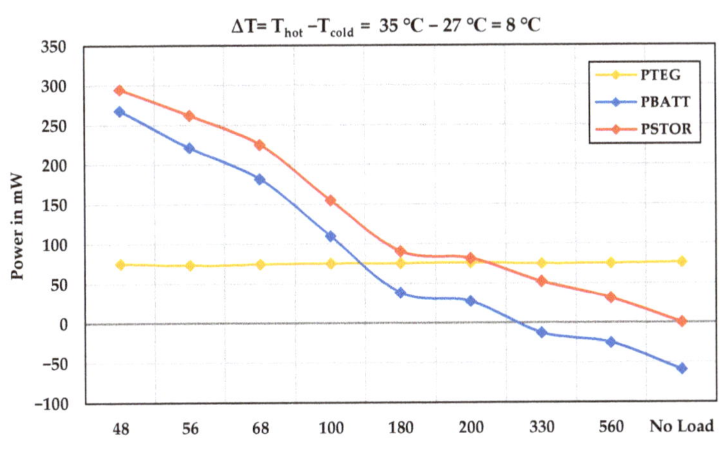

Figure 9. The power demand of the TEG energy harvesting system under the various resistive loads.

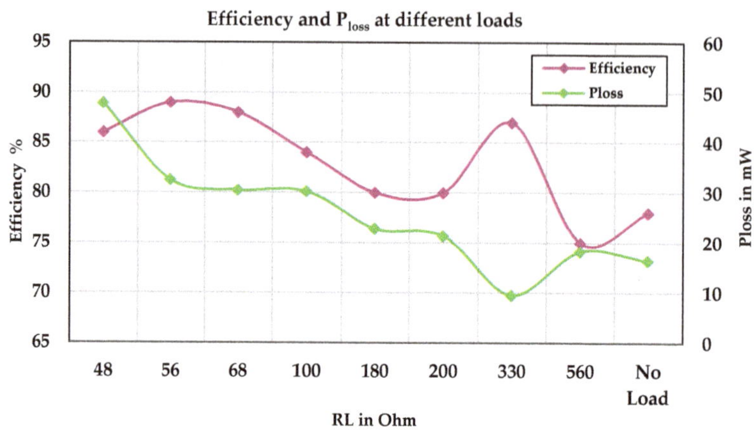

Figure 10. Efficiency and power losses of the TEG energy harvesting system under various resistive loads.

4.1.4. Fourth Experimental Stage: Hybrid Energy Harvesting

Figure 11 shows the contribution of hybrid energy harvesting sources under shade conditions and a temperature difference of 8 °C to supply a wearable sensor node and a 34 mW load. The battery was charged in addition to supplying the load, achieving a system efficiency of 92%. Furthermore, the system efficiency with the wearable sensor node was 95%.

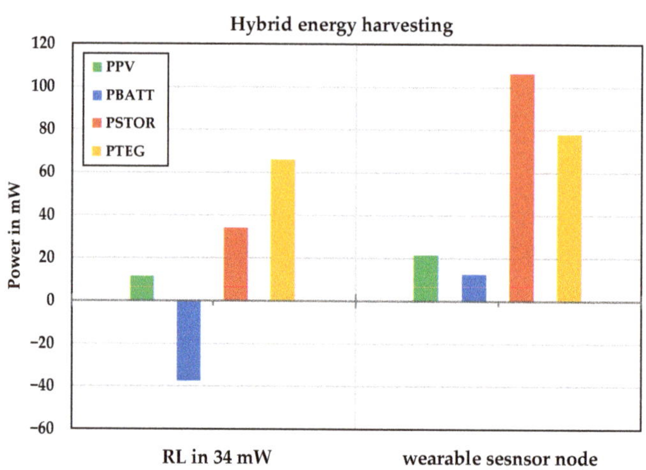

Figure 11. The contribution of hybrid energy harvesting resources in supplying the output load in shadow conditions.

Figure 12 compares the hybrid energy harvesting system under partly cloudy conditions and a temperature difference of 8 °C. According to this diagram, MPPT was achieved in the hybrid structure of both energy harvesting sources.

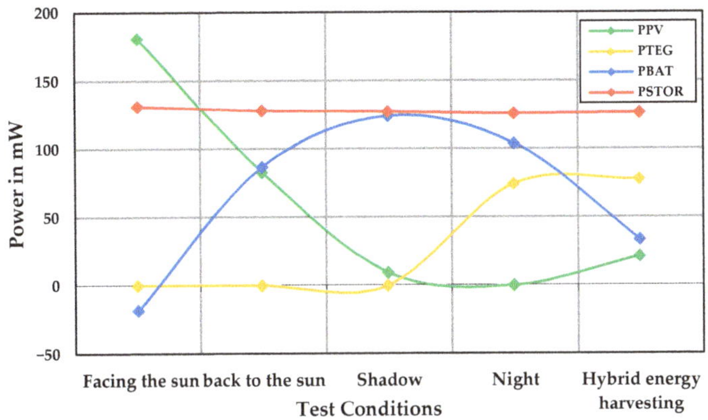

Figure 12. Power contribution of the hybrid energy harvesting system for wearable sensor node.

Figure 13 shows the efficiency and P_{loss} of the system and TEG module at a temperature difference of 8 °C, and the hybrid energy harvesting system under shadow conditions with a temperature difference of 8 °C. In the shadow condition, the battery provided most of the output power, resulting in high system efficiency. In the hybrid energy harvesting system, the system efficiency was 96%, with the battery providing only a small portion of the consumed power.

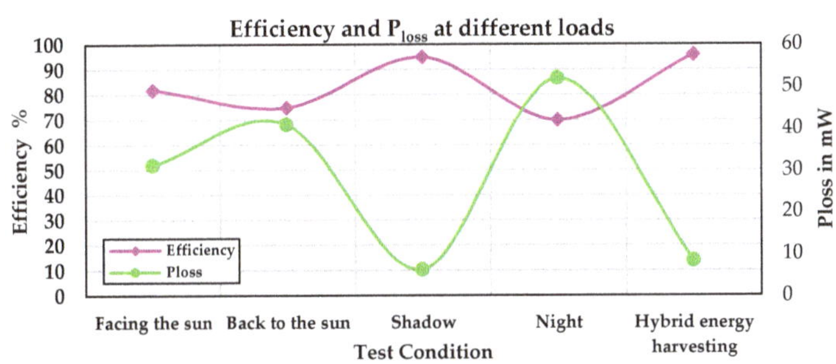

Figure 13. Energy harvesting system efficiency and P_{loss}.

5. Discussion and Comparison

We designed and developed a hybrid double-source energy harvesting system to support wearable devices for health monitoring in the head, face, and wrist regions, focusing on long-term continuous measurement. We primarily targeted individuals engaged in outdoor activities such as workers in the oil and gas industry. Given the high-power density of solar cells outdoors, a hybrid of PV and TEG energy was utilized to harvest energy quickly and sufficiently from sunlight and body heat. Since body heat is an inherent part of the human body, TEG can provide a useful energy source for the wearable device when sunlight is not available or when a person is indoors, considering the temperature difference of the available energy source. This improves the temporal-spatial stability and reliability of the harvesting system.

The system harvests input powers from PV and TEG sources simultaneously, utilizing two BQ25504 low-power DC/DC boost converters to supply the load and charge the battery. By eliminating Schottky diodes with high power losses and voltage drops, and only harvesting energy from the source with the higher voltage at any given moment, this structure can enhance the system's performance. Additionally, the PV panel and TEG modules support small form factor and high efficiency, which are essential in providing the ease of use, user experience, and unobtrusiveness of wearable devices.

Table 6 shows the comparison of the proposed hybrid energy harvesting system with the related previous works [28,30,50,51]. In any wearable device, the total dimension and form factor play pivotal roles in the user experience, useability, and practicability that lead to unobtrusiveness. Thus, careful circuit design and component selection are vital due to the limited space. Considering these points, we tried to reduce the overall area of the PV and TEG energy harvesting system, which was much lower compared to other works. Consequently, this can facilitate further wearing wearable devices by users and drive them toward the wear-and-forget. Although our careful design and component selection resulted in shrinking the size of the PV panel and TEG module, this reduction in dimensions did not decrease the output power of the PV panel and TEG module. The harvesting sources could achieve an appropriate output power from the solar sources and body heat, which can support powering the sensors and wearable devices and store excess harvested energy in batteries—in the ideal condition. The power consumption of the systems depends on various factors including the types of sensors. The other influencing factor in assessing the total energy consumption is the MCU/embedded system. For example, low-power MCUs such as nRF52840 (Nordic Semiconductor, Trondheim, Norway), MSP430FR5969 (Texas Instruments, Dallas, TX, USA), ADuCM302/ADuCM305 (Analog Devices, Wilmington, MA, USA), and STM32L4 (STMicroelectronics, Geneva, Switzerland) can extended the battery life. However, due to our focus on the energy harvesting system itself and its suitability for integration in wearable devices, along with considerations of the cost and availability, we used the NodeMCU microcontroller. Although this microcontroller has higher power consumption compared to other low-power options, the total energy consumption was not prioritized. This choice does not reflect the actual efficiency of the system, as it could simply be impacted by the microcontroller's higher power demands.

When harvesting energy from multiple sources, some kind of OR-ing structure is needed. This can be carried out before or after the voltage conversion step [40,41]. While using the first structure has the advantage of using a single voltage converter for both sources, it limits the power sources to having the same internal impedance, and only one source can be used at a time. Therefore, using two BQ25504 ultra-low power boost converters with the same configuration, we made it possible to achieve simultaneous energy harvesting from input sources and separate the MPPT for each source, which did not require an additional diode in the output part, and reduced the power loss.

Although the continuous measurement of physiological and non-physiological parameters is of concern to all groups of occupations and health, however, some of the targeting group could take priority due to several reasons such as safety and harsh environmental conditions that expose them to more frequent risks. Those in the oil and gas industry are such workers, and the continuous monitoring would provide them with several advantages. For instance, it is known for its hazardous working conditions including exposure to toxic chemicals, high-pressure equipment, extreme temperatures, and physically demanding tasks. Wearable health monitors can help identify potential health risks and provide real-time alerts in case of emergencies, enhancing worker safety. Furthermore, fatigue is a major concern in the industry—not only in oil and gas, but long working hours can also lead to decreased alertness and cognitive function. Wearable devices can track sleep patterns and activity levels to help employers and workers manage fatigue effectively, reducing the risk of accidents. Additionally, wearable devices can continuously monitor vital signs like HR, body temperature, and respiratory rate. This enables the early detection of health problems such as heat stress or cardiac issues, allowing for timely intervention and the prevention of more serious health events. Monitoring the health of workers in the oil and gas industry through wearables is crucial for enhancing safety, preventing accidents, complying with regulations, and improving overall worker well-being and productivity.

Table 6. Comparison of the current work with some previous studies.

Ref.	Energy Source	Sensors Deployed	Energy Storage	Area of Harvester (mm^2)	Power of Harvester (mW)	Mode of Device Wearability	Energy Management IC	MCU Unit	Circuit Techniques for Hybrid
This work	PV, TEG	PPG, Accelerometer	Battery, 300 mAh	Panel = 1840, TEG = 1024	Panel = 307, TEG = 78.2 at (ΔT = 8 °C)	Glasses, Wrist-worn	Two BQ25504 boost converters	NodeMCU ESP8266	Energy harvesting from both sources, without diode
[28]	PV, TEG	Temperature, Pulse oximeter, Accelerometer	Supercapacitor, 50 F	Panel = 4320, TEG = 1600	Panel = 207, TEG = 50 at (ΔT = 20 °C)	Wrist-worn	LTC3105 boost converter	ATmega-328p	Power OR-ing.
[30]	TEG	Powering a LED	N/A	TEG = 559	TEG = 0.023 at (ΔT = 10 °C)	Wrist-worn	LTC3108 boost converter	N/A	—
[50]	PV	N/A	Battery, CR2025-supercapacitor, 4 F	40,000	820	N/A	BQ25570 buck-boost converter	Atmel ATMEGA328P-AU	—
[51]	PV, TEG	Nano-power accelerometer, Temperature, Analog microphone	Battery, 40 mAh	Panel = 3892, TEG = 560	Panel = 4.42, TEG = 2.62 at (ΔT = 16 °C)	Bracelet	BQ25570 buck-boost converter and LTC3108 boost converter	MSP430FR5969	Energy harvesting from both sources, with diode

Therefore, considering the condition in which our proposed system was worn by outdoor workers exposed to sunlight for two hours (in the actual situation of workers in the oil and gas industry, this period is longer), the energy consumption of the wearable sensor node was turned into a self-powered system. In this situation, surplus energy was also stored in the battery for times when energy harvesting sources were not available. However, one of our main limitations was in evaluating the system in real conditions in the workplace, particularly in the oil and gas industry. Additionally, convincing individuals to wear these glasses during work poses a challenge. The application of this system can be considered not only for workers in the oil and gas industry, but also for other individuals such as mountaineers and those interested in health monitoring.

Wearable electronics, despite their negligible individual power consumption, significantly impact the global energy usage due to their sheer numbers. In 2022, 1.1 billion wearable healthcare devices, consuming an average of 656 mW each (including hub consumption), accounted for 727 MW of power and 1942 tons of CO_2 emissions, necessitating 11.34 million conifer trees to offset. Given their exponential growth, it is crucial to design more energy-efficient wearables. Implementing power harvesting techniques could mitigate their environmental footprint and reduce the burden on the energy sector [52].

In future work, it is possible to expand energy harvesting input sources and provide a multi-input hybrid structure (e.g., body motion energy or RF ambient energy). Furthermore, integrating health monitoring sensors such as those for checking the blood glucose levels of individuals with diabetes, skin temperature for detecting fever, and more. Additionally, the use of flexible components, which bring a lot of comfort to wearable devices, can be explored.

6. Conclusions

One of the main challenges in the continuous and unobtrusive measurement of health-related parameters by wearable biomedical sensors is the battery's capacity and form factor. Energy harvesting techniques (single/multisource) are widely used to extend the lifetime of wearable nodes. We designed and implemented an efficient hybrid PV/TEG energy harvesting system in a compact, low-cost form factor that is compatible with wearables worn on the face, head, and wrist. This system prolongs measurement and supports the continuous monitoring of physiological parameters.

Our meticulous design and component selection resulted in a smaller PV panel and TEG module without reducing their output power. The PV panel conversion efficiency of our proposed system was 16.68%, significantly higher than the previously reported 4.79%.

At the core of the system, we utilized two BQ25504 DC/DC boost converters, both active and employed simultaneously. Unlike previous studies using the OR-ing structure, our design allowed each power source to operate with MPPT, charging/supplying a common battery/load simultaneously or individually.

Experimental results demonstrated the feasibility of the overall design, doubling the sensor system's battery lifetime to 18 h. The efficiency of the hybrid energy harvesting system was 96%, enabling the system to become self-powered under direct sunlight for two hours. Our results indicate that the system could be further extended for use in outdoor occupations.

Author Contributions: Conceptualization, Z.T., M.H. and S.D.; Methodology, Z.T., M.H. and S.D.; Software, Z.T.; Validation, Z.T. and M.H.; Formal analysis, Z.T.; Investigation, Z.T.; Resources, Z.T.; Data curation, Z.T.; Writing—original draft preparation, Z.T., M.H. and S.D.; Writing—review and editing, Z.T., M.H., M.V., R.S., N.T. and S.D.; Visualization, Z.T.; Supervision, M.H. and S.D.; Project administration, S.D. and M.H. All authors have read and agreed to the published version of the manuscript.

Funding: This research received no external funding.

Institutional Review Board Statement: Not applicable.

Informed Consent Statement: Not applicable.

Data Availability Statement: No new data were created or analyzed in this study. Data sharing is not applicable to this article.

Conflicts of Interest: The authors declare no conflicts of interest.

References

1. Hasasneh, A.; Hijazi, H.; Talib, M.A.; Afadar, Y.; Nassif, A.B.; Nasir, Q. Wearable Devices and Explainable Unsupervised Learning for COVID-19 Detection and Monitoring. *Diagnostics* **2023**, *13*, 3071. [CrossRef]
2. Cheong, S.H.; Ng, Y.J.; Lau, Y.; Lau, S.T. Wearable technology for early detection of COVID-19: A systematic scoping review. *Prev. Med.* **2022**, *162*, 107170. [CrossRef]
3. Wang, W.H.; Hsu, W.S. Integrating artificial intelligence and wearable IoT system in long-term care environments. *Sensors* **2023**, *23*, 5913. [CrossRef]
4. Patil, V.; Singhal, D.K.; Naik, N.; Hameed, B.Z.; Shah, M.J.; Ibrahim, S.; Smriti, K.; Chatterjee, G.; Kale, A.; Sharma, A.; et al. Factors Affecting the Usage of Wearable Device Technology for Healthcare among Indian Adults: A Cross-Sectional Study. *J. Clin. Med.* **2022**, *11*, 7019. [CrossRef]
5. Popov, V.V.; Kudryavtseva, E.V.; Kumar Katiyar, N.; Shishkin, A.; Stepanov, S.I.; Goel, S. Industry 4.0 and digitalisation in healthcare. *Materials* **2022**, *15*, 2140. [CrossRef]
6. Chong, Y.W.; Ismail, W.; Ko, K.; Lee, C.Y. Energy harvesting for wearable devices: A review. *IEEE Sens. J.* **2019**, *19*, 9047–9062. [CrossRef]
7. Damre, S.S.; Shendkar, B.D.; Kulkarni, N.; Chandre, P.R.; Deshmukh, S. Smart Healthcare Wearable Device for Early Disease Detection Using Machine Learning. *Int. J. Intell. Syst. Appl. Eng.* **2024**, *12*, 158–166.
8. Guk, K.; Han, G.; Lim, J.; Jeong, K.; Kang, T.; Lim, E.K.; Jung, J. Evolution of wearable devices with real-time disease monitoring for personalized healthcare. *Nanomaterials* **2019**, *9*, 813. [CrossRef]
9. Zovko, K.; Šerić, L.; Perković, T.; Belani, H.; Šolić, P. IoT and health monitoring wearable devices as enabling technologies for sustainable enhancement of life quality in smart environments. *J. Clean. Prod.* **2023**, *413*, 137506. [CrossRef]
10. Davies, H.J.; Williams, I.; Peters, N.S.; Mandic, D.P. In-ear spo2: A tool for wearable, unobtrusive monitoring of core blood oxygen saturation. *Sensors* **2020**, *20*, 4879. [CrossRef]
11. Hussain, Z.; Sheng, Q.Z.; Zhang, W.E.; Ortiz, J.; Pouriyeh, S. Non-invasive techniques for monitoring different aspects of sleep: A comprehensive review. *ACM Trans. Comput. Healthc.* **2022**, *3*, 1–26. [CrossRef]
12. Haghi, M.; Ershadi, A.; Deserno, T.M. Recognizing Human Activity of Daily Living Using a Flexible Wearable for 3D Spine Pose Tracking. *Sensors* **2023**, *23*, 2066. [CrossRef]
13. Bellagente, P.; Crema, C.; Depari, A.; Ferrari, P.; Flammini, A.; Lanfranchi, G.; Lenzi, G.; Maddiona, M.; Rinaldi, S.; Sisinni, E.; et al. Remote and non-invasive monitoring of elderly in a smart city context. In Proceedings of the 2018 IEEE Sensors Applications Symposium (SAS), Seoul, Republic of Korea, 12–14 March 2018; pp. 1–6.
14. Razavi, M.; McDonald, A.; Mehta, R.; Sasangohar, F. Evaluating Mental Stress Among College Students Using Heart Rate and Hand Acceleration Data Collected from Wearable Sensors. *arXiv* **2023**, arXiv:2309.11097.

15. Kim, J.; Khan, S.; Wu, P.; Park, S.; Park, H.; Yu, C.; Kim, W. Self-charging wearables for continuous health monitoring. *Nano Energy* **2021**, *79*, 105419. [CrossRef]
16. Nozariasbmarz, A.; Collins, H.; Dsouza, K.; Polash, M.H.; Hosseini, M.; Hyland, M.; Liu, J.; Malhotra, A.; Ortiz, F.M.; Mohaddes, F.; et al. Review of wearable thermoelectric energy harvesting: From body temperature to electronic systems. *Appl. Energy* **2020**, *258*, 114069. [CrossRef]
17. Hesham, R.; Soltan, A.; Madian, A. Energy harvesting schemes for wearable devices. *AEU-Int. J. Electron. Commun.* **2021**, *138*, 153888. [CrossRef]
18. Davies, H.J.; Bachtiger, P.; Williams, I.; Molyneaux, P.L.; Peters, N.S.; Mandic, D.P. Wearable in-ear PPG: Detailed respiratory variations enable classification of COPD. *IEEE Trans. Biomed. Eng.* **2022**, *69*, 2390–2400. [CrossRef]
19. Haghi, M.; Danyali, S.; Thurow, K.; Warnecke, J.M.; Wang, J.; Deserno, T.M. Hardware prototype for wrist-worn simultaneous monitoring of environmental, behavioral, and physiological parameters. *Appl. Sci.* **2020**, *10*, 5470. [CrossRef]
20. Haghi, M.; Danyali, S.; Ayasseh, S.; Wang, J.; Aazami, R.; Deserno, T.M. Wearable devices in health monitoring from the environmental towards multiple domains: A survey. *Sensors* **2021**, *21*, 2130. [CrossRef] [PubMed]
21. He, Z.; Wang, K.; Zhao, Z.; Zhang, T.; Li, Y.; Wang, L. A Wearable Flexible Acceleration Sensor for Monitoring Human Motion. *Biosensors* **2022**, *12*, 620. [CrossRef] [PubMed]
22. Babar, M.; Rahman, A.; Arif, F.; Jeon, G. Energy-harvesting based on internet of things and big data analytics for smart health monitoring. *Sustain. Comput. Inform. Syst.* **2018**, *20*, 155–164. [CrossRef]
23. Páez-Montoro, A.; García-Valderas, M.; Olías-Ruíz, E.; López-Ongil, C. Solar energy harvesting to improve capabilities of wearable devices. *Sensors* **2022**, *22*, 3950. [CrossRef] [PubMed]
24. Proto, A.; Bibbo, D.; Cerny, M.; Vala, D.; Kasik, V.; Peter, L.; Conforto, S.; Schmid, M.; Penhaker, M. Thermal energy harvesting on the bodily surfaces of arms and legs through a wearable thermo-electric generator. *Sensors* **2018**, *18*, 1927. [CrossRef] [PubMed]
25. Huet, F.; Boitier, V.; Seguier, L. Tunable piezoelectric vibration energy harvester with supercapacitors for WSN in an industrial environment. *IEEE Sens. J.* **2022**, *22*, 15373–15384. [CrossRef]
26. Gljušćić, P.; Zelenika, S.; Blažević, D.; Kamenar, E. Kinetic energy harvesting for wearable medical sensors. *Sensors* **2019**, *19*, 4922. [CrossRef] [PubMed]
27. Sherazi, H.H.; Zorbas, D.; O'Flynn, B. A comprehensive survey on RF energy harvesting: Applications and performance determinants. *Sensors* **2022**, *22*, 2990. [CrossRef] [PubMed]
28. Mohsen, S.; Zekry, A.; Youssef, K.; Abouelatta, M. A self-powered wearable wireless sensor system powered by a hybrid energy harvester for healthcare applications. *Wirel. Pers. Commun.* **2021**, *116*, 3143–3164. [CrossRef]
29. Bai, Y.; Jantunen, H.; Juuti, J. Energy harvesting research: The road from single source to multisource. *Adv. Mater.* **2018**, *30*, 1707271. [CrossRef]
30. Shi, Y.; Wang, Y.; Mei, D.; Feng, B.; Chen, Z. Design and fabrication of wearable thermoelectric generator device for heat harvesting. *IEEE Robot. Autom. Lett.* **2017**, *3*, 373–378. [CrossRef]
31. Xiao, L.; Wu, K.; Tian, X.; Luo, J. Activity-specific caloric expenditure estimation from kinetic energy harvesting in wearable devices. *Pervasive Mob. Comput.* **2020**, *67*, 101185. [CrossRef]
32. Pillatsch, P.; Yeatman, E.M.; Holmes, A.S. Real world testing of a piezoelectric rotational energy harvester for human motion. *J. Phys. Conf. Ser.* **2013**, *476*, 012010. [CrossRef]
33. Delnavaz, A.; Voix, J. Energy harvesting for in-ear devices using ear canal dynamic motion. *IEEE Trans. Ind. Electron.* **2013**, *61*, 583–590. [CrossRef]
34. Li, X.; Sun, Y. WearETE: A scalable wearable e-textile triboelectric energy harvesting system for human motion scavenging. *Sensors* **2017**, *17*, 2649. [CrossRef]
35. Farooq, M.; Sazonov, E. Segmentation and characterization of chewing bouts by monitoring temporalis muscle using smart glasses with piezoelectric sensor. *IEEE J. Biomed. Health Inform.* **2016**, *21*, 1495–1503. [CrossRef]
36. Yu, B.Y.; Wang, Z.H.; Ju, L.; Zhang, C.; Liu, Z.G.; Tao, L.; Lu, W.B. Flexible and wearable hybrid RF and solar energy harvesting system. *IEEE Trans. Antennas Propag.* **2021**, *70*, 2223–2233. [CrossRef]
37. Veloo, S.G.; Tiang, J.J.; Muhammad, S.; Wong, S.K. A Hybrid Solar-RF Energy Harvesting System Based on an EM4325-Embedded RFID Tag. *Electronics* **2023**, *12*, 4045. [CrossRef]
38. Noh, Y.S.; Seo, J.I.; Kim, H.S.; Lee, S.G. A reconfigurable DC/DC converter for maximum thermoelectric energy harvesting in a battery-powered duty-cycling wireless sensor node. *IEEE J. Solid-State Circuits* **2022**, *57*, 2719–2730. [CrossRef]
39. Ali, A.; Shaukat, H.; Bibi, S.; Altabey, W.A.; Noori, M.; Kouritem, S.A. Recent progress in Energy Harvesting Systems for wearable technology. *Energy Strategy Rev.* **2023**, *49*, 101124. [CrossRef]
40. Tan, Y.K.; Panda, S.K. Energy harvesting from hybrid indoor ambient light and thermal energy sources for enhanced performance of wireless sensor nodes. *IEEE Trans. Ind. Electron.* **2010**, *58*, 4424–4435. [CrossRef]
41. Carli, D.; Brunelli, D.; Benini, L.; Ruggeri, M. An effective multi-source energy harvester for low power applications. In Proceedings of the IEEE 2011 Design, Automation & Test in Europe, Grenoble, France, 14–18 March 2011; pp. 1–6.
42. Mouser. AEM10941 Solar Energy Harvesting IC, e-peas. Available online: https://www.mouser.com/new/e-peas/e-peas-aem10941-solar-energy-harvesting-ic/ (accessed on 23 August 2023).
43. Mouser. AEM20940 Thermal Energy Harvesting IC, e-peas. Available online: https://www.mouser.com/new/e-peas/e-peas-aem20940-thermal-energy-harvesting-ic/ (accessed on 23 August 2023).

44. Analog Devices. ADP5091 Datasheet and Product Info. Available online: https://www.analog.com/en/products/adp5091.html (accessed on 23 August 2023).
45. TI.com. BQ25504 Data Sheet, Product Information and Support. Available online: https://www.ti.com/product/BQ25504 (accessed on 23 August 2023).
46. TI.com. BQ25570 Data Sheet, Product Information and Support. Available online: https://www.ti.com/product/BQ25570 (accessed on 23 August 2023).
47. SPV1050—STMicroelectronics, STMicroelectronics. Available online: https://www.st.com/en/power-management/spv1050.html (accessed on 23 August 2023).
48. Analog Devices. LTC3105 Datasheet and Product Info. Available online: https://www.analog.com/en/products/ltc3105.html (accessed on 23 August 2023).
49. Analog Devices. LTC3106 Datasheet and Product Info. Available online: https://www.analog.com/en/products/ltc3106.html (accessed on 23 August 2023).
50. Guragain, D.P.; Budhathoki, R.K.; Ghimire, P. Programmable timer triggered energy harvesting wireless sensor-node using long range radio access technology. *Int. J. Electr. Comput. Eng.* **2022**, *12*, 3869–3881.
51. Magno, M.; Brunelli, D.; Sigrist, L.; Andri, R.; Cavigelli, L.; Gomez, A.; Benini, L. InfiniTime: Multi-sensor wearable bracelet with human body harvesting. *Sustain. Comput. Inform. Syst.* **2016**, *11*, 38–49. [CrossRef]
52. TaheriNejad, N.; Perego, P.; Rahmani, A.M. Mobile Health Technology: From Daily Care and Pandemics to their Energy Consumption and Environmental Impact. *Mob. Netw. Appl.* **2022**, *27*, 652–656. [CrossRef]

sensors MDPI

Article

Validity and Test–Retest Reliability of Spatiotemporal Running Parameter Measurement Using Embedded Inertial Measurement Unit Insoles

Louis Riglet [1,2], Baptiste Orliac [1,2], Corentin Delphin [1,2], Audrey Leonard [3], Nicolas Eby [3], Paul Ornetti [1,2,4,5,6], Davy Laroche [1,2,4,6] and Mathieu Gueugnon [1,2,4,6,*]

1 CHU Dijon–Bourgogne, Centre d'Investigation Clinique, Module Plurithématique, Plateforme d'Investigation Technologique, 21000 Dijon, France; louis.riglet@chu-dijon.fr (L.R.); paul.ornetti@chu-dijon.fr (P.O.); davy.laroche@u-bourgogne.fr (D.L.)
2 INSERM, CIC 1432, Module Plurithématique, Plateforme d'Investigation Technologique, 21000 Dijon, France
3 Zhortech SAS, 54000 Nancy, France
4 INSERM, UMR1093-CAPS, Université Bourgogne Franche-Comté, UB, 21000 Dijon, France
5 Rheumatology Department, CHU Dijon–Bourgogne, 21000 Dijon, France
6 Collaborative Research Network STARTER (Innovative Strategies and Artificial Intelligence for Motor Function Rehabilitation and Autonomy Preservation), 21000 Dijon, France
* Correspondence: mathieu.gueugnon@chu-dijon.fr

Abstract: Running is the basis of many sports and has highly beneficial effects on health. To increase the understanding of running, DSPro® insoles were developed to collect running parameters during tasks. However, no validation has been carried out for running gait analysis. The aims of this study were to assess the test–retest reliability and criterion validity of running gait parameters from DSPro® insoles compared to a motion-capture system. Equipped with DSPro® insoles, a running gait analysis was performed on 30 healthy participants during overground and treadmill running using a motion-capture system. Using an intraclass correlation coefficient (ICC), the criterion validity and test–retest reliability of spatiotemporal parameters were calculated. The test–retest reliability shows moderate to excellent ICC values (ICC > 0.50) except for propulsion time during overground running at a fast speed with the motion-capture system. The criterion validity highlights a validation of running parameters regardless of speeds (ICC > 0.70). This present study validates the good criterion validity and test–retest reliability of DSPro® insoles for measuring spatiotemporal running gait parameters. Without the constraints of a 3D motion-capture system, such insoles seem to be helpful and relevant for improving the care management of active patients or following running performance in sports contexts.

Keywords: running gait analysis; insole; validation; repeatability; 3D motion analysis; running

1. Introduction

Running is among the most popular physical activities, and it is the basis of many sports or leisure activities. The practice of running, outside or on a treadmill, has greatly increased in recent years [1] with high health benefits [2,3]. To evaluate its effects and improve performance [4,5], analyzing running patterns is increasingly performed. Traditionally, rating scales and subjective observations are used for this analysis, but these approaches are less sensitive to performance changes in training or injury [6,7].

To provide objective, reliable, and reproducible outcomes, running gait analysis can be performed using 3D motion capture [8]. This non-invasive motion system is considered to be a gold standard for walking and running gait analysis [9–11] and helps to measure locomotion in depth through spatiotemporal, kinematic, and kinetic parameters. However, the principal disadvantages of this system are its long operation time, dedicated space requirements, technical expertise requirements, and high costs, preventing an assessment of running in an ecological context [9]. Significant progress has been made with the development of markerless systems and smartphone apps for performing gait analysis in clinical and sports contexts [12–14]. They have the advantage of reducing data collection and processing time [13]. However, these systems have limitations, including limited capture volume (based on the number of cameras) or partial accuracy (e.g., joint angle estimation) [12,13].

To overcome these 3D gait recording limitations, wearable devices are increasingly accepted and used by runners and clinicians [15–17]. Wearable systems (Inertial Measurement Unit—IMU) are based on a combination of accelerometers, gyroscopes, and magnetometers, allowing the description of normal or pathological human locomotion in a wide variety of environments [18]. In running gait analysis, IMUs help to measure running gait parameters, notably spatiotemporal parameters, and quantify running performance [19,20]. Despite their good feasibility and performance, there are major remaining technical disadvantages, such as sensor attachment errors, external signal noise, signal filtering errors, and integration drift [18].

To record foot movements and orientations, IMUs can be fixed on or under the foot. In particular, DSPro® insoles, developed by DigitSole SAS, have the advantage of being able to collect data during different tasks. For instance, this device has been validated for walking gait analysis compared to 3D motion analysis [21]. Recent studies of this device have highlighted good repeatability in the measurement of walking gait parameters such as cadence, walking speed, and stride length [21–23]. However, the validation of the extracted parameters during running is a prerequisite for extending their use in this activity.

For this reason, the aim of this present study was to evaluate the test–retest reliability and criterion validity (i.e., the extent to which scores for volunteers who have not changed are the same for repeated measurements over time and the degree to which the scores of Patient-Reported Outcome Measures (PROMs) are an adequate reflection of a "gold standard", respectively [24]), of the running parameters measured using embedded insoles in comparison to those obtained from the gold-standard system, i.e., a motion analysis system.

2. Materials and Methods

2.1. Population

A monocentric study was performed from October 2021 to February 2022 and included 30 healthy participants [21] who provided informed oral consent. The local institutional ethics committee and the French National Agency for Drug Safety approved and authorized the study protocol. The Clinical Trial registration reference is NCT05104645.

2.2. Procedure and Materials

The procedure and materials used in this study were described by Riglet et al. [21]. Briefly, the experiments were carried out at the INSERM U1093 laboratory (Dijon University, France). Each participant participated in two visits, with the second visit occurring seven days after the first. Motion analysis was carried out using Nexus software (Vicon System®, Oxford, UK, 2.12.1 version), and the volunteers were equipped with the Conventional Gait Model (version 2.5) markers set [25,26]. One marker was added at the extremity of the shoe, near the hallux. The 3D position of each marker was tracked with 18 optoelectronic cameras (11 VERO and 7 MX-T10 cameras, Vicon System®, Oxford, UK; 100 Hz). Two force plates (AMTI®, Watertown, MA, USA; 1000 Hz) were embedded in the floor to record ground reaction forces. The volume of interest was approximately 10 m × 2 m × 2.5 m, with an error threshold set to 0.5 mm in accordance with manufacturer data. Additionally, each participant wore identical shoes (Ekiden One, Kalenji®, Villeneuve d'Ascq, France) fitted with DSPro® insoles (Zhortech® algorithms, DigitSole SAS, Nancy, France, 104 Hz) (Figure 1). A sensor integrating accelerometers, gyroscopes and magnetometers was fixed onto the proximal part of each insole, allowing the calculation of linear accelerations and angular velocities in three dimensions.

Figure 1. DSPro® insole device.

Participation involved two running conditions (overground and treadmill running) at two different speeds. Participants were asked to run on a 10-m walkway at a self-selected speed (comfortable) and then at a fast speed. These running overground tasks were performed at least five back-and-forth runs for each speed. In the second condition, participants were asked to run on a treadmill at a self-selected speed (comfortable speed obtained after a 2-min familiarization trial), then at a fast speed (adding 2 km/h to the comfortable speed).

2.3. Data Analysis

The synchronization between the two systems (IMUs and motion capture) was synchronized using an analog button to trigger the beginning of each running trial. For each participant, a minimum of 5 trials for running overground and 1 trial of 2 min for treadmill running were collected (for comfortable and fast speed).

For motion capture, lower limb markers and ground reaction forces were post-processed following the method described in Riglet et al. [21]. Briefly, gait events (heel strike and toe off) were computed using the position of the foot velocity algorithm by O'Connor et al. [25]. The maximum vertical foot velocity corresponds to the toe off and the minimum to heel strike. Then, two other gait parameters were computed: flat foot in (FFI) and flat foot out (FFO), corresponding, respectively, to the sample in which the toe touches the ground after the heel strike and to the sample in which the heel takes off the ground after heel strike (please see Riglet et al. [21] for more details). For insoles, these events were also extracted (Zhortech properties). A comparison of the initial heel strike of each gait cycle between motion capture and insoles was performed and considered to be synchronized if the difference was less than 0.35 s. Custom-made scripts with Python (3.9) and Matlab (MathWorks®, Natick, MA, USA, R2023a) were used to perform all post-processing.

Regardless of running conditions, the parameters of interest were speed, stride cadence, flight time, stance time, stride time, swing time, stride length, stride height, and plantar flexion angle foot in. For overground running, loading time, propulsion time, impact force, and leg stiffness were also measured based on force plate data. Without an instrumented treadmill, these parameters were not measured for treadmill conditions. The definition of running parameters measured with motion capture is detailed in Table S1 (definition of parameters measured with insoles are not available: property of Zhortech company).

2.4. Statistical Analysis

Based on the comparison of each running parameter measured during two sessions, test–retest reliability was estimated. As suggested by Koo & Li [26], estimation was performed using intraclass correlation (ICC) and its 95% confidence interval (CI) based on a mean-rating, absolute agreement, 2-way mixed-effects model. For motion capture and insoles, ICC and p-value (p) were computed between the two sessions for each running condition. Additionally, absolute and relative errors between sessions were computed, and a comparison between motion capture and insoles was performed using a Student's t-test. Based on the standard error of measurement (SEM) and the standard deviation (SD), minimal detectable change (MDC) was also computed for each variable and condition [27] and was defined as follows:

$$\text{MDC} = 1.96.\text{SEM}.\sqrt{2} \text{ with SEM} = \text{SD}\sqrt{1 - \text{ICC}}.$$

Based on the comparison of each running parameter measured by motion capture and insoles, criterion validity was estimated. As suggested by Koo & Li [26], estimation was performed using intraclass correlation based on a mean-rating, consistency, 2-way mixed-effects model (regardless of the session and gait side (right and left) for both systems). Additionally, a Lin concordance coefficient (CCC) and Bland–Altman analysis were also carried out. Based on Bland–Altman plots, limits of agreement (LoA) were calculated and defined as the mean difference between two measurements ± 1.96 SD of the difference.

ICC values were classified into four categories: poor (lower than 0.5), moderate (between 0.5 and 0.75), good (between 0.75 and 0.9), and excellent (higher than 0.9) [26]. Using the Bonferroni correction, the threshold of significance was fixed at $p < 0.004$ (0.05 divided by the number of variables (13)). CCC values were considered to be excellent when the coefficient was higher than 0.8 [28]. All statistical analyses were performed using Matlab (MathWorks®, R2023a).

3. Results

3.1. Population Characteristics

The population was composed of 14 females and 16 males. Female population characteristics were age 27.6 ± 5.2 years, height 165.2 ± 5.3 cm, body mass 61.0 ± 9.2 kg and shoe size 39.2 ± 1.8. Male population characteristics were age 28.2 ± 6.1 years, height 180.3 ± 5.3 cm, body mass 74.4 ± 8.6 kg and shoe size 43.4 ± 1.5.

3.2. Running Gait Cycle

For overground running, 2896 gait cycles were recorded (comfortable speed: 1640 cycles and fast speed: 1256 cycles). For treadmill running, 18,945 gait cycles were recorded (comfortable: 9317 cycles and fast: 9628 cycles).

3.3. Test–Retest Reliability

For overground running, the mean and standard deviation of each parameter, ICC values, and MDC are presented in Table 1 for motion capture and Table 2 for insoles.

Table 1. Mean, standard deviation, ICC values (with 95% CI), and minimal detectable change (MDC) of running parameters were measured using a motion-capture system for overground conditions. Significant results are indicated as follows: * = $p < 0.001$, + = $p < 0.004$.

	OVERGROUND RUNNING—MOTION CAPTURE							
	Comfortable Speed				Fast Speed			
	Mean ± SD		ICC [95% CI]	MDC	Mean ± SD		ICC [95% CI]	MDC
	Session 1	Session 2			Session 1	Session 2		
Speed (m/s)	2.75 ± 0.44	2.87 ± 0.34	0.549 [0.238; 0.761] +	0.73	4.14 ± 0.61	4.17 ± 0.55	0.776 [0.581; 0.887] *	0.75
Stride Cadence (step/min)	79.89 ± 3.12	80.16 ± 3.07	0.850 [0.705; 0.928] *	3.29	87.81 ± 6.10	87.87 ± 5.19	0.720 [0.489; 0.857] *	8.23
Flight Time (s)	0.20 ± 0.06	0.22 ± 0.06	0.837 [0.637; 0.926] *	0.07	0.26 ± 0.04	0.26 ± 0.04	0.912 [0.824; 0.957] *	0.04
Stance Time (s)	0.28 ± 0.03	0.27 ± 0.03	0.723 [0.454; 0.866] *	0.04	0.21 ± 0.02	0.21 ± 0.02	0.746 [0.531; 0.871] *	0.03
Stride Time (s)	0.75 ± 0.03	0.75 ± 0.03	0.847 [0.699; 0.926] *	0.03	0.68 ± 0.05	0.68 ± 0.04	0.704 [0.463; 0.848] *	0.07
Swing Time (s)	0.48 ± 0.04	0.48 ± 0.04	0.903 [0.784; 0.956] *	0.03	0.47 ± 0.04	0.47 ± 0.04	0.839 [0.688; 0.920] *	0.04
Loading Time (s)	0.13 ± 0.01	0.13 ± 0.01	0.757 [0.539; 0.881] *	0.02	0.11 ± 0.01	0.11 ± 0.02	0.620 [0.316; 0.809] *	0.02
Propulsion Time (s)	0.15 ± 0.02	0.14 ± 0.02	0.524 [0.164; 0.756] +	0.04	0.11 ± 0.02	0.10 ± 0.02	0.499 [0.152; 0.738]	0.04
Stride Length (m)	2.08 ± 0.33	2.17 ± 0.29	0.657 [0.385; 0.825] *	0.51	2.85 ± 0.38	2.87 ± 0.35	0.904 [0.809; 0.953] *	0.31
Stride Height (m)	0.30 ± 0.09	0.33 ± 0.08	0.725 [0.473; 0.865] *	0.12	0.49 ± 0.11	0.50 ± 0.11	0.881 [0.766; 0.942] *	0.11
Plantar Flexion Foot In (°)	15.16 ± 4.60	15.43 ± 3.98	0.863 [0.726; 0.934] *	4.38	15.71 ± 4.88	15.37 ± 4.48	0.646 [0.376; 0.815] *	7.67
Impact Force (kN)	1.60 ± 0.31	1.63 ± 0.30	0.963 [0.909; 0.984] *	0.16	1.78 ± 0.32	1.78 ± 0.30	0.961 [0.914; 0.982] *	0.17
Leg Stiffness (kN/m)	9.87 ± 2.04	9.96 ± 1.82	0.952 [0.898; 0.978] *	1.17	10.14 ± 2.23	10.35 ± 2.34	0.717 [0.469; 0.862] *	3.38

Table 2. Mean, standard deviation, ICC values (with 95% CI), and minimal detectable change (MDC) of running parameters were measured using insoles for overground conditions. Significant results are indicated as follows: * = $p < 0.001$, $^{+}$ = $p < 0.004$.

	OVERGROUND RUNNING—INSOLES							
	Comfortable Speed				Fast Speed			
	Mean ± SD		ICC [95% CI]	MDC	Mean ± SD		ICC [95% CI]	MDC
	Session 1	Session 2			Session 1	Session 2		
Speed (m/s)	2.76 ± 0.41	2.88 ± 0.33	0.514 [0.192; 0.738] $^{+}$	0.73	3.98 ± 0.50	4.02 ± 0.42	0.687 [0.439; 0.838] *	0.71
Stride Cadence (step/min)	79.83 ± 3.08	80.31 ± 2.87	0.745 [0.524; 0.873] *	4.14	90.02 ± 6.74	90.27 ± 6.41	0.662 [0.399; 0.824] *	10.51
Flight Time (s)	0.21 ± 0.06	0.22 ± 0.05	0.859 [0.694; 0.935] *	0.06	0.25 ± 0.04	0.25 ± 0.04	0.928 [0.855; 0.965] *	0.03
Stance Time (s)	0.27 ± 0.03	0.27 ± 0.03	0.758 [0.512; 0.884] *	0.04	0.21 ± 0.02	0.21 ± 0.02	0.718 [0.485; 0.855] *	0.03
Stride Time (s)	0.75 ± 0.03	0.75 ± 0.03	0.745 [0.525; 0.873] *	0.04	0.67 ± 0.05	0.67 ± 0.05	0.685 [0.434; 0.837] *	0.07
Swing Time (s)	0.48 ± 0.03	0.48 ± 0.03	0.883 [0.764; 0.944] *	0.03	0.46 ± 0.04	0.46 ± 0.04	0.798 [0.617; 0.898] *	0.05
Loading Time (s)	0.12 ± 0.02	0.12 ± 0.01	0.645 [0.362; 0.820] *	0.02	0.10 ± 0.01	0.09 ± 0.01	0.677 [0.402; 0.840] *	0.02
Propulsion Time (s)	0.15 ± 0.02	0.15 ± 0.02	0.655 [0.379; 0.825] *	0.03	0.12 ± 0.01	0.12 ± 0.01	0.531 [0.190; 0.758] $^{+}$	0.02
Stride Length (m)	2.07 ± 0.29	2.16 ± 0.26	0.628 [0.341; 0.808] *	0.47	2.66 ± 0.25	2.67 ± 0.22	0.844 [0.700; 0.922] *	0.25
Stride Height (m)	0.27 ± 0.10	0.30 ± 0.09	0.700 [0.451; 0.848] *	0.14	0.46 ± 0.11	0.46 ± 0.12	0.887 [0.776; 0.944] *	0.11
Plantar Flexion Foot In (°)	21.59 ± 4.75	21.97 ± 4.64	0.885 [0.770; 0.945] *	4.37	21.17 ± 5.29	20.63 ± 4.87	0.612 [0.328; 0.794] *	8.72
Impact Force (kN)	1.46 ± 0.29	1.46 ± 0.26	0.935 [0.863; 0.970] *	0.27	1.69 ± 0.32	1.69 ± 0.30	0.958 [0.909; 0.981] *	0.17
Leg Stiffness (kN/m)	10.00 ± 2.15	9.74 ± 1.80	0.749 [0.524; 0.876] *	2.74	10.14 ± 1.99	10.55 ± 2.11	0.672 [0.396; 0.838] *	3.24

For motion capture, ICC values were considered excellent for swing time, impact force, and leg stiffness at a comfortable speed and for flight time, stride length, and impact force at a fast speed. Speed, stance time, propulsion time, stride length, and stride height at comfortable speed had a moderate ICC value. The ICC values for the other parameters were considered to be good. Concerning fast speed, the ICC values for speed, swing time, and stride height were good. A lower ICC value was found for propulsion time (ICC = 0.499, $p = 0.004$). The other parameters had moderate ICC values.

For insole devices, ICC values were considered excellent for impact force and leg stiffness at a comfortable speed and for flight time, impact force, and leg stiffness at a fast speed. For comfortable speed, the following parameters had ICC values higher than 0.75: flight time, stance time, swing time, and flexion angle foot in. Fast speed, swing time, stride length, and stride height also had ICC values higher than 0.75. Regardless of speed conditions, the other parameters had moderate ICC values.

Concerning treadmill running, the mean and standard deviation of each parameter, ICC values, and MDC are presented in Table 3 for motion capture and Table 4 for insoles. For motion capture, parameters were significantly correlated between sessions with good to excellent ICC values (0.853 to 0.999, $p < 0.001$). For insole devices, parameters were significantly correlated between sessions with good to excellent ICC values (0.875 to 0.995, $p < 0.001$) regardless of speed.

Table 3. Mean, standard deviation, ICC values (with 95% CI), and minimal detectable change (MDC) of running parameters measured using a motion-capture system for treadmill conditions. Significant results are indicated as follows: * = $p < 0.001$.

	TREADMILL RUNNING—MOTION CAPTURE							
	Comfortable Speed				Fast Speed			
	Mean ± SD		ICC [95% CI]	MDC	Mean ± SD		ICC [95% CI]	MDC
	Session 1	Session 2			Session 1	Session 2		
Speed (m/s)	2.35 ± 0.31	2.35 ± 0.32	0.999 [0.997; 0.999] *	0.03	2.88 ± 0.32	2.88 ± 0.32	0.997 [0.994; 0.999] *	0.05
Stride Cadence (step/min)	78.64 ± 4.31	79.53 ± 3.86	0.877 [0.728; 0.943] *	3.97	81.28 ± 3.97	81.40 ± 4.13	0.919 [0.837; 0.961] *	3.17
Flight Time (s)	0.14 ± 0.06	0.13 ± 0.07	0.954 [0.905; 0.978] *	0.04	0.20 ± 0.05	0.19 ± 0.06	0.924 [0.848; 0.963] *	0.04
Stance Time (s)	0.32 ± 0.04	0.31 ± 0.04	0.950 [0.899; 0.976] *	0.02	0.27 ± 0.03	0.27 ± 0.03	0.948 [0.895; 0.975] *	0.02
Stride Time (s)	0.77 ± 0.04	0.76 ± 0.04	0.884 [0.734; 0.947] *	0.04	0.74 ± 0.04	0.74 ± 0.04	0.934 [0.867; 0.968] *	0.03
Swing Time (s)	0.45 ± 0.04	0.44 ± 0.04	0.927 [0.822; 0.968] *	0.03	0.47 ± 0.04	0.47 ± 0.04	0.920 [0.840; 0.961] *	0.03
Stride Length (m)	1.69 ± 0.20	1.67 ± 0.21	0.982 [0.957; 0.992] *	0.08	1.98 ± 0.20	1.98 ± 0.21	0.985 [0.970; 0.993] *	0.07
Stride Height (m)	0.26 ± 0.06	0.26 ± 0.06	0.928 [0.855; 0.965] *	0.04	0.35 ± 0.07	0.34 ± 0.08	0.914 [0.828; 0.958] *	0.06
Plantar Flexion Foot In (°)	13.39 ± 3.39	13.28 ± 3.22	0.853 [0.715; 0.927] *	3.48	13.67 ± 4.03	14.09 ± 3.83	0.886 [0.777; 0.944] *	3.64

Table 4. Mean, standard deviation, ICC values (with 95% CI), and minimal detectable change (MDC) of running parameters measured using insoles for treadmill conditions. Significant results are indicated as follows: * = $p < 0.001$.

	TREADMILL RUNNING—INSOLES							
	Comfortable Speed				Fast Speed			
	Mean ± SD		ICC [95% CI]	MDC	Mean ± SD		ICC [95% CI]	MDC
	Session 1	Session 2			Session 1	Session 2		
Speed (m/s)	2.36 ± 0.30	2.35 ± 0.30	0.995 [0.989; 0.998] *	0.06	2.86 ± 0.30	2.86 ± 0.31	0.990 [0.979; 0.995] *	0.08
Stride Cadence (step/min)	78.66 ± 4.31	79.56 ± 3.85	0.875 [0.725; 0.942] *	3.99	81.26 ± 3.97	81.37 ± 4.12	0.919 [0.837; 0.961] *	3.17
Flight Time (s)	0.17 ± 0.05	0.16 ± 0.06	0.906 [0.814; 0.954] *	0.05	0.21 ± 0.05	0.21 ± 0.06	0.903 [0.807; 0.952] *	0.05
Stance Time (s)	0.30 ± 0.04	0.30 ± 0.04	0.897 [0.796; 0.949] *	0.03	0.26 ± 0.02	0.27 ± 0.03	0.931 [0.861; 0.967] *	0.02
Stride Time (s)	0.77 ± 0.04	0.76 ± 0.04	0.883 [0.736; 0.946] *	0.04	0.74 ± 0.04	0.74 ± 0.04	0.935 [0.867; 0.968] *	0.03
Swing Time (s)	0.46 ± 0.04	0.45 ± 0.04	0.881 [0.760; 0.942] *	0.04	0.48 ± 0.04	0.48 ± 0.04	0.907 [0.815; 0.955] *	0.04
Stride Length (m)	1.80 ± 0.21	1.78 ± 0.22	0.983 [0.948; 0.993] *	0.08	2.11 ± 0.21	2.11 ± 0.22	0.985 [0.969; 0.993] *	0.07
Stride Height (m)	0.23 ± 0.07	0.23 ± 0.07	0.920 [0.841; 0.961] *	0.05	0.34 ± 0.08	0.33 ± 0.09	0.910 [0.821; 0.956] *	0.07
Plantar Flexion Foot In (°)	19.53 ± 3.80	19.05 ± 3.96	0.888 [0.779; 0.945] *	3.57	19.17 ± 4.46	19.29 ± 4.47	0.922 [0.843; 0.962] *	3.42

The absolute and relative error between the two sessions of each parameter extracted with motion capture and with insoles are presented in Tables S2 and S3. No significant difference was found, except for stride height during treadmill running (fast condition, $p = 0.002$).

3.4. Criterion Validity

Mean, standard deviation, and ICC values for running parameters are presented in Table 5 for the overground and Table 6 for the treadmill. Additionally, Bland–Altman plots for running parameters are presented in Figure 2 for the overground and Figure 3 for the treadmill. CCC values and plots are presented in Table S4 and Figures S1 and S2.

Table 5. Mean, standard deviation, and ICC values (with 95% CI) of running parameters for overground conditions. Significant results are indicated as follows: * = $p < 0.001$.

	OVERGROUND RUNNING					
	Comfortable Speed			Fast Speed		
	Mean ± SD		ICC [95% CI]	Mean ± SD		ICC [95% CI]
	Motion Capture	Insoles		Motion Capture	Insoles	
Speed (m/s)	2.81 ± 0.34	2.81 ± 0.32	0.987 [0.972; 0.994] *	4.15 ± 0.55	4.00 ± 0.42	0.948 [0.894; 0.975] *
Stride Cadence (step/min)	79.96 ± 3.07	79.94 ± 2.85	0.968 [0.934; 0.985] *	87.89 ± 5.23	90.21 ± 5.99	0.908 [0.816; 0.955] *
Flight Time (s)	0.20 ± 0.06	0.21 ± 0.05	0.965 [0.928; 0.983] *	0.26 ± 0.04	0.25 ± 0.04	0.940 [0.878; 0.971] *
Stance Time (s)	0.27 ± 0.03	0.27 ± 0.03	0.971 [0.940; 0.986] *	0.21 ± 0.02	0.21 ± 0.02	0.953 [0.903; 0.977] *
Stride Time (s)	0.75 ± 0.03	0.75 ± 0.03	0.977 [0.952; 0.989] *	0.68 ± 0.04	0.67 ± 0.04	0.950 [0.898; 0.976] *
Swing Time (s)	0.48 ± 0.04	0.48 ± 0.03	0.952 [0.902; 0.977] *	0.47 ± 0.04	0.46 ± 0.04	0.918 [0.836; 0.960] *
Loading Time (s)	0.13 ± 0.01	0.12 ± 0.01	0.858 [0.723; 0.930] *	0.11 ± 0.01	0.09 ± 0.01	0.811 [0.640; 0.905] *
Propulsion Time (s)	0.15 ± 0.02	0.15 ± 0.02	0.910 [0.821; 0.956] *	0.10 ± 0.02	0.12 ± 0.01	0.760 [0.555; 0.878] *
Stride Length (m)	2.12 ± 0.27	2.11 ± 0.24	0.979 [0.956; 0.990] *	2.86 ± 0.35	2.66 ± 0.22	0.849 [0.707; 0.925] *
Stride Height (m)	0.31 ± 0.08	0.28 ± 0.09	0.974 [0.947; 0.988] *	0.49 ± 0.11	0.46 ± 0.11	0.967 [0.932; 0.984] *
Plantar Flexion Foot In (°)	15.53 ± 4.04	21.98 ± 4.36	0.922 [0.842; 0.962] *	15.47 ± 4.21	20.85 ± 4.53	0.891 [0.784; 0.947] *
Impact Force (kN)	1.62 ± 0.30	1.47 ± 0.27	0.973 [0.944; 0.987] *	1.78 ± 0.31	1.70 ± 0.30	0.957 [0.912; 0.979] *
Leg Stiffness (kN/m)	9.87 ± 1.86	9.81 ± 1.79	0.917 [0.833; 0.960] *	10.40 ± 2.28	10.47 ± 1.91	0.729 [0.504; 0.861] *

Concerning intraclass correlation, ICC values from 0.729 to 0.987 ($p < 0.001$) were found for all parameters in overground running (comfortable and fast speed). For fast speed, leg stiffness had a moderate intraclass correlation (ICC < 0.75, $p < 0.001$). The other parameters showed good and excellent intraclass correlations (ICC > 0.75, $p < 0.001$). For treadmill running, ICC values from 0.899 to 0.999 ($p < 0.001$) were found for all parameters (comfortable and fast speed).

Table 6. Mean, standard deviation, and ICC values (with 95% CI) of running parameters for treadmill conditions. Significant results are indicated as follows: * = $p < 0.001$.

	TREADMILL RUNNING					
	Comfortable Speed			Fast Speed		
	Mean ± SD		ICC [95% CI]	Mean ± SD		ICC [95% CI]
	Motion Capture	Insoles		Motion Capture	Insoles	
Speed (m/s)	2.35 ± 0.32	2.36 ± 0.30	0.996 [0.991; 0.998] *	2.88 ± 0.32	2.86 ± 0.30	0.986 [0.971; 0.993] *
Stride Cadence (step/min)	79.10 ± 3.98	79.12 ± 3.98	0.999 [0.999; 0.999] *	81.35 ± 3.97	81.32 ± 3.96	0.999 [0.999; 0.999] *
Flight Time (s)	0.13 ± 0.07	0.16 ± 0.06	0.915 [0.830; 0.959] *	0.19 ± 0.05	0.21 ± 0.05	0.948 [0.894; 0.975] *
Stance Time (s)	0.31 ± 0.04	0.30 ± 0.04	0.948 [0.893; 0.975] *	0.27 ± 0.02	0.26 ± 0.03	0.940 [0.878; 0.971] *
Stride Time (s)	0.76 ± 0.04	0.76 ± 0.04	0.999 [0.999; 0.999] *	0.74 ± 0.04	0.74 ± 0.04	0.999 [0.999; 0.999] *
Swing Time (s)	0.45 ± 0.04	0.46 ± 0.04	0.962 [0.923; 0.982] *	0.47 ± 0.04	0.48 ± 0.04	0.977 [0.952; 0.989] *
Stride Length (m)	1.68 ± 0.20	1.79 ± 0.22	0.994 [0.987; 0.997] *	1.98 ± 0.21	2.11 ± 0.22	0.990 [0.979; 0.995] *
Stride Height (m)	0.26 ± 0.06	0.23 ± 0.06	0.981 [0.961; 0.991] *	0.35 ± 0.07	0.33 ± 0.08	0.981 [0.961; 0.991] *
Plantar Flexion Foot In (°)	13.33 ± 3.18	19.29 ± 3.77	0.899 [0.799; 0.951] *	13.88 ± 3.82	19.23 ± 4.37	0.919 [0.836; 0.960] *

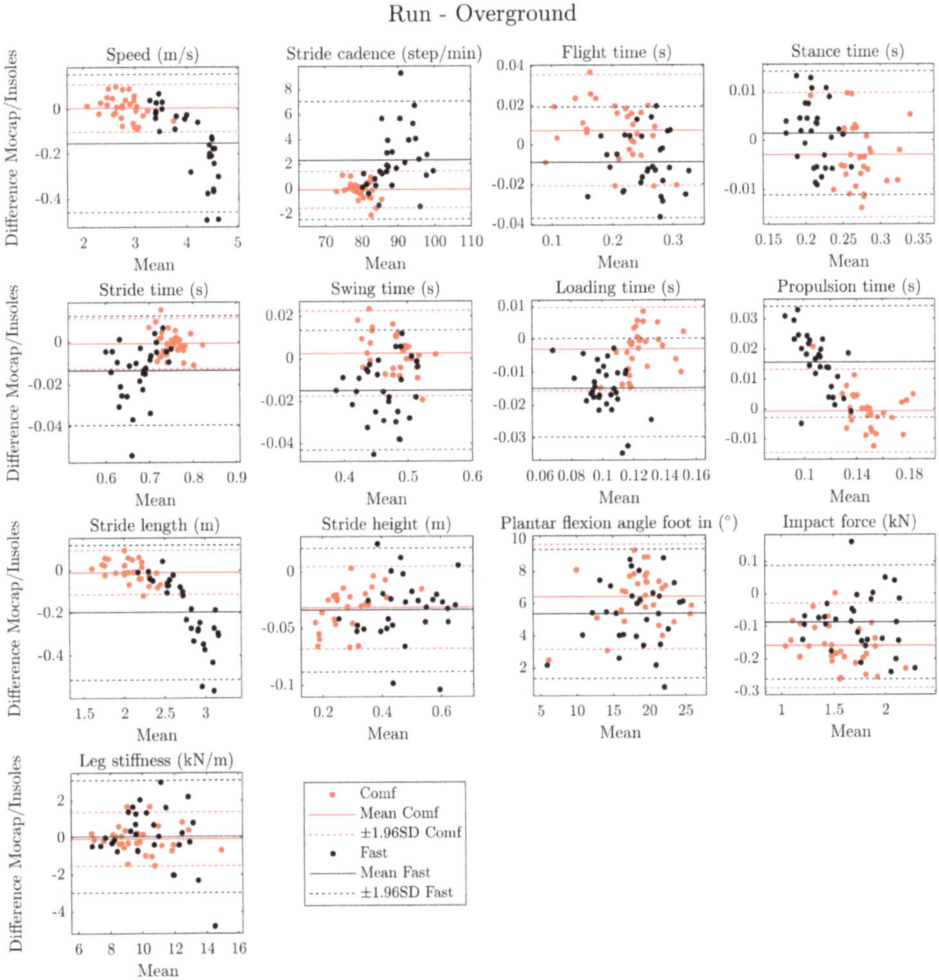

Figure 2. Bland–Altman plot for all running parameters during fast (black points) and comfortable (red points) overground running. Solid line = mean, dashed line = ±1.96 SD.

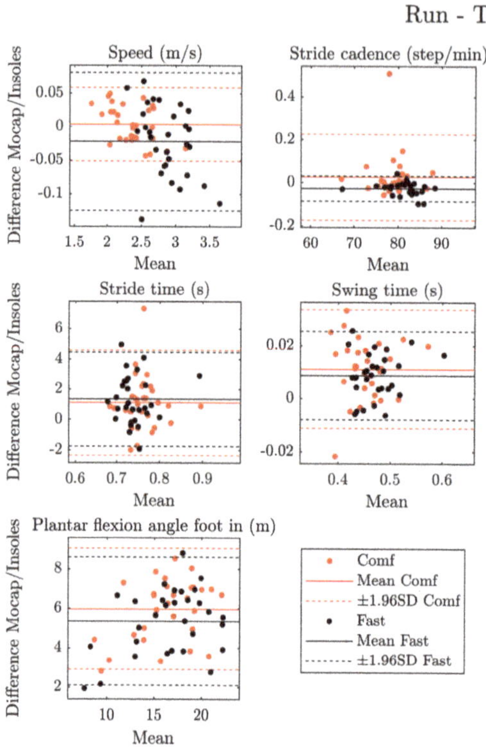

Figure 3. Bland–Altman plot for all running parameters during fast (black points) and comfortable (red points) treadmill running. Solid line = mean, dashed line = ±1.96 SD.

Concerning Lin concordance coefficient, CCC values were considered excellent (higher than 0.8) for all parameters except for plantar flexion foot during all speeds in overground and during treadmill running (CCC = [0.358–0.500]) as well as loading time (CCC = 0.467), propulsion time (CCC = 0.453) and stride length (CCC = 0.690) for fast speed during overground running.

Based on Bland–Altman plots, the following temporal parameters were found to have a mean bias close to zero seconds: flight time, stride time, stance time, swing time, loading time, and propulsion time. The following spatial parameters also have a mean bias close to zero: speed, stride cadence, stride length, and stride height. These spatial parameters also showed an increase in error heterogeneity with speed. The mean bias for plantar flexion angle foot was close to 6°. During overground running, the mean bias for dynamic parameters (impact force and leg stiffness) was close to −0.1 kN and 0 kN/m. Moreover, LoA was found to be lower than 15% of mean values for all parameters except plantar flexion foot in (20.5% and 25.2%, respectively, for comfortable and fast speed) and leg stiffness (27.9% for fast speed) for overground running. For treadmill running, flight time had an LoA of 26.6% for comfortable speed and 18.2% for fast speed. Plantar flexion foot in had an LoA of 20.6% for comfortable speed and 21.8% for fast speed.

4. Discussion

The aim of the present study was to evaluate the test–retest reliability and criterion validity of running parameters measured using embedded insoles in comparison to the values obtained from the gold standard motion-capture system. Based on the motion capture of 30 participants during overground and treadmill running, the present results highlight the relevance of DSPro® insoles in a sports context. Furthermore, the evaluation of two distinct speeds for both overground and treadmill running introduced a novel dataset, enhancing the precision and comprehensiveness of the validation process for these insoles.

Test–retest reliability was measured to quantify the degree of agreement of parameters for treadmill and overground running. For overground running, ICC values varied between moderate and excellent reliability. The moderate ICC values were explained by the high variability in running between two specific sessions overground because, unlike treadmill running, participants could choose their speed during the second session, thus modifying the gait parameters. However, the results obtained for insoles and motion capture on the treadmill highlight good to excellent reliability due to the identical speed setting between the two sessions. Nevertheless, propulsion time for motion capture shows a moderate and poor ICC value for comfortable and fast speed, respectively. Only a few previous studies have examined the reliability of measurements derived from IMUs during running [29], but our results are comparable to previous research, regardless of the conditions. For example, Deflandre et al. [30] reported good to excellent reliability for stance time and step frequency with a portable accelerometer during treadmill running. Using textile socks incorporating IMU, Mason et al. [31] also highlighted good to excellent test–retest agreement for stride and swing time during treadmill running and moderate to good during overground running. Additionally, the good results of this present study were reinforced by the absence of significant differences in absolute and relative errors between the two sessions (except for stride height during treadmill running in the fast condition), and all these values were lower than the MDC.

For criterion validity, good results were obtained with the ICC analysis. Spatiotemporal parameter measurements had good to excellent agreement between insoles and the motion-capture system regardless of speed (except for leg stiffness at fast speed). Additionally, these results were observed for treadmill conditions with ICC values higher than 0.80 for all parameters. The good reliability of DSPro® is consistent with other validation studies of IMU against a 3D motion analysis system during a running task. For instance, Uno et al. [32] found excellent relative validity (ICC > 0.900) for stride length, stride duration, stride frequency, stride speed, vertical height, stance phase duration, swing phase duration, and sagittal angle. Brahms et al. [33] found an ICC value of 0.955 for stride length during running. Moreover, in a recent meta-analysis, Horsley et al. [34] reported similar ICC values for stance phase time (ICC = 0.81–0.97), swing phase time (ICC = 0.56–0.81), stride duration (ICC = 0.55–0.99), stride frequency (ICC = 0.96–0.99), stride length (ICC = 0.75–0.99), and flight time (ICC = 0.81–0.86). Additionally, CCC values were in accordance with ICC values except for loading time (fast speed—overground running), propulsion time (fast speed—overground running), and plantar flexion foot in (comfortable and fast speed—overground and treadmill running). Differences between these two coefficients could be linked to deviations of the point cloud from the identity line. Finally, the Bland–Altman plots show that there is low heterogeneity in the measurement bias, with a mean bias close to 0 for spatiotemporal parameters, impact force, and leg stiffness in overground and treadmill conditions. Moreover, a bias of 6° was observed for the plantar flexion angle foot in but not influenced by the speed of running, and a bias of about 20 cm was observed for stride length during fast speed. Additionally, a linear error was observed for speed, stride cadence, stride time, propulsion time, and stride length during overground running. All these biases/errors could be partially explained by the hardware calibration and the algorithm used (fusion filters and calculation method) between motion capture and embedded IMU insoles. Additionally, LoA were not negligible but seem satisfactory with values lower than 15% compared to the mean. However, due to LoA higher than 20%, plantar flexion foot in was only partially validated for all conditions, flight time for treadmill condition, and leg stiffness for overground running at fast speed. However, these variables had moderate to excellent ICC values.

Furthermore, this study validates the use of IMU embedded insoles during overground and treadmill running and, since running was performed in different conditions and at different speeds, indicates that it is a clinically relevant means of assessing gait performance. IMUs are low-cost systems and could easily be used to follow different running parameters for professionals or individuals. Additionally, considering the limitations of motion capture (dedicated space requirements, technical expertise requirements, and high costs) [9], running analysis could be easily performed only with insoles in a clinical context.

This study has some limitations. First, running experiments were conducted in a laboratory, which can modify kinematics and temporal variables compared to outdoor running. Many studies have highlighted the importance of measuring running outside of the laboratory in natural training and/or competition environments [35]. Moreover, the treadmill is the most common system used to evaluate and quantify running gait [17]. This approach has the advantage of providing a standardized and reproducible environment. However, treadmill running only partially reflects natural running behavior, and there are differences in kinematic and kinetic patterns compared to overground running [15,17,36]. Additionally, without an instrumented treadmill and due to the lack of international consensus definition, the loading time parameter was not measured for this condition. However, depending on the attack of the foot during treadmill running, there is a potential for considerable variability, and comparison with data from insoles seems difficult. Additionally, a choice has been made not to estimate leg stiffness and impact force with literature models (e.g., presented by Morin et al. [37]) in order to validate insoles with real measurements. Second, experiments involved young (mean 27.6 $\pm$ 5.2 years) and healthy participants. Showing good accuracy for this population, these results could be modified for older or injured populations. Further investigations are warranted in patients with locomotor disabilities (e.g., with a higher risk of falls) to confirm these good psychometric properties of the insoles in care settings (neurologic rehabilitation, orthopedic surgery, etc.).

5. Conclusions

To conclude, by comparing running parameters measured with insoles and a motion-capture system, this present study confirms the good accuracy and repeatability of DSPro® insoles for most running parameters. Although a few parameters, such as plantar flexion foot in, flight time, and leg stiffness, were only partially validated under specific conditions, this wearable device seems to be a useful tool to improve the understanding of running performance. These insoles are a relevant way to overcome the constraints of the motion-capture system, and they have the potential to improve patient care and enhance the analysis of running performance in a sports context.

Supplementary Materials: The following supporting information can be downloaded at: https://www.mdpi.com/article/10.3390/s24165435/s1, Table S1: Calculation of running parameters for motion-capture system; Table S2: Comparison of absolute and relative error between the two sessions of each parameter extracted with motion capture and with insoles for overground running; Table S3: Comparison of absolute and relative error between the two sessions of each parameter extracted with motion capture and with insoles for treadmill running; Table S4: Comparison between intraclass and concordance correlation coefficient for overground and treadmill running; Figure S1: Lin's concordance correlation plot for all running parameters during fast and comfortable overground running; Figure S2: Lin's concordance correlation plot for all running parameters during fast and comfortable treadmill running. References [38–42] are cited in Supplementary Materials.

Author Contributions: Conceptualization, D.L., M.G. and P.O.; methodology, D.L., M.G. and P.O.; software, N.E., A.L. and B.O.; validation, M.G.; formal analysis, B.O., L.R. and M.G.; investigation, A.L., B.O., C.D. and N.E.; resources, A.L., B.O., C.D. and N.E.; data curation, B.O. and M.G.; writing—original draft preparation, L.R.; writing—review and editing, L.R., M.G., D.L. and P.O.; supervision, M.G.; project administration, M.G. All authors have read and agreed to the published version of the manuscript.

Funding: This research was funded (2020R217) by DigitSole SAS and the CHU Dijon–Bourgogne, FRANCE.

Institutional Review Board Statement: The study was conducted in accordance with the Declaration of Helsinki and approved by the Ethics Committee of CPP Est I Dijon (France).

Informed Consent Statement: Informed consent was obtained from all subjects involved in the study.

Data Availability Statement: Data are available on request due to restrictions (e.g., privacy and ethics).

Conflicts of Interest: The authors declare no conflicts of interest.

References

1. Gallow, A.; Heiderscheit, B. Clinical Aspects of Running Gait Analysis. In *Endurance Sports Medicine*; Miller, T.L., Ed.; Springer International Publishing: Cham, Switzerland, 2016; pp. 201–213. ISBN 978-3-319-32980-2.
2. Hespanhol Junior, L.C.; Pillay, J.D.; van Mechelen, W.; Verhagen, E. Meta-Analyses of the Effects of Habitual Running on Indices of Health in Physically Inactive Adults. *Sports Med.* **2015**, *45*, 1455–1468. [CrossRef] [PubMed]
3. van Mechelen, W. Running Injuries. *Sports Med.* **1992**, *14*, 320–335. [CrossRef]
4. Norris, M.; Anderson, R.; Kenny, I.C. Method Analysis of Accelerometers and Gyroscopes in Running Gait: A Systematic Review. *Proc. Inst. Mech. Eng. Part. P J. Sports Eng. Technol.* **2014**, *228*, 3–15. [CrossRef]
5. Chen, D.; Cai, Y.; Qian, X.; Ansari, R.; Xu, W.; Chu, K.-C.; Huang, M.-C. Bring Gait Lab to Everyday Life: Gait Analysis in Terms of Activities of Daily Living. *IEEE Internet Things J.* **2020**, *7*, 1298–1312. [CrossRef]
6. Pipkin, A.; Kotecki, K.; Hetzel, S.; Heiderscheit, B. Reliability of a Qualitative Video Analysis for Running. *J. Orthop. Sports Phys. Ther.* **2016**, *46*, 556–561. [CrossRef]
7. Dingenen, B.; Barton, C.; Janssen, T.; Benoit, A.; Malliaras, P. Test-Retest Reliability of Two-Dimensional Video Analysis during Running. *Phys. Ther. Sport* **2018**, *33*, 40–47. [CrossRef] [PubMed]
8. Nüesch, C.; Roos, E.; Pagenstert, G.; Mündermann, A. Measuring Joint Kinematics of Treadmill Walking and Running: Comparison between an Inertial Sensor Based System and a Camera-Based System. *J. Biomech.* **2017**, *57*, 32–38. [CrossRef]
9. Simon, S.R. Quantification of Human Motion: Gait Analysis—Benefits and Limitations to Its Application to Clinical Problems. *J. Biomech.* **2004**, *37*, 1869–1880. [CrossRef]
10. Cimolin, V.; Galli, M. Summary Measures for Clinical Gait Analysis: A Literature Review. *Gait Posture* **2014**, *39*, 1005–1010. [CrossRef]
11. Toro, B.; Nester, C.; Farren, P. A Review of Observational Gait Assessment in Clinical Practice. *Physiother. Theory Pract.* **2003**, *19*, 137–149. [CrossRef]
12. Uhlrich, S.D.; Falisse, A.; Kidziński, Ł.; Muccini, J.; Ko, M.; Chaudhari, A.S.; Hicks, J.L.; Delp, S.L. OpenCap: Human Movement Dynamics from Smartphone Videos. *PLoS Comput. Biol.* **2023**, *19*, e1011462. [CrossRef] [PubMed]
13. Wade, L.; Needham, L.; McGuigan, P.; Bilzon, J. Applications and Limitations of Current Markerless Motion Capture Methods for Clinical Gait Biomechanics. *PeerJ* **2022**, *10*, e12995. [CrossRef]
14. Mündermann, L.; Corazza, S.; Andriacchi, T.P. The Evolution of Methods for the Capture of Human Movement Leading to Markerless Motion Capture for Biomechanical Applications. *J. Neuroeng. Rehabil.* **2006**, *3*, 6. [CrossRef]
15. Willy, R.W. Innovations and Pitfalls in the Use of Wearable Devices in the Prevention and Rehabilitation of Running Related Injuries. *Phys. Ther. Sport* **2018**, *29*, 26–33. [CrossRef]
16. Benson, L.C.; Räisänen, A.M.; Clermont, C.A.; Ferber, R. Is This the Real Life, or Is This Just Laboratory? A Scoping Review of IMU-Based Running Gait Analysis. *Sensors* **2022**, *22*, 1722. [CrossRef]
17. Mason, R.; Pearson, L.T.; Barry, G.; Young, F.; Lennon, O.; Godfrey, A.; Stuart, S. Wearables for Running Gait Analysis: A Systematic Review. *Sports Med.* **2023**, *53*, 241–268. [CrossRef] [PubMed]
18. Ferrete Ribeiro, N.; Santos, C.P. Inertial Measurement Units: A Brief State of the Art on Gait Analysis. In Proceedings of the 2017 IEEE 5th Portuguese Meeting on Bioengineering (ENBENG), Coimbra, Portugal, 16–18 February 2017; pp. 1–4.
19. Reenalda, J.; Maartens, E.; Homan, L.; Buurke, J.H.J. Continuous Three Dimensional Analysis of Running Mechanics during a Marathon by Means of Inertial Magnetic Measurement Units to Objectify Changes in Running Mechanics. *J. Biomech.* **2016**, *49*, 3362–3367. [CrossRef] [PubMed]
20. Strohrmann, C.; Harms, H.; Kappeler-Setz, C.; Tröster, G. Monitoring Kinematic Changes with Fatigue in Running Using Body-Worn Sensors. *IEEE Trans. Inf. Technol. Biomed.* **2012**, *16*, 983–990. [CrossRef]
21. Riglet, L.; Nicol, F.; Leonard, A.; Eby, N.; Claquesin, L.; Orliac, B.; Ornetti, P.; Laroche, D.; Gueugnon, M. The Use of Embedded IMU Insoles to Assess Gait Parameters: A Validation and Test-Retest Reliability Study. *Sensors* **2023**, *23*, 8155. [CrossRef]
22. Ziagkas, E.; Loukovitis, A.; Zekakos, D.X.; Chau, T.D.-P.; Petrelis, A.; Grouios, G. A Novel Tool for Gait Analysis: Validation Study of the Smart Insole PODOSmart®. *Sensors* **2021**, *21*, 5972. [CrossRef]
23. Loukovitis, A.; Ziagkas, E.; Zekakos, D.X.; Petrelis, A.; Grouios, G. Test-Retest Reliability of PODOSmart® Gait Analysis Insoles. *Sensors* **2021**, *21*, 7532. [CrossRef] [PubMed]
24. Mokkink, L.; Prinsen, C.; Patrick, D.; Alonso, J.; Bouter, L.; De Vet, H.; Terwee, C.B.; Mokkink, L. COSMIN Methodology for Systematic Reviews of Patient-Reported Outcome Measures (PROMs). *User Man.* **2018**, *78*, 1–78.
25. O'Connor, C.M.; Thorpe, S.K.; O'Malley, M.J.; Vaughan, C.L. Automatic Detection of Gait Events Using Kinematic Data. *Gait Posture* **2007**, *25*, 469–474. [CrossRef] [PubMed]

26. Koo, T.K.; Li, M.Y. A Guideline of Selecting and Reporting Intraclass Correlation Coefficients for Reliability Research. *J. Chiropr. Med.* **2016**, *15*, 155–163. [CrossRef]
27. Haley, S.M.; Fragala-Pinkham, M.A. Interpreting Change Scores of Tests and Measures Used in Physical Therapy. *Phys. Ther.* **2006**, *86*, 735–743. [CrossRef]
28. Akoglu, H. User's Guide to Correlation Coefficients. *Turk. J. Emerg. Med.* **2018**, *18*, 91–93. [CrossRef] [PubMed]
29. Zeng, Z.; Liu, Y.; Hu, X.; Tang, M.; Wang, L. Validity and Reliability of Inertial Measurement Units on Lower Extremity Kinematics During Running: A Systematic Review and Meta-Analysis. *Sports Med.-Open* **2022**, *8*, 86. [CrossRef]
30. Deflandre, D.; Miny, K.; Bury, T. Myotest Efficiency in the Mechanical Analysis of the Stride. *Gazz. Medica Ital. Arch. Sci. Mediche* **2018**, *177*, 293–300. [CrossRef]
31. Mason, R.; Barry, G.; Robinson, H.; O'Callaghan, B.; Lennon, O.; Godfrey, A.; Stuart, S. Validity and Reliability of the DANU Sports System for Walking and Running Gait Assessment. *Physiol. Meas.* **2023**, *44*, 115001. [CrossRef]
32. Uno, Y.; Ogasawara, I.; Konda, S.; Yoshida, N.; Otsuka, N.; Kikukawa, Y.; Tsujii, A.; Nakata, K. Validity of Spatio-Temporal Gait Parameters in Healthy Young Adults Using a Motion-Sensor-Based Gait Analysis System (ORPHE ANALYTICS) during Walking and Running. *Sensors* **2022**, *23*, 331. [CrossRef]
33. Brahms, C.M.; Zhao, Y.; Gerhard, D.; Barden, J.M. Stride Length Determination during Overground Running Using a Single Foot-Mounted Inertial Measurement Unit. *J. Biomech.* **2018**, *71*, 302–305. [CrossRef] [PubMed]
34. Horsley, B.J.; Tofari, P.J.; Halson, S.L.; Kemp, J.G.; Dickson, J.; Maniar, N.; Cormack, S.J. Does Site Matter? Impact of Inertial Measurement Unit Placement on the Validity and Reliability of Stride Variables during Running: A Systematic Review and Meta-Analysis. *Sports Med.* **2021**, *51*, 1449–1489. [CrossRef] [PubMed]
35. Blades, S.; Jensen, M.; Stellingwerff, T.; Hundza, S.; Klimstra, M. Characterization of the Kinetyx SI Wireless Pressure-Measuring Insole during Benchtop Testing and Running Gait. *Sensors* **2023**, *23*, 2352. [CrossRef] [PubMed]
36. Hong, Y.; Wang, L.; Li, J.X.; Zhou, J.H. Comparison of Plantar Loads during Treadmill and Overground Running. *J. Sci. Med. Sport.* **2012**, *15*, 554–560. [CrossRef] [PubMed]
37. Morin, J.B.; Dalleau, G.; Kyröläinen, H.; Jeannin, T.; Belli, A. A Simple Method for Measuring Stiffness during Running. *J. Appl. Biomech.* **2005**, *21*, 167–180. [CrossRef]
38. Souza, G.S.D.S.E.; Rodrigues, F.B.; Andrade, A.O.; Vieira, M.F. A Simple, Reliable Method to Determine the Mean Gait Speed Using Heel Markers on a Treadmill. *Comput. Methods Biomech. Biomed. Eng.* **2017**, *20*, 901–904. [CrossRef]
39. Novacheck, T.F. The Biomechanics of Running. *Gait Posture* **1998**, *7*, 77–95. [CrossRef]
40. Supakkul, K. Using Positional Heel-Marker Data to More Accurately Calculate Stride Length for Treadmill Walking: A Step Length Approach. *arXiv* **2017**. [CrossRef]
41. Canete, S.; Jacobs, D.A. Novel Velocity Estimation for Symmetric and Asymmetric Self-Paced Treadmill Training. *J. Neuroeng. Rehabil.* **2021**, *18*, 27. [CrossRef]
42. Farley, C.T.; González, O. Leg Stiffness and Stride Frequency in Human Running. *J. Biomech.* **1996**, *29*, 181–186. [CrossRef]

sensors

MDPI

Article

Analyzing Optimal Wearable Motion Sensor Placement for Accurate Classification of Fall Directions

Sokea Teng [1,†], Jung-Yeon Kim [2], Seob Jeon [3,†], Hyo-Wook Gil [4], Jiwon Lyu [5], Euy Hyun Chung [6], Kwang Seock Kim [7] and Yunyoung Nam [8,*]

1 Department of ICT Convergence, Soonchunhyang University, Asan 31538, Republic of Korea; teng.sokea@sch.ac.kr
2 ICT Convergence Research Center, Soonchunhyang University, Asan 31538, Republic of Korea; betterwithme@sch.ac.kr
3 Department of Obstetrics and Gynecology, Soonchunhyang University Cheonan Hospital, Cheonan 31151, Republic of Korea; sjeon4595@gmail.com
4 Department of Internal Medicine, Soonchunhyang University Cheonan Hospital, Cheonan 31151, Republic of Korea; hwgil@schmc.ac.kr
5 Division of Respiratory Medicine, Department of Internal Medicine, Soonchunhyang University Cheonan Hospital, Cheonan 31151, Republic of Korea; 78214@schmc.ac.kr
6 Department of Dermatology, College of Medicine, Soonchunhyang University Cheonan Hospital, Cheonan 31151, Republic of Korea; djavan@naver.com
7 Future Innovation Medical Research Center, Soonchunhyang University Cheonan Hospital, Cheonan 31151, Republic of Korea; kimks5005gt@gmail.com
8 Department of Computer Science and Engineering, Soonchunhyang University, Asan 31538, Republic of Korea
* Correspondence: ynam@sch.ac.kr
† These authors contributed equally to this work.

Abstract: Falls represent a significant risk factor, necessitating accurate classification methods. This study aims to identify the optimal placement of wearable sensors—specifically accelerometers, gyroscopes, and magnetometers—for effective fall-direction classification. Although previous research identified optimal sensor locations for distinguishing falls from non-falls, limited attention has been given to the classification of fall direction across different body regions. This study assesses inertial measurement unit (IMU) sensors placed at 12 distinct body locations to determine the most effective positions for capturing fall-related data. The research was conducted in three phases: first, comparing classifiers across all sensor locations to identify the most effective; second, evaluating performance differences between sensors placed on the left and right sides of the body; and third, exploring the efficacy of combining sensors from the upper and lower body regions. Statistical analyses of the results for the most effective classifier model demonstrate that the support vector machine (SVM) is more effective than other classifiers across all sensor locations, with statistically significant differences in performance. At the same time, the comparison between the left and right sensor locations shows no significant performance differences within the same anatomical areas. Regarding optimal sensor placement, the findings indicate that sensors positioned on the pelvis and upper legs in the lower body, as well as on the shoulder and head in the upper body, were the most effective results for accurate fall-direction classification. The study concludes that the optimal sensor configuration for fall-direction classification involves strategically combining sensors placed on the pelvis, upper legs, and lower legs.

Keywords: falls; optimal sensor; fall direction; classification; machine learning; feature extraction; IMU sensors; daily life activity; optimal sensor; sensor location

1. Introduction

Falls are a major public health concern, ranking as the second leading cause of injuries and accidental death globally. They can lead to severe physical and emotional repercussions, including impairment, loss of autonomy, post-fall syndrome, depression, reduced activity levels, and possibly early mortality [1]. Wearable sensor systems significantly enhance the well-being of patients and older adults by assessing fall risk, detecting falls, and monitoring other health-related factors [2]. These systems typically place sensors on body areas like the waist, chest, and thighs to track daily activities, estimate energy expenditure, aid rehabilitation, and monitor heart activity [3,4].

The accuracy and efficiency of classification or detection systems are significantly influenced by sensor placement on the body. Previous studies have primarily focused on identifying optimal sensor locations for distinguishing between fall events and non-fall activities or activities of daily living (ADLs). Our study builds upon this by specifically classifying fall directions, including forward fall, backward fall, and sideways fall, in addition to non-fall activities such as sitting, standing, walking, and transitions between sitting and standing. While the waist is often recommended as an optimal location for fall detection due to its proximity to the body's center of mass, placement on the arms and legs is generally discouraged. This is because higher acceleration in these areas can impair the performance of detection systems, as noted in previous research [5].

Optimal performance varies by location, with several studies suggesting the waist, chest, thigh, or right pocket as effective sites for fall detection and classification [6–11]. Notably, the waist and thigh are often highlighted for their ability to capture comprehensive movement dynamics essential for accurate fall detection and classification [12,13]. In previous studies, fall detection focused on identifying whether a fall event occurred. It is important for systems that trigger alerts or further monitoring. However, the purpose of fall classification is to categorize the type or direction of the fall, such as forward, backward, or sideways. Classifying the direction of a fall is crucial because different injuries occur depending on the direction of impact, and this information is critical for tailoring medical interventions. For example, forward falls may result in wrist or arm fractures, while backward falls are more likely to cause head or hip injuries. By classifying fall direction, we can better understand the biomechanics of the fall and design more effective interventions.

Studies have shown that sensor location profoundly influences fall detection and classification accuracy. Identifying optimal body locations for IMUs is essential, particularly because different age groups prefer using the minimum number of wearable devices [14–16]. Common sensor locations for fall classification include the waist, chest, and thigh, with alternatives such as the forehead, neck, ear, shoulder, back, wrist, ankle, and foot, though these locations are generally less favorable [17–25]. For instance, Ponce et al. [5] reinforced the significant impact of sensor placement by determining that a combination of sensors at the waist and a lateral viewpoint provided the best results for fall detection, while Martínez-Villasenor et al. [6] found that waist sensors achieved 95.76% accuracy in fall classification using the UP-Fall dataset with various machine learning techniques, including random forest (RF), support vector machine (SVM), and multilayer perceptron (MLP). Santoyo et al. [7] found that sensors on the chest, waist, or both provided the best results using KNN, SVM, NB, and DT classifiers. Moreover, Özdemir et al. [12] demonstrated that a single waist sensor provided the best performance when analyzing 31 sensor combinations. Ntanasis et al. [13] highlighted that sensors placed on the waist and thigh were most effective for fall detection and daily activities monitoring. Table 1 summarizes key studies comparing sensor placements for optimal fall detection.

While several studies focus on fall detection, some studies specifically address fall-detection classification, a key method for determining fall direction. Accurate fall-direction classification requires comprehensive data, making it essential to identify the body regions that provide the most relevant information. For example, the authors in [26] used IMUs and eight machine learning classifiers with accelerometers, gyroscopes, and magnetometers placed on the waist for fall-direction classification. Despite the substantial body of research on fall detection, the optimal sensor locations for this purpose remain undetermined.

Proper sensor placement is crucial for accurately capturing falls and activities of daily living (ADLs), especially for distinguishing fall directions. Therefore, our study divided the body into upper and lower regions to identify the most effective sensor combination locations for fall-direction classification, which is crucial for understanding the biomechanics of falls and for implementing effective fall prevention strategies. Distinguishing the direction of a fall holds significant clinical and scientific importance, as the direction can directly affect the likelihood and severity of injuries, such as fractures or head trauma. For instance, forward falls often lead to wrist or arm fractures as individuals attempt to break their fall, whereas backward falls are more likely to result in head or hip injuries.

Classifying the direction of a fall is essential because different injuries occur depending on the direction of impact, and this information is critical for tailoring medical interventions. Clinicians often require detailed fall-direction data to guide treatment decisions, especially when the patient has lost consciousness and cannot describe how they fell. Knowing the fall direction can help medical professionals identify which body parts may have been impacted and adjust treatment accordingly. For example, a backward fall might pay more attention to the head and hips, whereas a forward fall might require focusing on the arms or wrists. Robinovitch et al. [27] have demonstrated that injury risk and protective responses vary significantly depending on fall direction. Thus, accurately identifying fall direction is key to developing personalized interventions and preventive measures, particularly for vulnerable populations such as older adults who are at higher risk of falls and injuries. This understanding can also inform the development of targeted fall prevention technologies and clinical interventions to reduce injury risks in high-risk populations.

This approach is driven by the distinct movement dynamics of the upper and lower body during fall events, which may influence the performance of fall-direction classification systems. For example, sensors on the lower body, such as the ankle or thigh, capture different fall aspects compared to sensors on the upper body, such as the chest or head. No more research has systematically explored sensor placement by dividing the body into upper and lower regions, leaving a gap in understanding that locations provide the most relevant data. This division also addresses the need for flexibility in sensor placement owing to discomfort, medical reasons, or physical limitations, catering to specific populations like the elderly or athletes. For instance, elderly individuals with lower-body weakness may benefit from lower-body sensors, while athletes might require upper-body sensors. This separation facilitates flexible sensor placement tailored to the specific needs of each target population.

This research addresses the gap by investigating the impact of sensor placements on the efficiency of fall-direction classification. We analyzed multiple body locations to identify the most effective sensor positions and employed advanced machine learning techniques to evaluate classification accuracy across various regions of the body.

The primary aim of this study was to determine the optimal sensor placements among 12 body locations and ascertain the critical body positions for recognizing and classifying fall directions using wearable IMU sensors. The key contributions of this study are as follows:

- Identify the most effective machine learning classifier for fall-direction analysis using data from 12 body locations;
- Examine how the left and right body sides affect fall-direction classification accuracy;
- Determine the most effective sensor location combinations for upper and lower body parts;
- Utilize a multiclass classification framework to evaluate and compare the effectiveness of sensor placement in identifying specific fall directions (e.g., forward, backward, and lateral) and non-falls.

The remainder of this paper is organized as follows: Section 2 explains the dataset and methodology for comparative analysis, Section 3 presents the results and analysis, Section 4 presents a discussion of results, and the conclusion summarizes the study.

Table 1. Summarizing relevant work on finding the best sensor placements for fall detection.

Ref	Dataset	Sensor	Sensor Location	Subject	Combination	Falls, ADLs	Algorithm	Best Performance
[5]	UP-Fall	5 IMUs, 1 helmet, 6 ambient, and 2 cameras	Neck, waist, left wrist, right pocket's trousers, and left ankle	17	5 IMU, 2 cameras	5/6	RF, SVM, MLP, KNN	- Waist and a lateral viewpoint - Acc: 98.72% (RF) - Right pocket acc: 98.57% (RF) - Wrist, acc: 98.32% (RF)
[6]	UP-Fall	5 IMUs, 1 helmet, 6 ambient, and 2 cameras	Neck, waist, left wrist, right pocket's trousers, and left ankle	17	Single	5/6	RF, SVM, MLP, KNN	Waist (RF) acc: 95.76%
[7]	UMA-Fall	1 smartphone and 4 IMU	Thigh pocket, chest, waist, right wrist, and ankle	19	31	3/11	SVM, KNN, NB, DT	Waist, chest: sen: 95% using SVM
[12]	Daily and Sports Activities	6MTw sensor (six three-DOF)	Right wrist, right thigh, right ankle, chest, waist, and head	14	63	20/16	KNN, BDM, SVM, LSM, DTW, ANNs	Waist: acc 99.87% using k-NN
[13]	Daily and Sports Activities	Acc + Gyr + Mag	Heads, chests, waists, right wrists, right thighs, and right ankles	14	Single	20/16	WEKA J48, DT, KNN, RF, RC, SVM	Waist: acc 99.28% using RF Thigh: acc 99.48% using SVM

2. Methodology

This section outlines the overall procedure for identifying the best sensor locations to classify fall directions. First, we collected IMU data from 12 sensor placements on the human body. The proposed methodology involves defining scenarios for selecting specific sensor locations and segmenting body parts. Then, we developed supervised learning models to assess these locations. Statistical analyses were employed to determine the most effective sensor placements for each body part and to compare different body regions for optimal and accurate fall-direction classification.

2.1. Acquisition

The study included 24 participants (13 females and 11 males) with ages ranging from 18 to 50 years and heights from 155 cm to 183 cm. The dataset was collected from various environments, including hospitals, homes, roads, and nursing homes. The inclusion criteria required participants to be physically healthy, with no musculoskeletal problems or neurological conditions problem that could impair their ability to perform falls. Participants with any injuries or mobility issues were excluded from the study to ensure safety during fall simulations. While participants were not at high risk of falls, they were trained to simulate realistic fall behaviors based on knowledge of fall risk dynamics. This allowed the participants to imitate falls typically experienced by individuals at a higher risk of falling, ensuring realistic and accurate fall simulations.

The data collection involved the use of three types of IMU sensors—a 3D accelerometer, 3D gyroscope, and 3D magnetometer—attached to 12 body locations, including the head, shoulders, upper arms, forearms, pelvis, upper legs, and lower legs. These sensors recorded movement data across three axes (X, Y, and Z). Participants were instructed to self-initiate falls in three different directions: backward, forward, and lateral (side falls). Non-fall activities, such as sitting, standing, walking, and transitioning between sitting and standing, were also recorded to provide control data for distinguishing between fall and non-fall events. Specifically, participants were asked to fall 1–10 times in each direction and each place depending on the participant's ability and comfort level. These activities were chosen to represent common daily movements that could potentially be misclassified as falls. The dataset was classified into four categories: non-fall, backward fall, forward fall, and lateral (side) fall. Table 2 provides a detailed breakdown of the sample counts for each category and the maximum row count per sample, ensuring a robust dataset for fall detection and classification analysis.

Table 2. The number of simples for each class and the maximum row count per sample.

Fall Type	Simple Each Class	Maximum of Row Each Simple
None fall	720	600
Backward	736	600
Forward	977	600
Side	438	600
Total	2871	600

2.2. Body Parts Division and Sensor Selection from Each Part

To identify the most effective sensor locations for fall classification, we divided the body into two main regions: upper and lower. This division was based on [28], which categorized the body into three parts and later into more than three parts. In our study, the upper body includes the head, shoulders, upper arms, and forearms, whereas the lower body consists of the pelvis, upper legs, and lower legs. This division is illustrated in Figure 1.

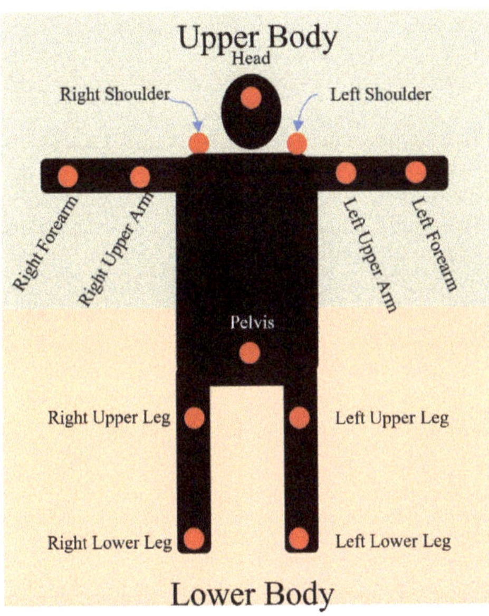

Figure 1. Description of wearable sensor placements and body region divisions.

The rationale behind dividing the body into upper and lower regions is to evaluate how each area contributes to fall-direction classification. This approach allows us to identify which sensors provide the most relevant information from each body region, enabling the selection of effective sensor configurations that minimize redundancy. By comparing the performance of upper- and lower-body sensors separately, we aimed to determine the optimal sensor locations for accurately capturing fall dynamics while also considering user comfort and practical placement considerations.

For each body part, we employed a factorial design approach, $N = 2^k - 1$, where k represents the number of selected sensor positions [7], to study the effect of sensor placement on detection accuracy. N denotes the total number of possible combinations for each body region. For the upper body, with four selected sensors, there are $2^4 - 1 = 15$ possible combinations, and for the lower body, with three sensors, there are $2^3 - 1 = 7$ combinations (for the reasons behind selecting the position of each one, see Section 3.2). This approach allowed us to systematically explore different sensor placements to capture fall-direction information and identify the optimal configuration to enhance fall-direction classification performance.

2.3. Materials and Methodologies

This section details the materials and methodologies used to identify the most effective sensor locations for fall classification using 12 wearable IMU sensors (Figure 1). Our analysis involved three main steps: (i) preprocessing and labeling, (ii) feature extraction, and (iii) building models to identify the most effective classifier, analyze differences between body sides, evaluate sensor combinations across upper and lower body regions, and validate the statistical significance of the results.

2.3.1. Preprocessing and Labeling

To improve model performance and reduce noise, we applied a 1D Gaussian filter during data processing and labeled the data for classification. Each dataset file contained up to 600 samples recorded at 60 Hz, representing 10 s of data per sequence. The data were classified into four categories for supervised learning: non-falls, backward falls, forward falls, and lateral falls. The IMU sensor data were synchronized with video recordings of each fall event and labeled by human experts on fall events. Trained observers reviewed the video footage to accurately classify the fall direction (e.g., forward, backward, or lateral). This method ensured objective and consistent labeling, minimizing potential bias that could arise from self-reporting. The labeled data were subsequently used to train and evaluate machine learning models for fall-direction classification.

Signal vector magnitude (M) was used to simplify the accelerometer, gyroscope, and magnetometer data by reducing the three-dimensional vectors (x, y, and z) into single scalar values, making pattern analysis and event detection more efficient [29]. This approach was chosen to minimize noise from individual axis variations, ensuring a more consistent representation of movement dynamics across different fall events. While analyzing individual axes (x, y, and z) could offer more detailed data for fall-direction classification, using M provided a more robust and simplified method for detecting fall-related patterns. In our study, M was integrated with the individual axes of the accelerometer, gyroscope, and magnetometer to improve accuracy for fall-direction classification. This section provides a detailed analysis of these equations.

$$Acc_M = \sqrt{{Acc_x}^2 + {Acc_y}^2 + {Acc_z}^2} \tag{1}$$

$$Gyr_M = \sqrt{{Gyr_x}^2 + {Gyr_y}^2 + {Gyr_z}^2} \tag{2}$$

$$Mag_M = \sqrt{{Mag_x}^2 + {Mag_y}^2 + {Mag_z}^2} \tag{3}$$

In total, we had 12 channels (9 raw data + 3 magnitudes data) per wearable sensor: three magnitudes (Acc_M, Gyr_M, and Mag_M), and the individual x, y, and z axes from the accelerometer, gyroscope, and magnetometer. Table 3 summarizes all raw data (R) and magnitudes (M) for feature extraction processing.

Table 3. Description of all raw data and magnitudes as input for feature extraction processing.

No,	Raw Data	Description
R_1–R_3	$Acc\ (x, y, z)$	3-axis of an accelerometer
M_1	Acc_M	1-scaler value from a 3-axis of an accelerometer
R_4–R_6	$Gyr\ (x, y, z)$	3-axis of a gyroscope
M_2	Gyr_M	1-scaler value from a 3-axis of a gyroscope
R_7–R_9	$Mag\ (x, y, z)$	3-axis of a magnetometer
M_3	Mag_M	1-scaler value from a 3-axis of a magnetometer

2.3.2. Feature Extraction

Feature extraction is crucial for machine learning classification in fall-detection systems. We used time- and frequency-domain techniques, applying a sliding-window approach with 600 samples (10 s of data) per window. In the time domain, we extracted 8 features: maximum, minimum, standard deviation, sum of absolute values, root mean square (RMS), mean, range, and maximum difference between consecutive values. In the frequency domain, we extracted 10 features, including the maximum, minimum, standard deviation, sum of absolute values, RMS, kurtosis, skewness, mean, range, and the maximum difference between consecutive fast Fourier transform (FFT) values. Each IMU sensor produced 12 channels multiplied by 18 features from both domains, so the total is 216 values (data point) per sensor.

2.3.3. Employing the Machine Learning Algorithm and Evaluation Method

We compared four commonly used supervised learning classification algorithms for fall-detection systems: RF [8], KNN [13], SVM [30], and MLP [10]. The optimal hyperparameters for each algorithm, identified through tuning, are listed in Table 4.

Table 4. Supervised machine learning models and parameters.

Models	Structure	Best Parameters
RF	Ensemble of decision trees	n_estimators=100, bootstrap=False, max_features=s'qrt'.
SVM	Nonlinear classification model	kernel=r'bf', C=10, gamma=s'cale'.
MLP	Feedforward neural network	hidden_layer_sizes=(100,), max_iter=1000, activation=r'elu', alpha=0.0001, solver=s'gd'.
KNN	Instance-based learning	n_neighbors=10, metric=m'anhattan', weights=u'niform'.

We evaluated our method using four key metrics: accuracy (4), precision (5), recall (sensitivity) (6), and F-score (7). Testing on a dataset partition showed exceptionally high accuracy and F-score. Testing on a dataset partition demonstrated exceptionally high accuracy and F-score, reflecting distinct patterns observed in each class—non-fall, backward fall, forward fall, and lateral fall. To ensure robustness, all results were derived using stratified 10-fold cross-validation to calculate the mean and standard deviation. This comprehensive analysis confirms that our system effectively distinguishes between different types of fall and non-fall activities, enhancing its overall performance.

$$Accuracy = \frac{TP + TN}{TP + TN + FP + FN} \tag{4}$$

$$Recall\ (sensitivity) = \frac{TP}{TP + FN} \tag{5}$$

$$Precision = \frac{TP}{TP + FP} \tag{6}$$

$$F - score = \frac{Recall \times Precision}{Recall + Precision} \times 2 \quad (7)$$

In the evaluation formulas, TP represents True Positives, FN denotes False Negatives, FP stands for False Positives, and TN signifies True Negatives.

We used the F-score (7) as a key metric for fall classification, offering a balanced and comprehensive evaluation of model performance [31]. This metric is particularly suitable for applications where both precision and recall are critical, especially in cases of class imbalance. We assessed performance using the mean of the F-score, validated statistically through the Kruskal–Wallis test to identify the most effective algorithm, and the *t*-test to check for significant differences in performance (f1_scores) between sensors on the left and right sides of the same area. We used the Kruskal–Wallis test to evaluate the statistical significance of performance differences across various sensor combinations. The *p*-values, being much larger than the typical significance threshold (e.g., 0.05), indicated no significant difference, supporting the null hypothesis. Post hoc testing with Tukey's Honest Significant Difference (HSD) was conducted using estimated means (circles within bars) with 95% confidence intervals to identify the most effective sensor combinations for each body part.

3. Results

This section evaluates how sensor quantity and placement affect fall-direction classification accuracy using the most effective sensor combinations. Our objectives are to (i) identify optimal machine learning algorithms, (ii) analyze differences between left and right sensor placements, and (iii) determine the best sensor combinations for each body part to identify the most effective placements.

3.1. Analysis of the Effect of Machine Learning Algorithm Selection on Fall Classification Performance

Our aim is to select a single classifier based on its overall performance across the 12 sensors, ensuring simplicity and efficiency in determining the optimal classifier for fall-direction classification. We evaluated four popular supervised learning algorithms—RF, SVM, MLP, and KNN—to determine the highest-performance single classifier for fall-direction classification across different sensor locations. A Kruskal–Wallis test was conducted to examine statistically significant differences ($p < 0.05$) in performance between these classifiers across 12 sensor locations. The results indicated that SVM consistently outperformed the other classifiers, including MLP, with statistically significant differences in performance for several sensor locations.

While the Kruskal–Wallis test revealed significant differences between the classifiers, we conducted a pairwise Tukey HSD test to specifically compare the F-scores between SVM and MLP across all sensor locations. The Tukey HSD test showed that the difference in F-scores between SVM and MLP was statistically significant ($p < 0.05$) for key sensor positions, including the right shoulder and left upper leg, where SVM demonstrated significant improvement over MLP. For the remaining sensor positions, although no significant differences were found ($p > 0.05$), SVM still maintained higher average F-scores compared to MLP. Additionally, the ranking of the classifiers based on their mean F-scores across all sensor locations consistently showed SVM as the top-performing classifier, with MLP following closely but not outperforming SVM significantly in most cases. These findings confirm that SVM is the most effective classifier for fall-direction classification across the 12 sensor locations, with statistically significant improvements over MLP for certain critical sensors. Table 5 summarizes the comparison of machine learning algorithms across the 12 sensor locations, highlighting the significant differences in performance between SVM and MLP and the overall ranking of the classifiers based on their mean F-scores.

Table 5. Results from comparing machine learning algorithms.

Sensor		A	B	C	D	E	F	G	H	I	J	K	L
Classifiers	SVM	75.74	87.24	76.53	77.40	56.14	57.16	66.37	63.49	85.69	84.30	79.01	80.21
	MLP	73.48	85.33	75.04	73.91	55.09	54.92	65.23	61.10	83.76	82.70	76.57	78.75
	p-value	0.112	0.082	0.364	0.023 *	0.364	0.326	0.406	0.131	0.096	0.049 *	0.082	0.199
	RF	66.93	82.02	70.85	69.66	51.68	52.63	59.13	58.78	78.69	78.55	74.50	75.93
	KNN	66.52	79.26	68.72	68.87	46.78	48.57	56.41	54.71	77.06	77.90	70.25	71.28
	p-value	0.705	0.028 *	0.174	0.597	0.008 **	0.003 **	0.174	0.082	0.257	0.290	0.000 ***	0.001 ***

Note: * $p < 0.05$, ** $p < 0.01$, *** $p < 0.001$, p-values below 0.05 indicate statistically significant differences in classifier performance across the respective sensor locations. The sensor locations are designated as follows: A (head), B (pelvis), C (left shoulder), D (right shoulder), E (left forearm), F (right forearm), G (right upper arm), H (left upper arm), I (right upper leg), J (left upper leg), K (left lower leg), and L (right lower leg).

3.2. Results for Analyzing Sensor Location of Different Sides (Left and Right) on the Body

We compared the sensor locations on the left and right sides of the body to determine the most effective representative for each anatomical area. This analysis aimed to streamline sensor placement while maintaining data accuracy. After identifying the SVM algorithm as the best-performing algorithm (Section 3.1), we used it to compare corresponding sensor locations across five areas: shoulders, upper arms, forearms, upper legs, and lower legs.

Table 6 shows that the t-test analysis found no significant performance differences between left and right sensor locations within the same anatomical areas (shoulders, upper arms, forearms, upper legs, and lower legs). All p-values were greater than 0.05, indicating that sensors on both sides perform similarly. The t-test results for these comparisons are summarized in Table 6.

Table 6. Statistical analysis results from comparing the left and right sides of the body.

Sensor Areas	Shoulders	Upper Arms	Forearms	Upper Legs	Lower Legs
T-statistic	−0.6534	−1.8727	−0.8059	−1.3912	−1.0864
p-value	0.5217	0.0774	0.4308	0.1811	0.2917

3.3. Results of Identifying Optimal Sensor Combinations and Comparing Body Parts for Effective Placement

We analyzed combinations of four upper-body sensors and three lower-body sensors for optimal fall-direction classification. After identifying SVM as the most effective model and finding no significant performance differences between sensors on the same anatomical area, we selected the best-performing sensors: head, right shoulder, right forearm, and left upper arm for the upper body; pelvis, left upper leg, and right lower leg for the lower body.

3.3.1. Results of Identifying Four Different Sensor Location Combinations on the Upper Body Part

We analyzed 17 sensor combinations on the upper body to identify the most effective locations, focusing on sensors positioned on the head, right shoulder, right forearm, and left upper arm. The Kruskal–Wallis test revealed a highly significant difference between the sensor combinations ($p < 0.0001$), indicating that sensor placement significantly affects the F1 score. A subsequent Tukey's HSD test pinpointed the specific combinations driving these differences, with the best sensor locations illustrated in Figure 2.

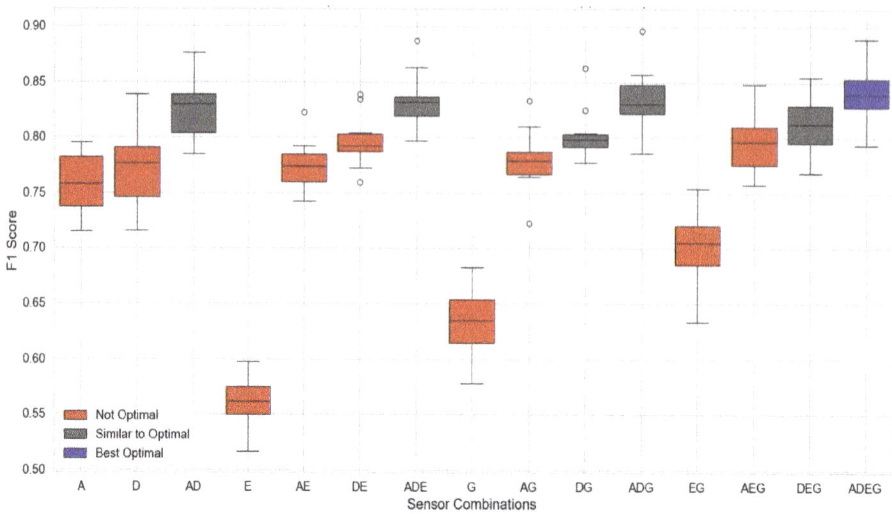

Figure 2. Results of identifying sensor combination on the upper body part, with positions indicated as follows: A: head, D: right shoulder, E: left forearm, and G: right upper arm.

Using these sensors, both individually and in combination, significantly improved algorithm performance compared with other configurations. As shown in Figure 2, six sensor setups—head–shoulder combination (AD), shoulder–upper arm (DG), shoulder–upper arm–forearm (DGE), head–shoulder–upper arm (ADG), head–shoulder–forearm (ADE), and all four sensors combined (ADEG, blue color)—achieved a mean F-score of approximately 0.82 and performance level, respectively. In contrast, sensors on the forearm and upper arm alone did not enhance system performance, indicating they may be less effective and could potentially reduce the overall accuracy of the fall-detection system.

The shoulder alone achieved a mean F-score of 77.40% (±3.68%), while the head achieved 75.74% (±2.86%). Performance improved when these positions were combined. For example, the shoulder–head combination yielded 82.64% (±2.83%), and the shoulder–upper arm combination reached 80.30% (±2.48%). Higher performances were observed when the three positions (shoulder, head, and upper arm) were combined, which resulted in a mean performance of 83.46% (±2.95%). The highest performance, 84.16% (±2.75%), was achieved when all four sensors were used together. These findings suggest that combining multiple sensor positions enhances the performance of fall classification systems. Figure 3 presents a confusion matrix based on the accuracy of each class, which evaluates the accuracy of various sensor combinations, helping to identify the most effective sensor placements for precise fall classification and making them more effective for real-world applications. While multiple sensor locations can enhance classification rates, focusing on key locations—such as the head, shoulder, and upper arm—proves most effective when used in combination. The confusion matrix (Figure 3) shows high precision and recall for non-fall and backward-fall events. However, slightly lower accuracy was observed for lateral falls, suggesting areas for improvement. Misclassifications primarily occurred between forward and lateral falls. Sensor combinations from the upper body, as described by the confusion matrix, are illustrated in Figure 3.

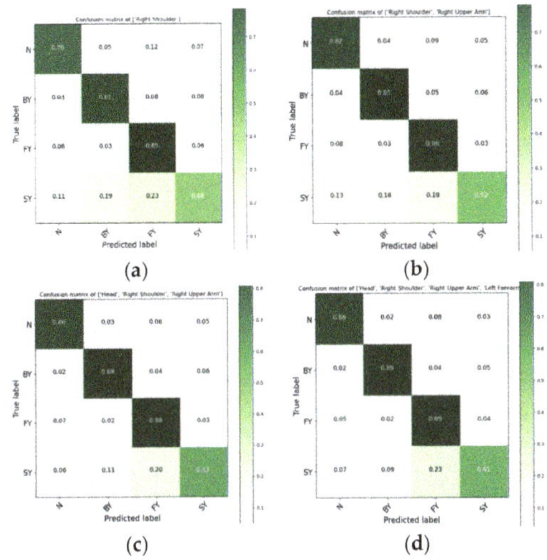

Figure 3. Confusion matrices for fall-direction classification based on optimal upper-body sensor locations: (**a**) shoulder-only sensor, (**b**) shoulder and upper arm combination, (**c**) head, shoulder, and upper arm combination, (**d**) head, shoulder, upper arm, and forearm combination. Classes are represented as "N" for non-fall, "BY" for backward fall, "FY" for forward fall, and "SY" for lateral fall.

3.3.2. Results of Three Different Sensor Location Combinations on the Lower Body

In this analysis, we focused on identifying the optimal single-sensor and sensor combinations for the lower body using an evaluation method similar to that described in the previous section. The results demonstrated that the best performance for single sensors in the lower body was achieved with sensors placed on the pelvis (B) and upper legs (J), as illustrated in Figure 4. These sensors, when used individually, significantly enhanced the effectiveness of the algorithm, outperforming other lower-body sensors. In Figure 5, three configurations—the pelvis–lower leg (BL), upper leg–lower leg (JL), and all three sensors combined (BJL)—achieved a geometric mean of approximately 0.915. Sensors on the lower leg alone did not improve performance, but when combined with sensors in other locations, they enhanced overall performance. This figure presents Tukey's HSD test results with a boxplot, comparing the mean F-score obtained using the SVM algorithm across these sensor positions. This analysis identifies the most effective sensor positions and combinations for accurate fall classification in the lower body parts, as shown in Figure 4.

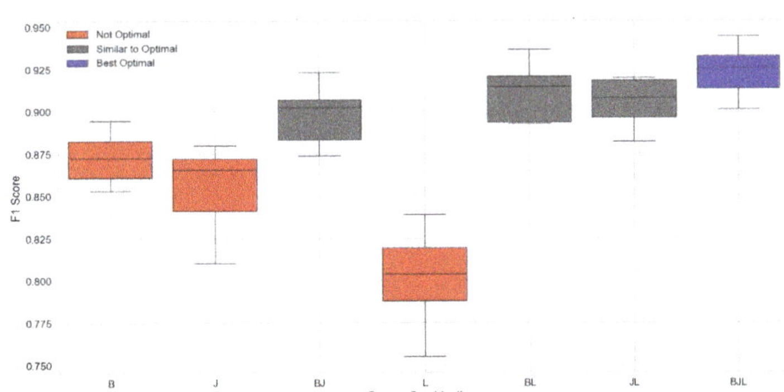

Figure 4. Results of identifying sensor combinations on the lower body, where positions are indicated as B: pelvis, J: left upper leg, and L: right lower leg.

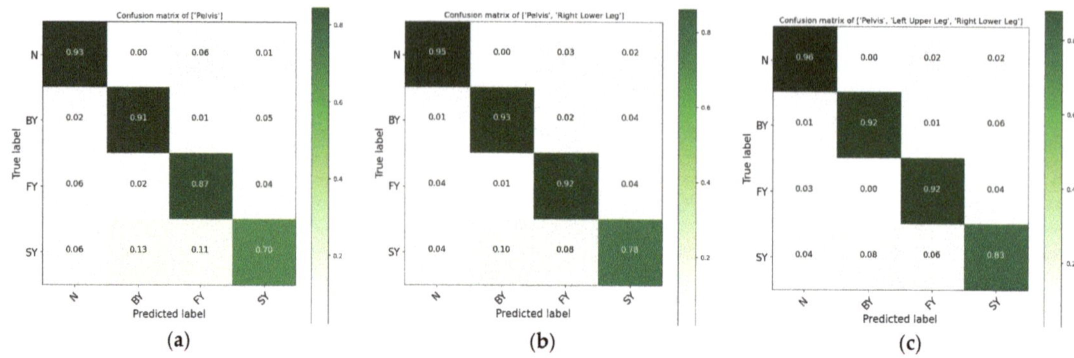

Figure 5. Confusion matrices for fall-direction classification based on optimal sensor locations on the lower body are presented as follows: (**a**) confusion matrix obtained from the pelvis location, (**b**) confusion matrix obtained from two combinations of pelvis–lower leg locations, (**c**) confusion matrix obtained from three combinations of pelvis–upper leg–lower leg locations. The classes are represented as "N" for "non-fall," "BY" for "backward-fall," "FY" for "forward-fall," and "SY" for "lateral-fall" or "side-fall".

For the lower body, the most effective areas were the pelvis at 87.24% (±1.38%) and the upper leg at 85.69% (±2.37%). Performance improved when these positions were combined. For example, combining the pelvis and lower legs resulted in a performance mean of 91.06% (±1.64%), while the combination of the upper and lower legs achieved 90.55% (±1.36%). The highest performance, 92.28% (±1.38%), was observed when all three locations—pelvis, upper leg, and lower leg—were combined. These findings indicate that combining multiple sensor positions generally enhances the performance of fall classification systems. Figure 5 presents the confusion matrix that highlights the performance of various sensor combinations for lower body parts. This analysis identifies the most efficient sensor locations and groupings for precise fall classification, enhancing their effectiveness for real-world applications. Combining multiple sensor locations, particularly the pelvis, upper leg, and lower leg, improves classification rates. The model performed exceptionally well in identifying non-fall, backward-fall, and forward-fall events with high precision and recall. However, lateral-fall classification showed slightly lower performance, indicating potential areas for model improvement. Misclassifications between forward and lateral falls were observed, likely due to similarities in fall patterns. The sensor combinations resulting from the lower body, as shown by the confusion matrices, are illustrated in Figure 5.

4. Discussion

Our results identified the shoulder and head as the most effective sensor locations for fall-direction classification on the upper body and the pelvis and upper leg for the lower body. These findings are consistent with previous studies that highlight the waist, chest, and thigh as key sensor locations (see Table 1) [5,7,15,16,31]. Although single-sensor placements are often sufficient for general fall detection, our study shows that they are inadequate for accurate fall-direction classification. Instead, combining sensors across both upper and lower body parts proved more effective. This approach enhances crucial data capture and simplifies the identification of the most effective sensor locations for multiclass fall classification and ADL systems.

We evaluated four classifiers across 12 sensor locations to select the best single classifier for fall-direction classification, with the SVM classifier emerging as the best classifier. This finding is consistent with previous studies that primarily focused on fall detection rather than fall-direction classification [7,16]. In those studies, SVM consistently outperformed other classifiers, including random forest (RF), multilayer perceptron (MLP), and K-Nearest Neighbors (KNN), in detecting falls from activities of daily living. Although some studies indicate that RF or MLP may perform better in specific fall-detection scenarios, our evaluation of fall-direction classification across various sensor locations demonstrated that SVM provided superior accuracy and consistency. Further research may explore the performance of alternative classifiers specifically for fall-direction classification.

We investigated whether sensor placement on the left versus right side of the body within the same anatomical area affected performance. Our analysis, covering five body areas (shoulders, upper arms, forearms, upper legs, and lower legs), found no significant performance differences between sides. This suggests that sensor placement within the same anatomical area can be flexible, allowing for random or alternating side selection without significantly affecting classification accuracy. This flexibility is beneficial for designing wearable systems as it can be adapted to user comfort or physical constraints.

While our results showed that combining sensors on multiple lower limb segments (pelvis, upper leg, and lower leg) provides the highest accuracy, we recognize that practical factors such as portability, affordability, and usability are just as important as accuracy in real-world applications of fall-detection systems. To maximize these practical aspects, we identified the pelvis as the single best sensor location for fall-direction classification. The pelvis offers a high classification performance due to its proximity to the center of mass, and it provides reliable data for detecting fall dynamics without requiring multiple sensors on the lower body. This makes it the most feasible choice for a single-sensor location, balancing accuracy with usability and cost-effectiveness.

Comparing the lower body to the upper body, the combination of the pelvis, upper leg, and lower leg sensors outperformed upper-body sensor combinations in classifying fall directions. The highest accuracy of 92.28% (±1.38%) was achieved with the three lower-body sensors, while upper-body sensors, such as the combination of shoulder, head, and upper arm, achieved a maximum accuracy of 83.46% (±2.95%). This suggests that lower-body sensors provide more accurate data for classifying fall direction, likely due to their closer proximity to the center of mass and the points of impact during a fall. These findings indicate that lower-body sensors should be prioritized in future wearable fall classification systems, as they significantly enhance classification accuracy for fall events.

However, we recognize the possibility that combining sensors from both upper and lower body regions could provide a more comprehensive view of body movement during a fall. This combination may improve the overall classification accuracy by leveraging complementary data from both areas. Therefore, future work could explore optimal combinations of upper and lower body sensors to maximize accuracy while minimizing the number of sensors required.

For the misclassifications between forward and lateral falls, contrary to initial assumptions about similarities in fall patterns, previous studies suggest that forward and lateral falls exhibit distinct biomechanical characteristics [32]. Forward falls typically involve different body segments and impact forces compared to lateral falls. Therefore, this first reason, the observed misclassifications may stem from sensor placement limitations or challenges in detecting lateral falls using the current setup. Lower-body sensor placements (pelvis, upper leg, and lower leg) also demonstrated superior classification performance compared to upper-body placements. This may indicate that lower-body sensors capture fall dynamics more effectively, especially in lateral falls, where the legs and pelvis may experience more pronounced movement and impact. And the second reason is the misclassification by a similar pattern. Further investigation into optimizing sensor configurations or applying more advanced machine learning models could help improve the detection of lateral falls.

Another observation is that lateral falls (SY) were misclassified as backward falls (BY) more frequently than as forward falls (FY), as shown in Figure 5. This is due to the similarities in body movement during backward and lateral falls, especially in the lower body. Both backward and lateral falls involve shifting of weight away from the center of mass, causing the legs and pelvis to experience similar patterns of acceleration and deceleration, particularly when using the current sensor placements. Forward falls, on the other hand, typically involve more pronounced upper body movement, which may make them easier to differentiate from lateral falls. The higher misclassification rate for lateral falls (SY) compared to other fall directions may also be attributed to the complex nature of lateral falls, where sideways movement of the body, particularly the legs and pelvis, produces sensor data that are more challenging to distinguish from other fall types, especially backward falls. This could indicate the need for further optimization of sensor placement or the addition of sensors specifically tailored to capture the unique dynamics of lateral falls.

Our results showed that lower-body sensor placements are more effective than upper-body sensor placements for fall-direction classification. The optimal combination involves three locations on the lower body: pelvis, upper leg, and lower leg, that significantly improved classification accuracy. Therefore, we recommend prioritizing the lower body for configuring sensor placements to collect essential data for accurate fall-direction classification.

This study has limitations. It focused on 12 specific sensor locations, excluding other potentially valuable locations such as the chest and neck. Although the SVM was the most effective model for this dataset, other machine learning or deep learning models not explored here may yield different results. The dataset may not represent the full range of daily activities, potentially limiting the robustness of the system in real-world scenarios. Lastly, although no significant differences were found between left and right sensor placements, this may not apply universally, as individual variations were not fully explored.

Additionally, the study population was not specifically at high risk for falls. The participants, while capable of imitating fall behaviors, were not elderly or mobility-impaired individuals, which may affect the applicability of our results to populations with a higher risk of falling. Falls in real-life scenarios are often unintentional and accidental, and these events may exhibit different biomechanical characteristics compared to the self-initiated falls simulated in our study. Real-life falls may involve higher levels of unpredictability, acceleration, and dynamic movement, which could influence the sensor data, particularly acceleration and gyroscope values. Therefore, while the falls in this study were designed to mimic high-risk behaviors, we acknowledge that self-initiated falls may not fully capture the complexities of accidental falls, limiting the generalizability of our findings. Future research should explore fall detection and classification in populations at higher risk and in more naturalistic, real-world settings to better understand the optimal sensor configurations for real-life fall scenarios.

One key limitation is that we did not account for changes in fall direction during the event. In real-life scenarios, a person may begin falling in one direction, such as forward, but rotate mid-fall and land on their side [27]. This dynamic shift in fall direction was not considered in our study, as our simulated falls followed a single, consistent trajectory. Future research should explore this aspect to better replicate real-world falls and improve the robustness of fall-direction classification systems.

5. Conclusions

This study aimed to optimize sensor placement on the human body for fall-direction classification while minimizing the number of sensors used. We focused on three main objectives: identifying the most effective machine learning classifier, analyzing whether sensor placement on different sides of the body impacts performance, and determining the optimal combination of upper- and lower-body sensors for accurate fall-direction classification.

Our findings highlight that while multiple sensors generally improve classification accuracy, focusing on two or three key sensor locations can provide robust performance. Specifically, the pelvis, upper leg, and lower leg emerged as the most effective lower-body locations, while the head, shoulder, and upper arm were identified as the best-performing upper-body locations. Although a single sensor can reliably distinguish between fall and non-fall events, it is less effective for identifying specific fall directions. Combining two or three sensors significantly enhances directional classification, achieving accuracies of 96% for non-fall events, 92% for backward and forward falls, and 83% for lateral falls. This demonstrates the importance of sensor placement, especially in areas like the pelvis and lower limbs, which are closer to the center of mass and impact points during a fall.

Future research should explore additional sensor placements beyond the 12 anatomical locations analyzed in this study to further improve fall-direction classification. Additionally, the application of advanced machine learning techniques, including deep learning models, could provide enhanced accuracy by capturing more complex movement patterns. Expanding the dataset to include a wider range of daily activities and fall types will be critical for improving the real-world applicability of fall-detection systems. Furthermore, individual variations—such as body type, age, and fall risk—should be explored to develop more personalized and effective fall classification solutions.

These future directions are essential for advancing the field of fall-direction classification, ensuring that the systems developed are not only highly accurate but also practical and adaptable, ensuring their effectiveness in diverse real-world scenarios, especially for populations at higher risk of falls.

Author Contributions: S.T. contributed to developing the proposed scenario and preparing the manuscript; J.-Y.K. advised and reviewed the manuscript; S.J., H.-W.G., J.L., E.H.C. and K.S.K. contributed to revising the manuscript; Y.N. supervised and reviewed the manuscript. All authors have read and agreed to the published version of the manuscript.

Funding: This work was supported by the National Research Foundation of Korea (NRF) grant funded by the Korea government (MSIT) (No. RS-2023-00218176), a grant of the Korea Health Technology R&D Project through the Korea Health Industry Development Institute (KHIDI), funded by the Ministry of Health & Welfare, Republic of Korea (grant number: HI21C1831) and the Soonchunhyang University Research Fund.

Institutional Review Board Statement: The study was conducted according to the guidelines of the Institutional Review Board on Human Subjects Research and Ethics Committees, Soonchunhyang University, Cheonan, Korea.

Informed Consent Statement: Informed consent was obtained from all subjects involved in the study.

Data Availability Statement: The study's data are available upon request from the corresponding author.

Conflicts of Interest: The authors declare no conflicts of interest.

References

1. World Health Organization. *Step Safely: Strategies for Preventing and Managing Falls across the Life-Course*; World Health Organization: Geneva, Switzerland, 2021. Available online: https://iris.who.int/handle/10665/340962 (accessed on 13 July 2024).
2. Pannurat, N.; Thiemjarus, S.; Nantajeewarawat, E.; Anantavrasilp, I. Analysis of Optimal Sensor Positions for Activity Classification and Application on a Different Data Collection Scenario. *Sensors* **2017**, *17*, 774. [CrossRef] [PubMed]
3. Davoudi, A.; Mardini, M.T.; Nelson, D.; Albinali, F.; Ranka, S.; Rashidi, P.; Manini, T.M. The Effect of Sensor Placement and Number on Physical Activity Recognition and Energy Expenditure Estimation in Older Adults: Validation Study. *JMIR mHealth uHealth* **2021**, *9*, e23681. [CrossRef] [PubMed]
4. Kulurkar, P.; Preethi, P. AI based elderly fall prediction system using wearable sensors: A smart home-care technology with IOT. *Sensors* **2023**, *25*, 100614. [CrossRef]
5. Ponce, H.; Martínez-Villaseñor, L.; Nuñez-Martínez, J. Sensor Location Analysis and Minimal Deployment for Fall Detection System. *IEEE Access* **2020**, *8*, 166678–166691. [CrossRef]
6. Martínez-Villaseñor, L.; Ponce, H.; Brieva, J.; Moya-Albor, E.; Núñez-Martínez, J.; Peñafort-Asturiano, C. UP-Fall Detection Dataset: A Multimodal Approach. *Sensors* **2019**, *19*, 1988. [CrossRef]
7. Santoyo-Ramón, J.A.; Casilari, E.; Cano-García, J.M. Analysis of a Smartphone-Based Architecture with Multiple Mobility Sensors for Fall Detection with Supervised Learning. *Sensors* **2018**, *18*, 1155. [CrossRef]
8. Sun, Z.; Wang, G.; Li, P.; Wang, H.; Zhang, M.; Liang, X. An improved random forest based on the classification accuracy and correlation measurement of decision trees. *Expert Syst. Appl.* **2024**, *237*, 121549. [CrossRef]
9. Mitchell, T.M. *The Discipline of Machine Learning*; School of Computer Science, Carnegie Mellon University: Pittsburgh, PA, USA, 2006; Volume 9.
10. Dang, X.; Li, W.; Zou, J.; Cong, B.; Guan, Y. Assessing the impact of body location on the accuracy of detecting daily activities with accelerometer data. *iScience* **2024**, *27*, 108626. [CrossRef]

11. Casilari, E.; Oviedo-Jiménez, M.A. Automatic Fall Detection System Based on the Combined Use of a Smartphone and a Smartwatch. *PLoS ONE* **2015**, *10*, e0140929. [CrossRef]
12. Özdemir, A.T. An Analysis on Sensor Locations of the Human Body for Wearable Fall Detection Devices: Principles and Practice. *Sensors* **2016**, *16*, 1161. [CrossRef]
13. Ntanasis, P.; Pippa, E.; Özdemir, A.T.; Barshan, B.; Megalooikonomou, V. Investigation of Sensor Placement for Accurate Fall Detection. In *Wireless Mobile Communication and Healthcare (MobiHealth 2016)*; Springer: Cham, Switzerland, 2017; pp. 225–232. [CrossRef]
14. Sikandar, T.; Rabbi, M.F.; Ghazali, K.H.; Altwijri, O.; Almijalli, M.; Ahamed, N.U. Minimum number of inertial measurement units needed to identify significant variations in walk patterns of overweight individuals walking on irregular surfaces. *Sci. Rep.* **2023**, *13*, 16177. [CrossRef]
15. Ahamed, N.U.; Kobsar, D.; Benson, L.C.; Clermont, C.A.; Osis, S.T.; Ferber, R. Subject-specific and group-based running pattern classification using a single wearable sensor. *J. Biomech.* **2019**, *84*, 227–233. [CrossRef] [PubMed]
16. Sikandar, T.; Rabbi, M.F.; Ghazali, K.H.; Altwijri, O.; Almijalli, M.; Ahamed, N.U. Evaluating the difference in walk patterns among normal-weight and overweight/obese individuals in real-world surfaces using statistical analysis and deep learning methods with inertial measurement unit data. *Phys. Eng. Sci. Med.* **2022**, *45*, 1289–1300. [CrossRef] [PubMed]
17. Ionut-Cristian, S.; Dan-Marius, D. Using Inertial Sensors to Determine Head Motion—A Review. *J. Imaging* **2021**, *7*, 265. [CrossRef] [PubMed]
18. Lin, F.-Y.; Lee, P.-T.; Ho, Y.-H.; Sung, P.-S.; Chen, P.-T.; Lin, C.-L. Fall Prediction Based on Head-Mounted IMU Sensor System. In Proceedings of the 2022 IEEE 11th Global Conference on Consumer Electronics (GCCE), Osaka, Japan, 18–21 October 2022; pp. 341–343. [CrossRef]
19. Martinez-Villaseñor, L.; Ponce, H. Design and Analysis for Fall Detection System Simplification. *J. Vis. Exp.* **2020**, e60361. [CrossRef] [PubMed]
20. Atallah, L.; Lo, B.; King, R.; Yang, G.-Z. Sensor Positioning for Activity Recognition Using Wearable Accelerometers. In *IEEE Transactions on Biomedical Circuits and Systems*; IEEE: New York, NY, USA, 2011; Volume 5, pp. 320–329. [CrossRef]
21. Yao, L.; Min, W.; Lu, K. A New Approach to Fall Detection Based on the Human Torso Motion Model. *Appl. Sci.* **2017**, *7*, 993. [CrossRef]
22. Shi, G.; Zhang, J.; Dong, C.; Han, P.; Jin, Y.; Wang, J. Fall detection system based on inertial mems sensors: Analysis design and realization. In Proceedings of the 2015 IEEE International Conference on Cyber Technology in Automation, Control, and Intelligent Systems (CYBER), Shenyang, China, 8–12 June 2015; pp. 1834–1839. [CrossRef]
23. Albert, M.V.; Kording, K.; Herrmann, M.; Jayaraman, A. Fall Classification by Machine Learning Using Mobile Phones. *PLoS ONE* **2012**, *7*, e36556. [CrossRef]
24. Kraft, D.; Srinivasan, K.; Bieber, G. Deep Learning Based Fall Detection Algorithms for Embedded Systems, Smartwatches, and IoT Devices Using Accelerometers. *Technologies* **2020**, *8*, 72. [CrossRef]
25. Montanini, L.; Del Campo, A.; Perla, D.; Spinsante, S.; Gambi, E. A Footwear-Based Methodology for Fall Detection. *IEEE Sens. J.* **2018**, *18*, 1233–1242. [CrossRef]
26. Turan, M.; Barshan, B. Classification of fall directions via wearable motion sensors. *Digit. Signal Process.* **2022**, *125*, 103129. [CrossRef]
27. Robinovitch, S.N.; Dojnov, A.; Komisar, V.; Yang, Y.; Shishov, N.; Yu, Y.; Bercovitz, I.; Cusimano, M.D.; Becker, C.; Mackey, D.C.; et al. Protective responses of older adults for avoiding injury during falls: Evidence from video capture of real-life falls in long-term care. *Age Ageing* **2022**, *51*, afac273. [CrossRef] [PubMed]
28. Pinto, A.B.A.; de Assis, G.A.; Torres, L.C.B.; Beltrame, T.; Domingues, D.M.G. Wearables and Detection of Falls: A Comparison of Machine Learning Methods and Sensors Positioning. *Neural Process. Lett.* **2022**, *54*, 2165–2179. [CrossRef]
29. Nandy, A.; Saha, J.; Chowdhury, C.; Singh, K.P.D. Detailed Human Activity Recognition using Wearable Sensor and Smartphones. In Proceedings of the 2019 International Conference on Opto-Electronics and Applied Optics (Optronix), Kolkata, India, 18–20 March 2019; pp. 1–6. [CrossRef]
30. Rabbi, M.F.; Diamond, L.E.; Carty, C.P.; Lloyd, D.G.; Davico, G.; Pizzolato, C. A muscle synergy-based method to estimate muscle activation patterns of children with cerebral palsy using data collected from typically developing children. *Sci. Rep.* **2022**, *14*, 3599. [CrossRef] [PubMed]
31. Liu, S.-H.; Cheng, W.-C. Fall Detection with the Support Vector Machine during Scripted and Continuous Unscripted Activities. *Sensors* **2012**, *12*, 12301–12316. [CrossRef] [PubMed]
32. Moon, Y.; Sosnoff, J.J. Safe Landing Strategies During a Fall: Systematic Review and Meta-Analysis. *Arch. Phys. Med. Rehabil.* **2017**, *98*, 783–794. [CrossRef]

sensors

MDPI

Article

Design and Development of a Smart Fidget Toy Using Blockchain Technology to Improve Health Data Control

Polina Bobrova *, Paolo Perego and Raffaele Boiano

Department of Design, Politecnico di Milano, Via Candiani 72, 20158 Milano, Italy; paolo.perego@polimi.it (P.P.); raffaele.boiano@polimi.it (R.B.)

* Correspondence: polina.bobrova@polimi.it

Abstract: This study explores the integration of blockchain technology in wearable health devices through the design and development of a Smart Fidget Toy. We aimed to investigate design challenges and opportunities of blockchain-based health devices, examine the impact of blockchain integration user experience, and assess its potential to improve data control and user trust. Using an iterative user-centered design approach, we developed a mid-fidelity prototype of a physical fidget device with a blockchain-based web application. Our key contributions include the design of a fidget toy using blockchain for secure health data management, an iterative development process balancing user needs with blockchain integration challenges, and insights into user perceptions of blockchain wearables for health. We conducted user studies, including a survey ($n = 28$), focus group ($n = 6$), interactive wireframe testing ($n = 7$), and prototype testing ($n = 10$). Our study revealed high user interest (70%) in blockchain-based data control and sharing features and improved perceived security of data (90% of users) with blockchain integration. However, we also identified challenges in user understanding of blockchain concepts, necessitating additional support. Our smart contract, deployed on the Polygon zkEVM testnet, efficiently manages data storage and retrieval while maintaining user privacy. This research advances the understanding of blockchain applications in health wearables, offering valuable insights for the future development of this field.

Keywords: blockchain; wearable devices; user-centered design; data privacy and control; health data collection

1. Introduction

1.1. Background

Wearable technologies, like fitness trackers and smartwatches, do more than count steps—they constantly track how much we move, how fast our hearts beat, and even how well we sleep. Wearable devices offer non-intrusive means of gathering data in real time on a range of behavioral and physiological markers associated with mental health [1–3].

However, using these wearables also brings several challenges, particularly related to data privacy, data ownership [4], user acceptance, and adoption [2]. To address these challenges, there is growing interest in exploring the potential of blockchain technology [5,6].

Blockchain offers a decentralized and secure way to store and manage data, providing transparency while ensuring privacy and control [7–11]. Blockchain employs advanced cryptographic methods such as hash chaining, anonymous signatures, and non-interactive zero-knowledge proofs to ensure data integrity and privacy [9,12].

Integrating blockchain with IoT systems can address security and privacy challenges by providing decentralized authentication, data integrity, and secure data sharing [13,14]. Blockchain can secure electronic health records by enabling decentralized access control and data confusion, balancing privacy with accessibility [14].

The immutability of blockchain records ensures that once data is recorded, it cannot be altered or deleted, which is crucial for maintaining the integrity of medical records [15–17].

Its potential applications in mental health data collection [18–20] present an opportunity to overcome the existing barriers and lay the foundation for the efficient use of wearable devices. Distributed ledger technology allows for efficient and secure sharing of electronic health records (EHRs) across different healthcare providers, improving the accuracy and timeliness of diagnoses and treatments [7,15,21].

Blockchain can address interoperability issues by providing a standardized framework for data sharing among different healthcare systems and providers, facilitating seamless access to patient records [7,11,22].

This research explores blockchain technology's potential to shape wearable devices and UX while enhancing trust. It lies in the overlap of design, technology, and health. The study uses human-centered design to create wearables that focus on developing a smart fidget toy that is conscious of patients' mental health needs while ensuring that the data gathered is secure, private, and valuable for doctors and therapists.

1.2. Motivation

Building upon the exploration of the capabilities of blockchain technology to improve wearable devices for healthcare, our research was narrowed down to focus on a specific device that embodies it: a smart fidget toy. Integrating smart technology into fidget devices offers a unique opportunity to collect mood and health-related data.

A fidget toy is a handheld object designed to help individuals focus, relieve stress, provide sensory experience, or keep their hands busy through repetitive movements [23]. These are used for various applications, notably in aiding individuals with ADHD, autism, and Asperger syndrome [24]. Traditional fidget toys like spinners have been widely adopted for simplicity and playfulness. No peer-reviewed scientific evidence showed that fidgets are effective treatments for mental health conditions [25]. An attempt to address the problem through the use of smart fidgets was proposed by Liang et al. [26] to provide real-time feedback and data tracking for individuals.

However, the potential of integrating smart technology into these toys remains not widely explored, offering opportunities to enhance their functionality and user engagement.

Smart fidget toys have the potential to collect valuable data on user interactions, which can be analyzed to help them build correlations. This data can be used not only to personalize the user experience, making the toys more effective for stress management and focus enhancement, but would also allow to build validated reports for third parties such as doctors, therapists, and researchers and improve collective health information [27].

Data collected from the smart fidget toy can be analyzed by the user and third parties to identify correlations between fidgeting patterns and mood states.

Given the sensitive nature of this data, blockchain technology can be proposed. Blockchain is increasingly being explored for its potential applications in healthcare, offering a decentralized and secure way to manage health-related data. This technology promises to enhance data sharing, maintain patient privacy, and improve the overall efficiency of healthcare systems [28], in particular:

1. Enhanced Data Security: The decentralized nature of blockchain ensures that data is not stored in a single location, reducing the risk of data breaches. Data recorded by the fidget toy can be securely encrypted and stored on the blockchain, making it tamper proof and ensuring that user data remains confidential and unaltered [29].
2. User Control and Privacy: Blockchain empowers users by giving them control over their data. Users can grant or revoke access to their data anytime, ensuring they maintain privacy and have a say in how their information is used. This is particularly important for sensitive health-related data [22,28].
3. Transparency and Trust: The transparency inherent in blockchain technology fosters trust among users. Every transaction is recorded on a public ledger, allowing users to verify the authenticity of their data. This transparency ensures that third parties accessing the data do so with the user's explicit consent, and any data manipulation attempts can be easily detected [30].

4. Immutable Health Records: Once data is recorded on the blockchain, it cannot be altered or deleted. This immutability is crucial for maintaining accurate health records over time. Users and healthcare providers can rely on the integrity of the data, knowing it reflects accurate historical metrics [31].
5. Efficient Data Sharing: Blockchain facilitates seamless and secure data sharing between users and authorized third parties. For instance, users can share their data with healthcare professionals, researchers, or wellness apps without the need for intermediaries, enhancing the efficiency and speed of data exchange [32].
6. Improved User Engagement: By leveraging blockchain technology, users can be motivated to engage more with the fidget toy and the associated web app. For example, they could earn tokens or rewards for contributing their data to research studies or maintaining consistent usage, adding a layer of gamification and motivation [33]. Moreover, blockchain can improve user engagement by providing security, privacy, transparency, trust, and traceability, as analyzed through user-generated content on Twitter [34].
7. Interoperability: Blockchain can support interoperability with other healthcare systems and applications, ensuring that data stored can be integrated with other health records and analytics platforms. This enhances the value of the collected data and provides a comprehensive view of the user's health [35].

Despite the promising potential of blockchain-powered fidget toys, their development poses several challenges. To provide users with valuable data, it is necessary to collect quantitative data, such as frequency and duration of fidget toy usage, and qualitative data that captures the user's emotional state or stress levels by implementing sensor technologies. While integrating blockchain, it is essential to identify the limits of the technology to list the restrictions and requirements for the product, service, and user experience design. Designing a toy that effectively balances traditional tactile satisfaction with digital capabilities requires a thorough understanding of user needs and preferences. The iterative design approach employed in the development of the smart fidget toy was essential in addressing these challenges and ensuring user-centered design.

In this study, we focus on a specific target audience—individuals aged 24 to 32 who are highly interested in well-being practices and are considered technology enthusiasts. This demographic is particularly receptive to new technology, making them an ideal group for testing and refining new smart wearable devices.

Our smart fidget toy is designed for on-demand play, with data collection occurring only when in use. This approach minimizes the computational power required and optimizes memory usage on both the device and the blockchain network. Limiting data collection to active usage periods ensures the device remains efficient and responsive, enhancing the overall user experience.

We assume that the moment the user interacts with our Toy, it is a moment of stress or when they are seeking concentration. Therefore, in the preliminary prototypes, we incorporated only one sensor—a photoplethysmography (PPG) that is effective for sensing the HR and HRV values of healthy subjects during rest [36].

1.3. Research Questions and Objectives

This study is part of a broader PhD research project titled User-Centric Design for Health Wearables: Exploring Blockchain Adoption for Data Privacy and Control. The overall aim of the doctoral research is to contribute to the developmental and ethical issues of health wearables and the adoption of blockchain technology. Within this framework, the current study addresses the following primary research questions:

- What are the design challenges in the design of blockchain-based devices for health?
- What are the design opportunities in the design of blockchain-based devices for health?

Additionally, we explore secondary questions to provide a comprehensive understanding:

- How can implementing blockchain technology shape the wearable device, user experience, and related applications?
- How can blockchain enhance data control, awareness of data value, and user trust?

This study employs a systematic approach that combines user feedback, iterative design processes, technology integration, and validation through user testing and serves as an evaluative single-case study. Through this investigation, we aim to contribute to the interdisciplinary area between design, technology, and health by addressing the design challenges and opportunities of blockchain-based devices for health.

To provide a comprehensive overview of our research framework, Figure 1 illustrates the key components of this study. The primary objective of this research is to explore how iterative design and user-centered techniques can be employed to develop a device collecting health-related data. Through focus group discussions, surveys, and user testing sessions, we seek to understand the key features and functionalities users desire and the most effective ways to implement these features. By documenting our design and development process, we aim to provide valuable insights and guidelines for future design projects incorporating blockchain technology and smart wearables.

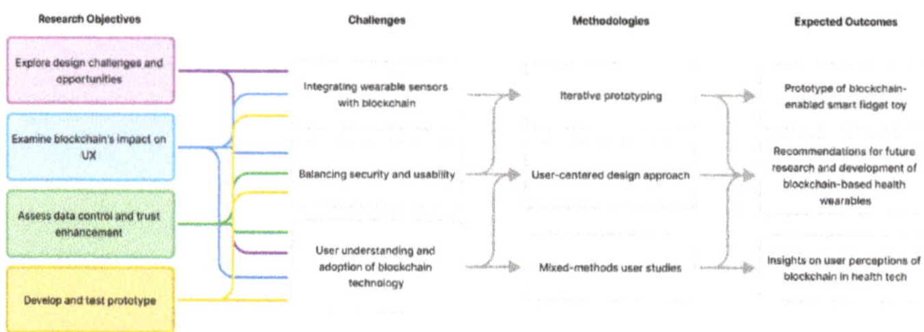

Figure 1. Overview of Research Objectives, Challenges, Methodology, and Expected Outcomes.

1.4. Paper Organization

This paper is organized as follows: Section 2 reviews the literature on wearable devices, blockchain in healthcare, and user-centered design. Section 3 describes our methodology, including the iterative design process, user involvement, and development of the prototypes. Section 4 presents results from user studies. Section 5 discusses key findings and challenges in integrating blockchain with wearable health devices. Section 6 concludes the paper, summarizing contributions and future directions. Throughout, we address the research questions and objectives outlined in Section 1.3, focusing on the design and development of our blockchain-based smart fidget toy.

2. Literature Review

2.1. Wearable Devices in Healthcare

Wearable devices have gained significant interest from experts in healthcare for their ability to monitor physiological and behavioral data continuously. They effectively monitor physiological parameters such as heart rate, blood pressure, and glucose levels, aiding in early diagnosis and treatment [37]. Lu et al. still underline that despite the potential, wearable devices face challenges such as user-friendliness, privacy, and security [37].

Advances in miniaturization, flexible electronics, and biosensors have significantly improved the functionality and reliability of wearable devices [38,39].

Piwek et al. highlight the potential of consumer health wearables to revolutionize healthcare, especially by assisting patients in "self-tracking" [4]. They also underline challenges related to data accuracy and privacy [4].

In the context of mental health, Onyeaka et al. report on the increasing use of smartphones and wearables for health promotion among individuals with anxiety or depression [1]. Schecter et al. discuss the purported benefits and potential drawbacks of fidget spinners, noting the lack of peer-reviewed evidence supporting their effectiveness for mental health conditions [25]. However, Liang et al. proposed an augmented fidget spinner for biofeedback and respiration training, demonstrating the potential for integrating smart technologies into fidget devices [26].

2.2. Blockchain Technology in Healthcare

Numerous researchers have explored the application of blockchain technology in healthcare. Hasselgren et al. provide a comprehensive scoping review of blockchain applications in healthcare and health sciences, highlighting potential benefits in areas such as electronic health records management and clinical trials [18].

Blockchain in healthcare is primarily used for secure data sharing, managing health records, and access control [18,28,40,41]. It addresses the challenges of electronic health records by providing a decentralized and secure method for data exchange. It improves interoperability between disparate healthcare systems, facilitating seamless data exchange and access control [40,41].

Fan et al. proposed MedBlock, a blockchain-based information management system that aims to efficiently and securely share medical data [7]. Their work demonstrates the potential of blockchain in improving the security of data and interoperability in healthcare settings.

Koumpounis and Perry proposed a blockchain-based electronic health record system with patient-centered data access control, highlighting the importance of user empowerment in managing health data [19].

Chowdhury et al. developed a blockchain-based wearable data marketplace to address privacy and trust issues in health data sharing [42], proving blockchain's potential to improve data security, user control, and privacy in health data sharing from wearable devices.

Dwivedi et al. presented a decentralized privacy-preserving healthcare blockchain for IoT devices, addressing some of the security and privacy challenges associated with health data collection from wearables [5].

2.3. User-Centered Design and Technology in Healthcare

The importance of user-centered design in health technologies is well-established. Genaro Motti and Caine provide an overview of wearable applications for healthcare, emphasizing the need for user-centered design approaches to address challenges in user acceptance and adoption [43].

Effective user-centered design involves end-users from the early stages of project development through to final deployment, ensuring their needs and requirements are met [44]. Involving patients in the design and testing phases ensures that health technologies are customized to meet their needs, enhancing functionality and usability [45].

Blockchain technology can be considered challenging from a design perspective [46]. Moniruzzaman et al. underline that the design process of blockchain-based products often lacks systemized practice guidelines [47]. Moreover, designers struggle to understand the particular challenges users face in blockchain-based products [48,49], and developers often cover design-related issues [46].

Despite the growing body of research on blockchain applications in healthcare, several gaps remain. There is limited empirical evidence on the user acceptance of blockchain-based health data-sharing systems. Furthermore, while user-centered design is recognized as important, there is a lack of specific design guidelines for blockchain-based health applications that balance technical requirements with user needs and regulatory compliance.

3. Materials and Methods

3.1. Iterative Design Process

The development of the smart fidget toy followed an iterative, user-centered design process consisting of multiple phases. Each phase incorporated user feedback to refine and improve the prototype. This approach ensures that the outcome will be closely aligned with user needs and preferences.

Each phase of the design process involved the participation of potential users in various formats (interviews, testing sessions, etc.) and iterative prototyping to refine the device based on direct user feedback.

3.1.1. Initial Concept

The initial concept for the Smart Fidget Toy (Figure 2) was developed based on preliminary research into existing fidget toys and wearable devices for stress management. The goal was to integrate blockchain technology into a device that provides secure data storage while maintaining the tactile satisfaction of traditional fidget toys.

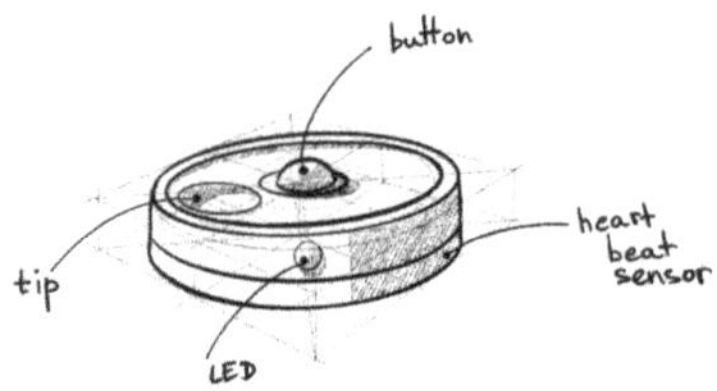

Figure 2. Sketch of Initial Concept of the Device.

Smart Fidget Toy is proposed as a device to collect data when a stressful moment happens to the user and to correlate it with some health parameters.

The product concept includes the following aspects: shape and ergonomics, sensors and fidgeting elements, possible use cases, and used sensors. The product will be in a round, flat shape to fit comfortably in the user's palm and enable it to be used as jewelry.

By leveraging blockchain technology, the wearable device aims to provide a holistic approach to mood monitoring, empowering individuals to take proactive steps in managing their mood state and assisting them in their rituals associated with OCD, ADHD, panic attacks, and depression. Smart Fidget Toy, however, is not a medical device and collects biomedical data solely for monitoring and research purposes.

3.1.2. Project Development

To define better the user interactions with the product, we conducted the following studies and exercises (Figure 3):

1. Development of the Storyboard
2. Persona Canvas Development
3. Development of the Empathy Map
4. Comparison of the User Flows of similar products
5. Development of User Flows
6. Development of Low-fidelity prototypes
7. App Map Development
8. User Journey Map Development
9. Service Blueprint Development
10. Focus Group
11. Anonymous Survey
12. Development and User Testing of Interactive Wireframes
13. Development and User Testing of Mid-fidelity prototype

That approach allowed us to continuously refine the product while maintaining its multidisciplinary development while focusing on a user-centered design approach.

Four of the activities involved potential users. In these studies, the number of participants varied, reflecting both methodological considerations and practical constraints. Our sample sizes are consistent with common practices in user-centered design research, particularly for qualitative data collection and iterative design processes.

For the focus group ($n = 6$), we aimed for a size that would facilitate a discussion while ensuring all participants had an opportunity to contribute. This aligns with recommendations for focus group sizes in design research, which typically range from 4 to 6 participants [50].

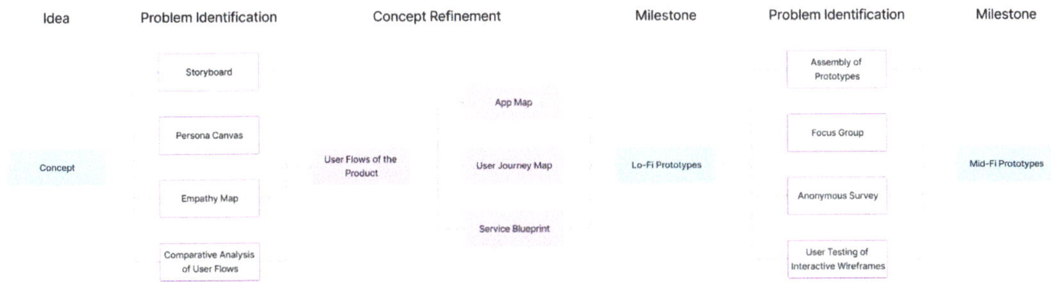

Figure 3. The Development Process of the Smart Fidget Toy.

The survey ($n = 28$) allowed us to gather a broader range of perspectives, providing quantitative data to complement our qualitative insights. While larger sample sizes are ideal for surveys, this number was sufficient for our exploratory purposes [51].

For the wireframe testing ($n = 7$) and mid-fidelity prototype testing ($n = 10$), these sample sizes are typical for usability studies, where 5 to 10 participants often uncover the majority of usability issues [52,53]. These numbers allowed for in-depth, qualitative feedback on user experience while remaining feasible.

It's important to note that our study faced time limitations and recruitment challenges, which influenced our final participant numbers. Additionally, to maximize insights from our participant pool, some individuals participated in multiple study phases: two participants were involved in both the focus group and mid-fidelity prototype testing, and two others participated in both wireframe testing and mid-fidelity prototype testing. This overlap allowed us to gather longitudinal insights on user perceptions throughout the design process. While larger sample sizes could potentially offer more detailed results, our approach prioritized deep, qualitative insights at each stage of the design process.

3.1.3. Development of Prototypes

To begin the development of the smart fidget, we created several low-fidelity prototypes focusing either on the technological aspects of the product or on the physical appearance of the device, incorporating basic interaction features.

Those prototypes were developed to validate that the model can be produced, the electronic elements fit inside, and we can test it with users.

The technology-focused prototype was not shown to the audience, but it served as an experimental exercise to define the technologies used for the project. Since the technological aspects significantly influence the design, it was essential to start developing the hardware and software components of the wearable device early in the design process. The draft software to transfer and store the data in the blockchain was developed and connected to a physical low-fi prototype.

The physical artifact was re-developed several times. The 3D model of the prototype was prepared using Grasshopper inside of Rhinoceros [54]. The Grasshopper script describes the design of several separate prototype elements parametrically to allow easier future editing. The following components were developed: the rotary disc, the closing cap to fix the disk, the button cap, and the base. The size of the artifact was influenced by the electronics selected, especially by the height of the rotary encoder and the size of the microcontroller.

The base has supportive elements for the closing cap and rotary disc, a hole for the PPG sensor, another one for the charging cable, and one more for LED. The rotary element is presented as a flat disc with a spherical tip on the top surface and a full-height hole in the center for the rotary encoder. The button cap has a special border to assemble it to the rotary disc and a little tip to fit in the rotary encoder rotating part. The closing cap of the device has a little border to lock the rotary disc inside the device and two holes aligned with the base for the LED and charging cable.

All the elements were fabricated using 3D printing technology, and all elements except for the button cover were printed with white PLA, while for the button, we used different mixes of rubber-like materials.

At this stage, the only electronic element placed inside was the rotary encoder since it provides tangible user feedback.

Those prototypes were developed to verify the selection of electronics and validate their design with potential users.

3.2. User Experience Design

Blockchain technology is considered a highly safe data storage method but has a complex user experience [55]. Moreover, most of the microcontrollers used for wearables cannot directly interact with the blockchain, which means that an intermediate layer or bridge solution is required to facilitate communication between the wearable device and the blockchain.

This can introduce additional complexities in designing and implementing blockchain-based wearables for health-related monitoring.

To define the user experience of our product, an analysis of User Flows of the products available on the market was conducted with a focus on user interactions, use of blockchain technology, data collection, and data storage.

We selected several products (Table 1): three blockchain-based products (Patientory is a health app [56,57], and two others are IoT devices—IoTeX Ucam [58] and IoTeX Pebble [59]). We also selected one symptom-tracking app (Bearable [60]), three mental health devices (Apollo [61], TouchPoints [62], and Lief [63]), and one mood companion phygital game (InTempo [64]).

Table 1. User Flow Analysis of Selected Products.

Product	Product Category	Use of Blockchain	Data Entry
Patientory [38,39]	Health App	For User Authentication, Data Storage, Data Access, User Reward, Subscription Payment	User Entry, Healthcare Providers, Third-Party Products
IoTeX Ucam [40]	IoT	For User Authentication, Data Access	From the device
IoTeX Pebble [41]	IoT	For Data Storage, Data Distribution	From the device
Bearable [42]	Mood Tracking App	No	User Entry, Third-Party Products
Apollo [43]	Mental Health Device	No	From the use of the product
Touchpoints [44]	Mental Health Device	No	No
Lief [45]	Mental Health Wearable	No	From the wearable
InTempo [46]	Mood Companion (Device + App)	No	From the user interactions with the device and/or app

From the analysis of blockchain-based products, we learned that Patientory uses blockchain technology for user authentication, data storage, data access, user rewards, and payments for storage subscriptions [56,57]. IoTeX Ucam uses blockchain for user authentication and data access only [58], while IoTeX Pebble utilizes blockchain for data storage and data distribution [59].

None of the mood/mental health-related products we selected for our analysis utilize blockchain. The physical devices of those products are collecting the data in a different manner. For example, Apollo does not collect the data using the device but rather from the fact of interactions with the product [61], while InTempo tracks user interactions with the device while it is connected to a smartphone and the user plays a game [64]. Lief collects the data from the wearable about users and provides users with biofeedback [63].

This analysis was used to define user flows overall and specifically at what steps of the user flows best to integrate interactions with blockchain technology and how to pass the data from the device to the blockchain.

For our web app, we implemented a similar experience of mood tracking apps for login, records retrieval, and uploading new records interactions. To share the data entries from the device to the blockchain, we decided to use the same web app with a connection to a cryptocurrency wallet for blockchain transactions as commonly implemented in blockchain-based apps.

To limit the efforts of the user, we decided to utilize wired communication between the device and the web app so that the user can charge the device and upload the new data collected by the device simultaneously.

The results influenced the definitions of User Flows, App Map, User Journey Map, and Service Blueprint. The regular user interaction is defined as follows:

1. The user uses the device on demand. The device collects data about user interactions and the user's heartbeat.
2. The user can connect the device to a web app. If new data is found on the device, the user is advised to upload it to the blockchain network, providing their notes and comments if they want to. Once the data is on the blockchain network, it is permanently deleted from the device.
3. Using the web app, the user can retrieve their previous records from the blockchain. They can only see them if they are logged in to the wallet they use for interactions with the blockchain network.
4. The user can grant or revoke access to their data to doctors and/or researchers. They can get monetary benefits, if applicable, from sharing their data.

3.3. Focus Group Study

We conducted a focus group discussion with six individuals from our target demographic (24–32 years old, interested in well-being practices and technology). Participants interacted with the lo-fi prototype and provided feedback on design, usability, and desired features. The feedback collected was used to inform the next iteration of the prototype.

The prototype was presented to a focus group to validate affordances, tactile properties and feedback, ergonomics, the influence of peers on the use of the device, and to adjust the position of the sensor.

The participants were asked to discuss the presented artifact, try using it, and provide their opinions. Then, they were described in more detail about the product and asked more precise questions about tactile feedback, shape, materials, and overall playfulness.

The audio transcript of the study was later coded to identify key issues, key potentials, opinions on material and tactile feedback, and the overall opinion of participants.

To validate the ergonomics and the sensor position, participants were invited to apply a transmittable colored element (eyeshades) on their hands and try to use the device again (Figure 4). Later, we could see the most and least touched spots on the prototype.

Feedback from the focus group highlighted the importance of tactile feedback and ergonomic design, which led to adjustments in the shape and materials used in the next prototype iteration.

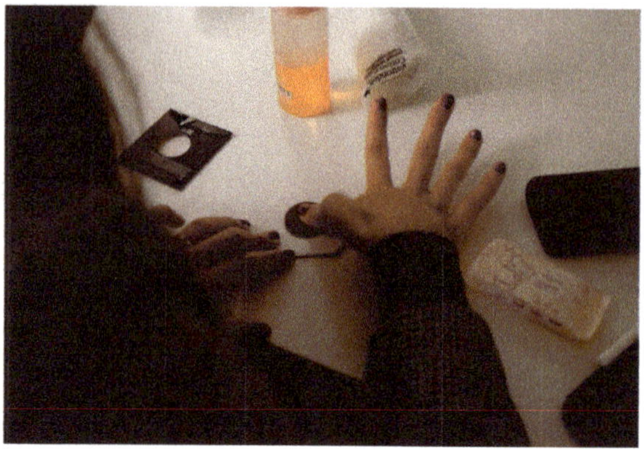

Figure 4. Participant Applying Transmittable Material on their Hands for an Experiment.

3.4. Anonymous Survey

An anonymous survey was designed to gather information on the types of data users would find valuable for their well-being practices while using the smart fidget toy. The survey had an exploratory nature and included general demographic questions, questions on user experience with wearables and fidgets, and suggestions for additional features.

The survey was developed as an Airtable [65] form and distributed online, targeting the same demographic as the focus group. We aimed to collect about 30 responses that were analyzed to identify common themes and preferences, which guided the selection of data points to be collected by the smart fidget toy and the web app linked to the device and blockchain network.

The survey had a strict structure; it included both quantitative and qualitative questions, with the quantitative data revealing trends in user preferences and the qualitative data providing insights into desired features. The qualitative data was coded by the themes of the questions in several groups, such as "fidgeting elements" or "important knowledge for the individual".

3.5. Development and User Testing of Interactive Wireframes

Based on the user flows, app map technologies selected, and insights from the focus group and survey, wireframes for the accompanying web application were developed. These wireframes illustrated all key user interactions with the web app:

1. Wallet Setup
2. Linking the Device to the Web App
3. Upload the Data Collected by the Device to the Blockchain
4. Shared Access

Each wireframe was initially developed using Whimsical [66], and later, we added interactivity to it on Figma [67].

To set up the user testing sessions, we created a project on Useberry [68]. The user testing was made of a few main sections:

1. Informed Consent
2. Demography and Experience Questionnaire
3. Test Tasks
4. Post-testing questionnaire

Seven participants were recruited to test the wireframes through a series of usability tasks. We invited young professionals with experience working in the tech sector or research experience related to blockchain and design to participate. All tests were conducted remotely using Useberry platform [68] in asynchronous modality. Participants were asked to navigate the web application and simulate all four main user flows. They were instructed to provide feedback on the usability, intuitiveness, and overall user experience of the wireframes.

To improve our design decisions in the next stages, we collected the data about interactions user completed and how they navigated through the app to complete the tasks (clicks, misclicks, flows, and time needed) along with their replies to questionnaires.

3.6. Development and Testing of the Mid-Fidelity Prototype

At this stage, we developed a mid-fidelity (mid-fi) prototype of the entire product: the device and the web app with the link between them.

3.6.1. Device

The physical device was developed incorporating electronic components to simulate the smart functionalities of the fidget toy. This prototype included Adafruit Qt Py -SAMD21 microcontroller (Adafruit Industries, New York City, NYUSA) with GD25Q16—2 MB SPI Flash in 8-Pin SOIC package (GigaDevice, Beijing, China) soldered on the back side of it, Pulse Sensor (World Famous Electronics LLC, Brooklyn, NY, USA) at the bottom, TSWA-3N-C LFS (C&K Switches, Waltham, MA, USA) rotary encoder with the button, Adafruit Li-Poly 3.7 V 150 mAh battery (Adafruit Industries, New York City, NY, USA) and Adafruit LiIon or LiPoly Charger BFF Add-On for QT Py (Adafruit Industries, New York City, NY, USA).

The 3D model of the device was designed using the Grasshopper plugin inside Rhinoceros [54]. To ensure the precision of the design, we used the drawing of the rotary encoder and PPG sensor provided by manufacturers in the datasheets and the 3D models of all other elements published online by Adafruit [69–72].

Most of the elements were 3D printed using PLA material, while the tips for fingers were manufactured using a flexible transparent material to provide a comfortable tactile experience for the user, ensuring the transparency of it for the PPG sensor and simplifying the manufacturing process.

The electronic elements were assembled inside of the smart fidget, ensuring that all elements function properly while still providing access to repair its elements.

3.6.2. Software of the Device

The designed code integrates the usage of various components, such as a Pulse Sensor, a Flash Memory module, and a Rotary Encoder. The code is designed to collect data about user interaction and their heartbeat and manage this data through flash memory. The collected data can be recorded, transmitted, or cleared based on the user interaction. The full code can be found on GitHub [73].

The setup() function initializes all hardware components. The device uses Serial Communication for debugging purposes but also for managing the communication with the web app.

The loop() function continuously checks for serial input and manages the system state (COLLECT, SEND, DELETE) based on received commands on a serial port.

The collect() function handles the core data collection logic:

- Tracks the user interactions with the rotary encoder elements.
- Reads the heart rate from the pulse sensor, updating the minimum, maximum, and average heart rate values.
- Monitors the activity timeout to determine the end of a session, storing the data to flash memory when the session ends.

The writeFile() function writes the data as a new session in the JSON file if there was no interaction for the past 5 min. The use of JSON for data formatting ensures that the collected data can be easily transmitted.

3.6.3. Web App

The web application was developed to enable users to upload data collected by the smart fidget toy to a blockchain network. The complete code can be found on GitHub [74].

The web app can be seen in 2 parts:

1. Frontend Application
2. Smart Contract

The frontend application is a user-friendly interface that allows users to interact with their data and the blockchain network. The smart contract is a script deployed on the blockchain network that defines the rules and interactions of the data-uploading process and its storage and retrieval. The overall user-blockchain interaction happens as described (Figure 5):

1. The user uses the fidget as a physical device.
2. The user connects the fidget to their device and opens the web app in the browser.
3. Frontend communicates with a smart contract on the blockchain.

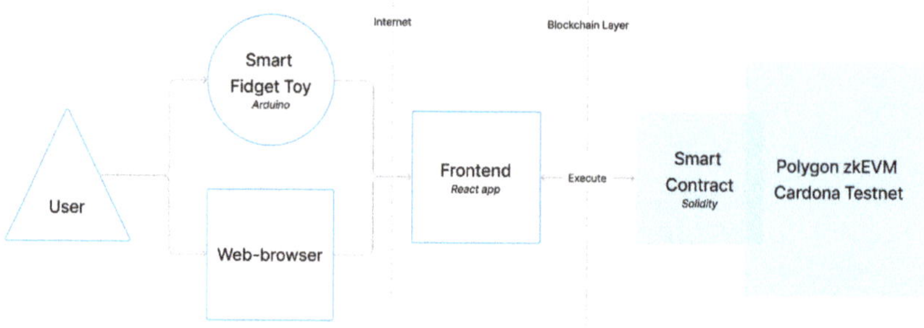

Figure 5. Architecture of the Web app.

The frontend application was developed using React framework for Javascript [75] and Tailwind CSS [76] for the visual configuration. The app also uses dayjs library [77] and Heroicons [78]. The frontend app followed the styles and design patterns defined by the user testing of the wireframes. We used MetaMask Wallet [79] as a blockchain provider for simplified development and user interaction. The app was deployed using the Netlify platform [80].

The smart contract was implemented using Solidity, a programming language specifically designed for creating smart contracts on the Ethereum blockchain [81], and deployed in Polygon zkEVM Cardona Testnet [82], ensuring the anonymization of the data uploaded. The smart contract manages the data to be written in the blockchain and its format, as well as allowing our app to read the records previously submitted and their details.

The blockchain integration was designed to securely store user data and provide a transparent record of all interactions.

The web app is made of several pages dedicated to different interactions that use smaller components for UI groups and elements to simplify the development and allow the reuse of the components.

To connect the device to the web app, the user has to connect it with the wire to the PC, open the web app in the browser, and ensure that they already have the wallet connected to the selected blockchain network.

The web app uses serial communication to retrieve and delete the data from the device. The UI notifies the user if new data is available on the device. User can decide to upload the data to the blockchain by submitting a form, where they can see the preview of their records and add more notes and comments if needed. Once the data is uploaded to the blockchain, it is permanently deleted from the device.

User can see their previously uploaded records in the calendar. Users can easily see if there were any records for each day, and by clicking on it, they can see more detailed information. The calendar was developed using the tutorial [83]. The app additionally emulates that a researcher requested access to the user's data.

For the mid-fi prototype, we limited the functionality of the product to the key features: data collection, data uploading, and mocking the data sharing interaction.

3.6.4. Blockchain Integration

To address the challenges of secure data storage and user-controlled health data management, we implemented a blockchain solution using the Polygon zkEVM Cardona Testnet [82]. This choice was motivated by its reduced transaction costs, improved scalability compared to the Ethereum mainnet, and enhanced privacy features, making it particularly suitable for our application.

Using Polygon zkEVM allows our app to benefit from zero-knowledge proofs that significantly increase user privacy [84,85]. zkEVM allows for the execution of smart contracts and transactions without revealing the underlying data, providing a higher level of anonymization than traditional blockchain solutions. This is particularly crucial for health-related data.

The core of our blockchain integration is a Solidity smart contract. This contract manages the storage and retrieval of data collected by Smart Fidget Toy. The contract's structure is designed to handle both daily summary records and detailed session data efficiently.

The contract utilizes two main data structures:

1. Record: stores daily summaries of fidgeting sessions, including average, minimum, and maximum heart rate data, number of sessions, main fidgeting activity, and total duration.
2. Session: stores detailed information about individual fidgeting sessions, including precise heart rate measurements, duration, timestamps, and user comments and tags.

The data flow from the Toy to the blockchain follows these steps:

1. The device collects session data during user interactions.
2. Data is temporarily stored on the device.
3. A user enters the web app on their device using a browser and connects the fidget toy to the device with a wire.
4. When connected to our web application, the device transfers the collected data.
5. The web application interacts with the smart contract to upload the data to the blockchain.
6. Daily summaries are stored on the blockchain.
7. Individual sessions are stored on the blockchain.
8. Users retrieve data.

Users interact with their data on the blockchain through our web application, which connects to their Ethereum MetaMask wallet. When adding new data, the application calls the appropriate smart contract functions, triggering a transaction that the user must sign. This process ensures that users maintain control over their data uploads.

3.6.5. User Testing of the Mid-Fidelity Prototype

The user testing of the mid-fi prototype was developed to evaluate the usability and effectiveness of the Smart Fidget Toy and its integration with the web app, overall interest in the product, and to understand the value provided by it to the users.

The mid-fi prototype was tested by a group of participants from the target demographic. Users were given the Toy to use in a lab setting, and data was collected during active usage periods. Afterward, participants were asked to upload their data to the blockchain, check their previous records, explore the shared access section, and provide or reject the request via the web app on the laptop provided by the researcher. Users would access the app in the browser and connect the device with the wire to the computer.

Each session was finalized by a semi-structured closing interview, where participants were asked to provide feedback on their experience with the device and web app and their feelings about the data sharing feature and the data control management within the app.

Each session was recorded. The user interaction with the web app was additionally recorded with screen recording.

The audio files were later transcribed and coded into the groups of "errors", "confusions", "liked/enjoyed", and "got curious". The coded groups were additionally supported with key screens or video recording frames. Metrics such as task completion time, error rates, and user satisfaction were analyzed.

4. Results

4.1. Focus Group Feedback on Lo-Fi Prototypes

The focus group discussions provided valuable insights into the design, usability, and desired features of the Smart Fidget Toy (Figure 6). Six individuals from our target demographic (24–32 years old, interested in well-being practices and technology) participated.

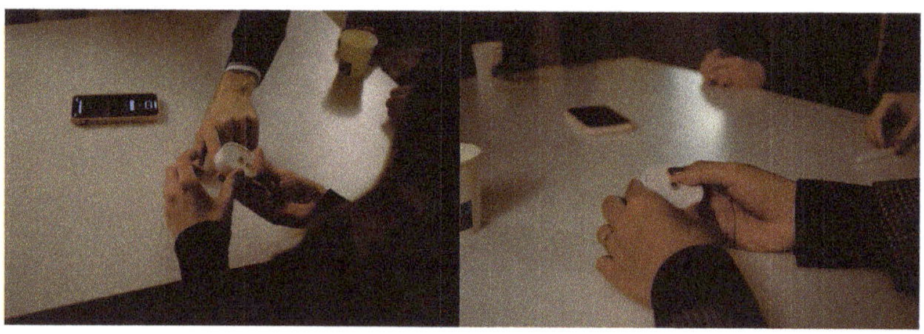

(a) (b)

Figure 6. Low-Fidelity Prototype during Focus Group: (**a**) participants discussing the low-fidelity prototype; (**b**) participants trying to use the low-fidelity prototype.

Key topics covered include perceived usage, ideal size and shape, button placement, texture preferences, data tracking features, and potential stigma around a visible fidget device.

Overall, the prototype received positive feedback, but suggestions were made to smooth sharp edges, add grips, hide seams, and collect additional physiological data. No explicit decisions were made, but action items centered on iterating the design based on ergonomic feedback and testing variations in size, shape, materials, and data tracking capabilities.

Participants appreciated the tactile and interactive nature of the prototype, noting that it provided a satisfying sensory experience. They especially liked the clicking feedback of the button, but they emphasized the lack of feedback of the rotary element.

Users easily understand the purpose of the device. The affordance of the rotating element was clear, though they needed more tips to understand that they also can use the button to fidget it. The item seemed to be a bit too tall. The item also needs rounded edges.

Users noted that the device is quite noticeable and might be subject to stigma. Nevertheless, they admit that people use other items at workplaces without any shame, e.g., softballs. Users became very curious about playing with the fidget. They were trying to get it from the hands of other participants and play more.

Based on the experiment with transmittable material on the palms of participants (Figure 7a), we can see four main contact points between their hands and the device: the tip, "upper part of the surface on the side", and 40° CW and CCW from it (Figure 7b). We also noticed that few participants naturally placed their fingers at the central point of the bottom of the device. That means that the sensor should be relocated, and additional affordance must be provided.

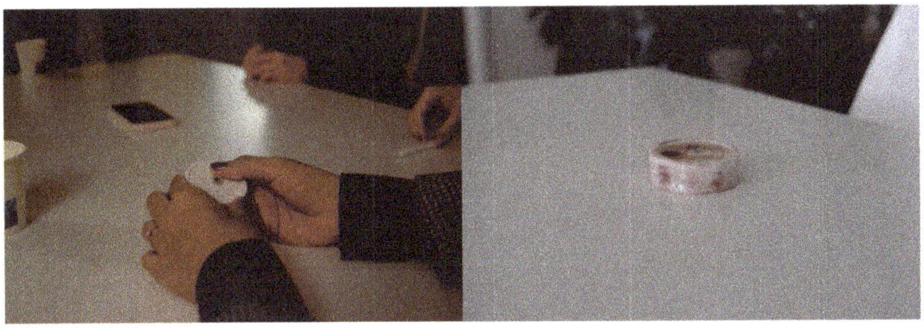

(a) (b)

Figure 7. Experiment during Focus Group with Transmittable Material on Hands of Participants: (**a**) participant trying the prototype with transmittable material on their hand; (**b**) low-fidelity prototype with marks left by participants' trials after the completion of the study.

4.2. Survey Results

We collected 28 responses. Most (over 95%) of the survey participants fall in the 25–34-year-old range group, which is our target audience. Most of them work in the creative (over 60%) or tech industries (over 55%). Most of them are quite familiar with the wearables and even use smartwatches daily. 1/3 of participants fidget a few times a day, while an equal number almost never do it.

Most users do not find the heartbeat value crucial while fidgeting, though people who fidget quite often find it more valuable than others (Table 2).

Table 2. Survey Responses to "How important is it for you that the smart fidget device tracks your heart rate?".

Option	Number of Participants Selected	% of Participants Selected
Very important	1	3.6
Somewhat important	13	46.4
Not very important	8	28.6
Not important at all	6	21.4

When analyzing the data interests of participants (Table 3), we notice that about 60% of respondents would like to see the data about the duration of the session and the emotional state before and after the session. Many participants also found it relevant to know the intensity of fidgeting, the date and time, specific triggers, and their main activity while fidgeting. At the same time, a tiny number (7.1%) notably expressed interest in the data collection about people around them while fidgeting and environmental factors.

Table 3. Survey Responses to "What other metrics or data would you find valuable for the smart fidget to track and analyze?".

Metrics	Number of Participants Selected	% of Participants Selected
Duration of fidgeting sessions	16	57.1
Intensity of fidgeting sessions	15	53.8
The date and time of the fidgeting session	13	46.4
Environmental factors	9	32.1
Emotional state before and after fidgeting	16	57.1
Specific triggers for fidgeting sessions	13	46.4
People around you while fidgeting	2	7.1
Your main activity while fidgeting	13	46.4
The location while fidgeting	10	35.7

From the qualitative data of the survey, we could collect information on two main topics: what and how people use to fidget and what they would expect from the product.

Most respondents mentioned pens and pencils as their most common fidget; many respondents mentioned using jewelry or simply anything in their proximity. Also, many emphasized that they use their hair or body parts, especially hands/palms/fingers.

4.3. User Testing of Wireframes

We invited 8 participants from different countries; 6 of them are 25 to 34 years old with diverse levels of knowledge about blockchain technology.

Most participants managed to complete all tasks and felt confident that they did it correctly. Most users found Shared Access functionality very valuable. 3 out of 4 tasks took relatively little time to complete (up to 40 s) with few clicks required by users to accomplish (Table 4).

Table 4. Overall Quantitative Results of User Testing of Interactive Wireframes.

Test Task	Average Time	Median Number of Clicks	Min Number of Clicks	Max Number of Clicks
Wallet Setup	64.3 s	10	10	11
Link Device	12.4 s	2	2	4
Upload Data	36.9 s	8	6	16
Shared Access	14.5	3	2	7

4.3.1. Wallet Setup Task

Even though our web app does not include the wallet creation functionality, we wanted to provide our participants with a more immersive experience by asking them to complete the creation of a MetaMask Wallet account and installation of its plugin for the Google Chrome browser.

This process was relatively easy to complete for all users but quite time-consuming (>1 min), with 10–11 clicks to complete.

4.3.2. Link Device Task

Since we have the interaction between the physical and digital worlds, we needed to test if this connection is clear to our users. We asked our participants to imagine that they have a Smart Fidget Toy and to complete a task where they associate their physical device with their MetaMask Wallet. Most users confidently finished this task without any wrong clicks.

4.3.3. Upload Data Task

In this task, users are asked to imagine that it has already collected some data. Users are required to try uploading the data to the blockchain network.

This task seemed much more confusing for the users. From the positions and number of user clicks on the Upload Page, we can see that many participants did not understand why they landed at the data uploading form or where they landed at all after connecting the device to the web app (Figure 8). Based on mouse movement and clicks, we can conclude that the users were trying to reach the form to upload the data while they were already there. The form also seemed complicated to understand, and the steps provided to users were confusing and did not reflect the expected mental model.

Nevertheless, all users completed the task and uploaded their data to the blockchain.

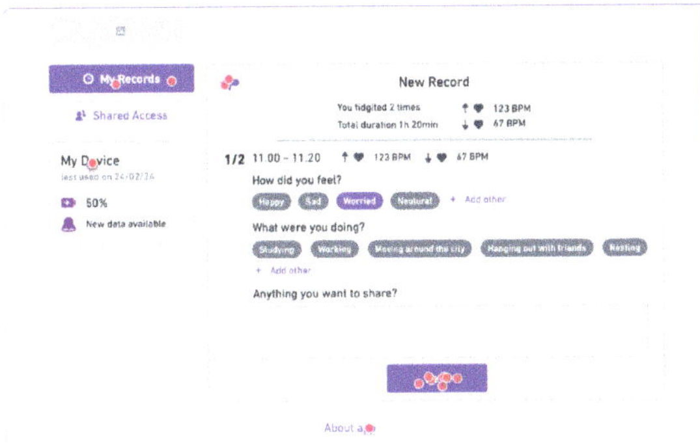

Figure 8. User Clicks on Upload Page on the Web App.

4.3.4. Shared Access Task

This user flow is about a feature provided by the platform that allows others to access user data with their permission. It requires the user to decide how to reach the Shared Access page and reject or decline new requests.

For most users, this task was very straightforward and took the least time to complete, and they took the expected user flow of 2 clicks. Most used the notice message to check shared access, while one used the dashboard navigation. None used an icon in the dashboard.

From this experiment, the overall user experience meets user needs and expectations, but some adjustments are still needed. The Upload Data user flow needs to be redefined and simplified.

From qualitative replies, we can see that potential users would love to know more about the benefits of using them.

4.4. Mid-Fi Prototype Testing

We conducted ten user testing sessions among five females and five males. All participants fall in the 25–36 years old age group. They were asked to try to interact with the device (Figure 9) and connect it to a web app (Figures 10 and 11).

Figure 9. Mid-fidelity Prototype of Smart Fidget Toy.

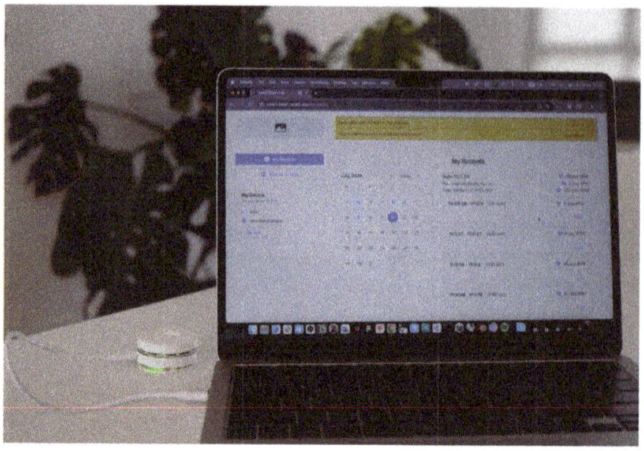

Figure 10. Mid-fidelity prototype of Smart Fidget Toy connected to a Web App.

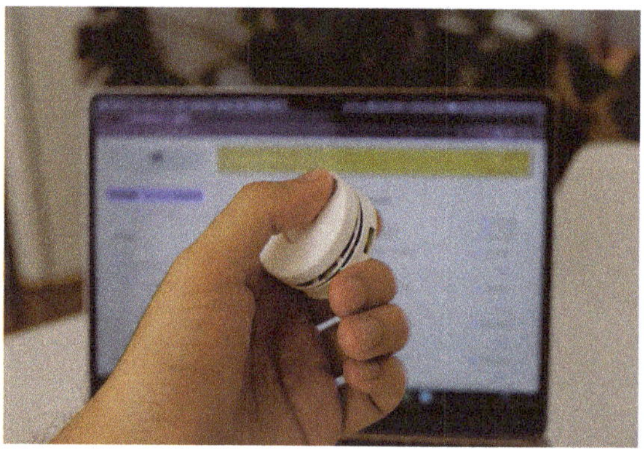

Figure 11. User Testing Session of Mid-fidelity Prototype.

All participants were excited to try the device after we demonstrated it to them. When users tried the device, it was noticeable that for some users, the device should be bigger to fit comfortably, but for users with smaller hand sizes, it was not the case. Even though the device itself felt "fragile" to users, they enjoyed the fidgeting interactions with the device, notably the "click" during the rotary fidgeting, which was perceived as "very satisfying". However, the clicking interaction could be more explicit.

Users also mentioned that they "wish for more feedforward and feedback" while interacting with the device. The affordances on the top part of the device seemed evident to them, while more affordances are needed for the base part.

After the device trial, users were asked to use the companion web app on a laptop. They easily connected the device to the computer using the wire, entering the web app and allowing serial communication between them.

On the interface, they were notified about new data availability for upload, and they followed the procedure. The updated upload form was more intuitive even though some users did not provide further details about their mood state, etc.

The interaction with MetaMask Wallet was easy for users to complete, thanks to the supportive texts in the app. Nevertheless, all participants confirmed they needed more explanations about the elements shown in the MetaMask Wallet popup. Then, users easily explored previous records.

The users found the Shared Access feature very interesting and were motivated to share their data for monetary benefits or easier data sharing with their doctors.

Overall, users expressed interest in the device and the web app. Many of them mentioned that they needed more explanations or supportive materials. For most of them, it was easy to complete all the tasks. From the post-testing interview, all users confirmed unfamiliarity with the blockchain technology. About 30% of participants claimed concern about how their data is collected, stored, and used. Still, most participants in everyday life do not take care of the data they produce or share online. Regarding health-related data, all participants highlighted its importance and their wish to have more control over it. Moreover, one participant mentioned that "considering the amount of data we give to [online platform], when I saw the data it collects since I started to use it a long time ago and underage, I regret there was no one to educate me" (P9) about their data.

5. Discussion

5.1. Key Findings

The study shows that integrating more technologies into fidget toys can significantly improve their functionality and user engagement. The smart fidget toy developed in this research successfully collected data on user interactions and health-related data, such as heart rate, during usage. The proposed web application helps users to interact with the data and blockchain network. This data can provide actionable insights for users to manage their moods and for healthcare professionals to offer more informed support.

5.2. User-Centered Design and Iterative Prototyping

The iterative design process, which included multiple phases of user feedback, was crucial in the development process of the device. Now, Smart Fidget Toy closely aligns with user needs and preferences. Focus group discussions and surveys provided important insights into the desired features and functionalities, such as the importance of tactile feedback, ergonomic design, and the ability to track specific data points like session duration and emotional state. The user testing sessions with interactive wireframes and the mid-fi prototype improved overall control over data and acceptability of the technologies used.

Participants found the device engaging and easy to use but suggested improvements in the feedback mechanisms and additional support materials to enhance the user experience.

5.3. Impact of Blockchain Integration

5.3.1. Enhanced Data Security

The decentralized nature of blockchain significantly reduces the risk of data breaches. Unlike traditional centralized systems, where data is stored in a single location vulnerable to attacks, blockchain distributes data across multiple nodes. This makes it more difficult for unauthorized parties to alter or access the data. During our testing, users expressed increased confidence in the security of their health-related data (90%), knowing it was protected by advanced cryptographic techniques inherent to blockchain and anonymization provided by it.

5.3.2. User Control and Privacy

Blockchain technology empowers users by granting them complete control over their data. Users can selectively grant or revoke access to their data, ensuring they maintain privacy and autonomy over their personal information. This feature was particularly well-received by participants, who valued the ability to manage their data-sharing preferences directly. The transparent and user-driven access control model aligns with the growing demand for personalized data privacy solutions.

Our design requires users to manually initiate data transfers by connecting the device to a computer. This approach gives users additional control over when their data is uploaded. It aligns with our goal of enhancing user empowerment and privacy.

5.3.3. User Engagement

The integration of blockchain also suggested improvements to user engagement. For example, users could earn tokens for contributing their data to research studies. 70% of participants found this feature not only easy to use but also beneficial and motivating. This aspect increased user motivation and promoted more sustained interaction with the smart fidget toy, making it a more valuable tool.

5.3.4. Computational Challenges

While blockchain technology offers significant benefits for data security and user control in our Smart Fidget Toy project, it also presents several computational challenges:

1. The Smart Fidget Toy, as a small wearable device, has limited computational power and storage capacity. This constraint necessitated the implementation of a hybrid approach where data is temporarily stored on the device and later uploaded to the blockchain via a web application.
2. We implemented an intermediate web application that bridges the smart fidget toy and the blockchain network. The web application handles all blockchain-related computations when users connect the device to a computer via a wired connection. This includes data serialization, encryption, transaction creation, and communication with the blockchain network.
3. Data collection is optimized to occur only during active user interaction with the device. Limiting data capture to these periods reduces unnecessary processing and conserves the device's memory and battery life. The collected data is formatted into lightweight JSON files, facilitating efficient storage and quick transfer to the web application upon connection.
4. We chose the Polygon zkEVM Cardano Testnet as our blockchain platform due to its compatibility with Ethereum smart contracts and its efficiency in processing transactions. This network offers lower transaction fees and faster confirmation times than other platforms, reducing the computational load and cost associated with data uploads.
5. In smart contract design, minimizing gas costs and improving overall system performance was crucial. Our iterative development process included multiple rounds of code optimization and gas usage analysis to ensure efficient blockchain interactions.

These challenges required careful consideration in designing and implementing our blockchain-based solution for the smart fidget toy. By acknowledging and addressing these computational aspects, we aimed to utilize the benefits of blockchain technology without compromising the user experience or the practical functionality of the device.

5.4. *Challenges and Limitations*

The development process highlighted several challenges, particularly related to integrating sensor technologies and the complexity of user interactions with blockchain-based systems. The requirement for an intermediate layer to facilitate communication between the wearable device and the blockchain added complexity to the design and significantly influenced design and development decisions. Additionally, the physical design of the fidget toy needed continuous adjustments to balance the inclusion of electronic components with user comfort and tactile satisfaction.

This study included only limited interactions with the device and web application. User testing sessions were conducted with a task list provided to the participants. Future research should explore the long-term effects of using blockchain technology for health data. Additionally, further refinements in the device's design and functionality, including more advanced sensor integration and enhanced user interaction features, will be necessary to improve its effectiveness and user satisfaction.

6. Conclusions

The Smart Fidget Toy developed in this study represents an advancement in wearable technology for the health sector using blockchain. Combining user-centered design techniques and the security features of blockchain technology, the device offers a novel approach to mood, stress, focus management, data collection, storage, and sharing. The iterative design process and continuous user feedback ensured that the product met the needs and expectations of its target audience. Moreover, it helped to improve user acceptance of the new technologies, making it not only a valuable tool for users, healthcare providers, and researchers but also providing insights on the design decisions to be taken.

To sum up, similar projects can follow the development process as described:

1. Concept Development
    - Initial Research: Investigate existing products and identify gaps where technology can improve functionality.
    - Define Objectives: Outline the specific goals of the product, focusing on user needs and potential user benefits.
    - Comparative Analysis: Analyze similar products to refine user flows, data interaction, and overall experience.
    - Technology Integration: Research security and implementation challenges of the suggested technologies.
2. Iterative Design Process
    - Low-Fidelity Prototyping: Start with basic prototypes involving stakeholders to shape the initial concept and get early feedback on the physical and technological aspects.
    - User Feedback: Conduct focus groups, user testing sessions, and surveys to gather early insights on usability, desired features, and ergonomic considerations.
    - Refinement: Continuously improve the design based on user feedback.
3. Blockchain Integration
    - Data Security: Make sure that the user data is securely stored and transmitted in the most secure manner.
    - User Control and Transparency: Enable user-controlled data access and transparency in data transactions and access history to build trust and ensure privacy.

Key Considerations:

- User-Centered Focus: Maintain a strong focus on the user's needs and preferences throughout the project to ensure the product is intuitive and meets the intended needs effectively.
- Technology Suitability: Evaluate and select technologies for their innovation and practical benefits in terms of security, privacy, and user engagement.
- Iterative Development: Leverage iterative design and testing processes to continually improve the product based on actual user interactions and feedback.
- Stakeholder Engagement: Engage with all stakeholders, including potential users, healthcare professionals, and technology experts, throughout the development process to align the product with broader system requirements and user expectations.

Author Contributions: Conceptualization, P.B.; Methodology, P.B.; Software, P.B. and P.P.; Validation, P.B. and P.P.; Formal Analysis, P.B.; Investigation, P.B.; Resources, P.B. and P.P.; Data Curation, P.B. and P.P.; Writing—Original Draft Preparation, P.B.; Writing—Review & Editing, P.P.; Visualization, P.B.; Supervision, P.P. and R.B.; Project Administration, P.B. and P.P. All authors have read and agreed to the published version of the manuscript.

Funding: This research was supported by a PhD scholarship funded under Ministerial Decree No. 352/2022, through the Italian National Recovery and Resilience Plan (PNRR), with partial co-financing from Fifth Beat SRL.

Institutional Review Board Statement: The study was conducted according to the guidelines of the Declaration of Helsinki and approved by the Ethics Committee of Politecnico di Milano (n.29/2024 21 May 2024).

Informed Consent Statement: Informed consent was obtained from all subjects involved in the study.

Data Availability Statement: The data presented in this study are available at the request of the corresponding author (for privacy, legal, and ethical reasons).

Acknowledgments: We would like to thank all the volunteers who participated in our user research and evaluation studies for their insightful feedback.

Conflicts of Interest: A co-author (Raffaele Boiano) is affiliated with Fifth Beat SRL, which provided partial financial support for this research. However, the company had no influence over the study design, data collection, analysis, or interpretation.

References

1. Onyeaka, H.; Firth, J.; Kessler, R.C.; Lovell, K.; Torous, J. Use of Smartphones, Mobile Apps and Wearables for Health Promotion by People with Anxiety or Depression: An Analysis of a Nationally Representative Survey Data. *Psychiatry Res.* **2021**, *304*, 114120. [CrossRef] [PubMed]
2. Genaro Motti, V.; Caine, K. An Overview of Wearable Applications for Healthcare: Requirements and Challenges. In Proceedings of the Adjunct Proceedings of 2015 ACM International Joint Conference on Pervasive and Ubiquitous Computing and Proceedings of 2015 ACM International Symposium on Wearable Computers, Osaka Japan, 7–11 September 2015; Association for Computing Machinery: New York, NY, USA, 2015; pp. 635–641.
3. Ajayi, K.V.; Wachira, E.; Onyeaka, H.K.; Montour, T.; Olowolaju, S.; Garney, W. The Use of Digital Health Tools for Health Promotion among Women with and without Chronic Diseases: Insights from the 2017–2020 Health Information National Trends Survey. *JMIR mHealth uHealth* **2022**, *10*, e39520. [CrossRef] [PubMed]
4. Piwek, L.; Ellis, D.A.; Andrews, S.; Joinson, A. The Rise of Consumer Health Wearables: Promises and Barriers. *PLoS Med.* **2016**, *13*, e1001953. [CrossRef] [PubMed]
5. Dwivedi, A.D.; Srivastava, G.; Dhar, S.; Singh, R. A Decentralized Privacy-Preserving Healthcare Blockchain for IoT. *Sensors* **2019**, *19*, 326. [CrossRef]
6. Dong, G.; Chen, Y.; Fan, J.; Liu, D.; Hao, Y.; Wang, Z. A Privacy-User-Friendly Scheme for Wearable Smart Sensing Devices Based on Blockchain. In Proceedings of the 2018 IEEE 15th International Conference on Mobile Ad Hoc and Sensor Systems (MASS), Chengdu, China, 9–12 October 2018. [CrossRef]
7. Fan, K.; Wang, S.; Ren, Y.; Li, H.; Yang, Y. MedBlock: Efficient and Secure Medical Data Sharing via Blockchain. *J. Med. Syst.* **2018**, *42*, 136. [CrossRef] [PubMed]
8. Karame, G.; Čapkun, S. Blockchain Security and Privacy. *IEEE Secur. Priv.* **2018**, *16*, 11–12. [CrossRef]
9. Zhang, R.; Xue, R.; Liu, L. Security and Privacy on Blockchain. *ACM Comput. Surv.* **2019**, *52*, 1–34. [CrossRef]
10. Jesse, Y.-H.; Ko, D.; Choi, S.; Park, S.; Smolander, K. Where Is Current Research on Blockchain Technology?—A Systematic Review. *PLoS ONE* **2016**, *11*, e0163477. [CrossRef]
11. Hussien, H.M.; Yasin, S.M.; Udzir, N.I.; Ninggal, M.I.H.; Salman, S. Blockchain Technology in the Healthcare Industry: Trends and Opportunities. *J. Ind. Inf. Integr.* **2021**, *22*, 100217. [CrossRef]
12. Feng, Q.; He, D.; Zeadally, S.; Khan, M.K.; Kumar, N. A Survey on Privacy Protection in Blockchain System. *J. Netw. Comput. Appl.* **2019**, *126*, 45–58. [CrossRef]
13. Zaghloul, E.; Li, T.; Mutka, M.W.; Ren, J. Bitcoin and Blockchain: Security and Privacy. *IEEE Internet Things J.* **2020**, *7*, 10288–10313. [CrossRef]
14. Yu, Y.; Li, Y.; Tian, J.; Liu, J. Blockchain-Based Solutions to Security and Privacy Issues in the Internet of Things. *IEEE Wirel. Commun.* **2018**, *25*, 12–18. [CrossRef]
15. Shamshad, S.; Minahil; Mahmood, K.; Kumari, S.; Chen, C.-M. A Secure Blockchain-Based e-Health Records Storage and Sharing Scheme. *J. Inf. Secur. Appl.* **2020**, *55*, 102590. [CrossRef]
16. Benchoufi, M.; Ravaud, P. Blockchain Technology for Improving Clinical Research Quality. *Trials* **2017**, *18*, 1–5. [CrossRef] [PubMed]
17. Lin, H.; Zhang, H.; Yan, H.; Wang, H.; Shi, Y.; Gao, F.; Wen, Q. A Secure Online Treatment Blockchain Service. *Wirel. Pers. Commun.* **2020**, *117*, 1773–1795. [CrossRef]
18. Hasselgren, A.; Kralevska, K.; Gligoroski, D.; Pedersen, S.A.; Faxvaag, A. Blockchain in Healthcare and Health Sciences—A Scoping Review. *Int. J. Med. Inform.* **2020**, *134*, 104040. [CrossRef]
19. Koumpounis, S.; Perry, M. Blockchain-Based Electronic Health Record System with Patient-Centred Data Access Control. In Proceedings of the 2023 IEEE/ACM 6th International Workshop on Emerging Trends in Software Engineering for Blockchain (WETSEB), Melbourne, Australia, 14 May 2023; pp. 17–24.

20. Ettaloui, N.; Arezki, S.; Gadi, T. A Blockchain-Based Electronic Mental Health Records Model. In Proceedings of the 2023 10th International Conference on Information Technology, Computer, and Electrical Engineering (ICITACEE), Semarang, Indonesia, 31 August–1 September2023; pp. 320–325.
21. Xia, Q.; Sifah, E.B.; Asamoah, K.O.; Gao, J.; Du, X.; Guizani, M. MeDShare: Trust-Less Medical Data Sharing among Cloud Service Providers via Blockchain. *IEEE Access* **2017**, *5*, 14757–14767. [CrossRef]
22. Dagher, G.G.; Mohler, J.; Milojkovic, M.; Marella, P.B. Ancile: Privacy-Preserving Framework for Access Control and Interoperability of Electronic Health Records Using Blockchain Technology. *Sustain. Cities Soc.* **2018**, *39*, 283–297. [CrossRef]
23. Fenton, C. Exploring the Content of Self-help Toolkits as an Intervention for Non-suicidal Self-injury in Young People in England. *Int. J. Ment. Heal. Nurs.* **2023**, *32*, 1326–1334. [CrossRef]
24. Graziano, P.A.; Garcia, A.; Landis, T.D. To Fidget or Not to Fidget, That Is the Question: A Systematic Classroom Evaluation of Fidget Spinners Among Young Children with ADHD. *J. Atten. Disord.* **2018**, *24*(1), 163–171. [CrossRef]
25. Schecter, R.A.; Shah, J.; Fruitman, K.; Milanaik, R.L. Fidget Spinners: Purported Benefits, Adverse Effects and Accepted Alternatives. *Curr. Opin. Pediatr.* **2017**, *29*, 616. [CrossRef] [PubMed]
26. Liang, R.-H.; Yu, B.; Xue, M.; Hu, J.; Feijs, L.M.G. BioFidget: Biofeedback for Respiration Training Using an Augmented Fidget Spinner. In Proceedings of the 2018 CHI Conference on Human Factors in Computing Systems, Montreal, QC, Canada, 21–26 April 2018; Association for Computing Machinery: New York, NY, USA, 2018; pp. 1–12.
27. Länsisalmi, H.; Peíró, J.M.; Kivimäki, M. Collective Stress and Coping in the Context of Organizational Culture. *Eur. J. Work. Organ. Psychol.* **2000**, *9*, 527–559. [CrossRef]
28. Angraal, S.; Krumholz, H.M.; Schulz, W.L. Blockchain Technology: Blockchain Technology: Applications in Health Care. *Circ. Cardiovasc. Qual. Outcomes* **2017**, *10*, e003800. [CrossRef]
29. Abu-elezz, I.; Hassan, A.; Nazeemudeen, A.; Househ, M.; Abd-alrazaq, A. The Benefits and Threats of Blockchain Technology in Healthcare: A Scoping Review. *Int. J. Med. Inform.* **2020**, *142*, 104246. [CrossRef] [PubMed]
30. Agbo, C.C.; Mahmoud, Q.H.; Eklund, J.M. Blockchain Technology in Healthcare: A Systematic Review. *Healthcare* **2019**, *7*, 56. [CrossRef]
31. McGhin, T.; Choo, K.-K.R.; Liu, C.Z.; He, D. Blockchain in Healthcare Applications: Research Challenges and Opportunities. *J. Netw. Comput. Appl.* **2019**, *135*, 62–75. [CrossRef]
32. Prokofieva, M.; Miah, S.J. Blockchain in Healthcare. *Australas. J. Inf. Syst.* **2019**, *23*, 1–22. [CrossRef]
33. Stallone, V.; Wetzels, M.; Mahr, D.; Klaas, M.I. Enhancing Digital Advertising with Blockchain Technology. *J. Interact. Mark.* **2023**, *59*, 76–98. [CrossRef]
34. Grover, P.; Kar, A.K.; Janssen, M.; Ilavarasan, P.V. Perceived Usefulness, Ease of Use and User Acceptance of Blockchain Technology for Digital Transactions—Insights from User-Generated Content on Twitter. *Enterp. Inf. Syst.* **2019**, *13*, 771–800. [CrossRef]
35. Shahnaz, A.; Qamar, U.; Khalid, A. Using Blockchain for Electronic Health Records. *IEEE Access* **2019**, *7*, 147782–147795. [CrossRef]
36. Lin, W.-H.; Wu, D.; Li, C.; Zhang, H.; Zhang, Y.-T. Comparison of Heart Rate Variability from PPG with That from ECG. In Proceedings of the International Conference on Health Informatics: ICHI 2013, Vilamoura, Portugal, 7–9 November 2013; Springer Nature: Berlin/Heidelberg, Germany, 2014; pp. 213–215. [CrossRef]
37. Lu, L.; Zhang, J.; Xie, Y.; Gao, F.; Xu, S.; Wu, X.; Ye, Z. Wearable Health Devices in Health Care: Narrative Systematic Review. *JMIR mHealth and uHealth* **2020**, *8*, e18907. [CrossRef] [PubMed]
38. Yetisen, A.K.; Martinez-Hurtado, J.L.; Ünal, B.; Khademhosseini, A.; Butt, H. Wearables in Medicine. *Adv. Mater.* **2018**, *30*, 1706910. [CrossRef]
39. Gao, Y.; Li, H.; Luo, Y. An Empirical Study of Wearable Technology Acceptance in Healthcare. *Ind. Manag. Data Syst.* **2015**, *115*, 1704–1723. [CrossRef]
40. Hölbl, M.; Kompara, M.; Kamišalić, A.; Nemec Zlatolas, L. A Systematic Review of the Use of Blockchain in Healthcare. *Symmetry* **2018**, *10*, 470. [CrossRef]
41. Chukwu, E.; Garg, L. A Systematic Review of Blockchain in Healthcare: Frameworks, Prototypes, and Implementations. *IEEE Access* **2020**, *8*, 21196–21214. [CrossRef]
42. Chowdhury, M.J.M.; Ferdous, M.S.; Biswas, K.; Chowdhury, N.; Kayes, A.; Watters, P.; Ng, A. Trust Modeling for Blockchain-Based Wearable Data Market. In Proceedings of the 2019 IEEE International Conference on Cloud Computing Technology and Science (CloudCom), Sydney, Australia, 11–13 December 2019; pp. 411–417.
43. Motti, V.G.; Caine, K. Human Factors Considerations in the Design of Wearable Devices. In Proceedings of the Human Factors and Ergonomics Society Annual Meeting, Adelaide, Australia, 17–19 November 2014; Volume 58, pp. 1820–1824. [CrossRef]
44. Smaradottir, B.; Fensli, R. User-Centred Design of Health Information Technology. *Int. J. Integr. Care* **2016**, *16*, 38. [CrossRef]
45. Dabbs, A.D.V.; Myers, B.A.; Mc Curry, K.R.; Dunbar-Jacob, J.; Hawkins, R.P.; Begey, A.; Dew, M.A. User-Centered Design and Interactive Health Technologies for Patients. *CIN Comput. Inform. Nurs.* **2009**, *27*, 175. [CrossRef]
46. Glomann, L.; Schmid, M.; Kitajewa, N. Improving the Blockchain User Experience—An Approach to Address Blockchain Mass Adoption Issues from a Human-Centred Perspective. In Proceedings of the Advances in Artificial Intelligence, Software and Systems Engineering, Virtual, 16–20 July 2020; Ahram, T., Ed.; Springer International Publishing: Cham, Switzerland, 2020; pp. 608–616.

47. Moniruzzaman, M.; Chowdhury, F.; Ferdous, M.S. Examining Usability Issues in Blockchain-Based Cryptocurrency Wallets. In Proceedings of the Cyber Security and Computer Science, Dhaka, Bangladesh, 15–16 February 2020; Bhuiyan, T., Rahman, M.M., Ali, M.A., Eds.; Springer International Publishing: Cham, Switzerland, 2020; pp. 631–643.
48. Pschetz, L.; Dixon, B.; Pothong, K.; Bailey, A.; Glean, A.; Soares, L.L.; Enright, J.A. Designing Distributed Ledger Technologies for Social Change: The Case of CariCrop. In Proceedings of the 2020 CHI Conference on Human Factors in Computing Systems, Honolulu, HI, USA, 25–30 April 2020; Association for Computing Machinery: New York, NY, USA, 2020; pp. 1–12.
49. Elsden, C.; Trotter, L.; Harding, M.; Davies, N.; Speed, C.; Vines, J. Programmable Donations: Exploring Escrow-Based Conditional Giving. In Proceedings of the 2019 CHI Conference on Human Factors in Computing Systems, Glasgow, UK, 4–9 May 2019; Association for Computing Machinery: New York, NY, USA, 2019; pp. 1–13.
50. Breen, R.L. A Practical Guide to Focus-Group Research. *J. Geogr. High. Educ.* **2006**, *30*, 463–475. [CrossRef]
51. Burmeister, E.; Aitken, L.M. Sample Size: How Many Is Enough? *Aust. Crit. Care* **2012**, *25*, 271–274. [CrossRef]
52. Lewis, J.R. Sample Sizes for Usability Studies: Additional Considerations. *Hum. Factors* **1994**, *36*, 368–378. [CrossRef]
53. Faulkner, L. Beyond the Five-User Assumption: Benefits of Increased Sample Sizes in Usability Testing. *Behav. Res. Methods Instrum. Comput.* **2003**, *35*, 379–383. [CrossRef] [PubMed]
54. Robert McNeel & Associates. *Rhinoceros 8. Version 8.0*; Robert McNeel & Associates: Seattle, WA, USA. Available online: https://www.rhino3d.com/8/new/ (accessed on 10 June 2024).
55. UX Challenges in Blockchain | Avark. Available online: https://avark.agency/learn/article/ux-challenges-in-blockchain/ (accessed on 6 November 2023).
56. Patientory Inc. Available online: https://patientory.com (accessed on 13 February 2024).
57. Patientory: Blockchain-Based App for Managing Health Data and Providing Incentives to Improve Health Outcomes. Available online: https://pixelplex.io/work/blockchain-mobile-app-for-maintaning-health/ (accessed on 13 February 2024).
58. Ucam. Available online: https://iotex.io/blog/ucam-security-camera-information/ (accessed on 13 February 2024).
59. Take Control of Your Data with Pebble. Available online: https://iotex.io/pebble (accessed on 13 February 2024).
60. Bearable. Available online: https://bearable.app (accessed on 13 February 2024).
61. How to Use: Let Apollo Help You Discover a New You. Available online: https://apolloneuro.com/en-intl/pages/how-to-use?ref=nav (accessed on 13 February 2024).
62. TouchPoints—Reduces Stress—Invented by Dr. Serin. Available online: https://serincenter.com/product/touchpoints-stress-reducing-wearables-invented-by-dr-amy-serin/ (accessed on 13 February 2024).
63. Lief: The Wearable for Mental Health. Available online: https://getlief.com/ (accessed on 13 February 2024).
64. InTempo: Safe Space in Your Hand. Available online: https://intempoapp.com (accessed on 13 February 2024).
65. Airtable, Inc. Available online: https://airtable.com (accessed on 10 June 2024).
66. Whimsical, Inc. Available online: https://whimsical.com (accessed on 13 February 2024).
67. Figma, Inc. Available online: https://www.figma.com/ (accessed on 13 February 2024).
68. Useberry. Available online: https://www.useberry.com (accessed on 13 February 2024).
69. Adafruit Adafruit_CAD_Parts—1317 150mAh Lipo Battery. Available online: https://github.com/adafruit/Adafruit_CAD_Parts/tree/main/1317%20150mAh%20Lipo%20Battery (accessed on 5 March 2024).
70. Adafruit Adafruit_CAD_Parts—4600 QT Py. Available online: https://github.com/adafruit/Adafruit_CAD_Parts/tree/main/4600%20QT%20Py (accessed on 5 March 2024).
71. Adafruit_CAD_Parts—5397 Lipo Charger BFF. Available online: https://github.com/adafruit/Adafruit_CAD_Parts/tree/main/5397%20Lipo%20Charger%20BFF (accessed on 5 March 2024).
72. Adafruit Pulse Sensor. Available online: https://cdn-shop.adafruit.com/product-files/1093/Pulse+Sensor+Data+Sheet-2018.pdf (accessed on 5 March 2024).
73. Bobrova, P. Smart Fidget Toy Device Software (Heartbeat and Activity Tracker). Available online: https://github.com/pbdes/smart-fidget-toy-device-midfi (accessed on 11 July 2024).
74. Bobrova, P. Smart Fidget Toy App. Available online: https://github.com/pbdes/smart-fidget-app-midfi (accessed on 11 July 2024).
75. Meta Platforms, Inc. *React*; Version 18.2.0; Meta Platforms, Inc.: Menlo Park, CA, USA; Available online: https://react.dev (accessed on 10 June 2024).
76. Tailwind Labs. Tailwind CSS. Available online: https://tailwindcss.com (accessed on 10 June 2024).
77. Dayjs. Version 1.11.10. Available online: https://day.js.org (accessed on 10 June 2024).
78. Tailwindcss, Heroicons. Available online: https://heroicons.com (accessed on 10 April 2024).
79. ConsenSys. *MetaMask Wallet*; ConsenSys: Brooklyn, NY, USA; Available online: https://metamask.io (accessed on 10 June 2024).
80. Netlify Inc. *Netlify*; Netlify: San Francisco, CA, USA; Available online: https://www.netlify.com (accessed on 5 May 2024).
81. Monrat, A.A.; Schelén, O.; Andersson, K. A Survey of Blockchain From the Perspectives of Applications, Challenges, and Opportunities. *IEEE Access* **2019**, *7*, 117134–117151. [CrossRef]
82. Polygon PoS and Polygon zkEVM: New Testnets for Polygon Protocols. Available online: https://polygon.technology/blog/polygon-pos-and-polygon-zkevm-new-testnets-for-polygon-protocols (accessed on 10 March 2024).
83. Daily Web Coding Build a React Tailwind Calendar Ui from Scratch. Available online: https://www.youtube.com/watch?v=s9-K02CP8hw (accessed on 1 April 2024).

84. Sun, X.; Yu, F.R.; Zhang, P.; Sun, Z.; Xie, W.; Peng, X. A Survey on Zero-Knowledge Proof in Blockchain. *IEEE Netw.* **2021**, *35*, 198–205. [CrossRef]
85. Partala, J.; Nguyen, T.H.; Pirttikangas, S. Non-Interactive Zero-Knowledge for Blockchain: A Survey. *IEEE Access* **2020**, *8*, 227945–227961. [CrossRef]

sensors

MDPI

Article

A Multi-Scale CNN for Transfer Learning in sEMG-Based Hand Gesture Recognition for Prosthetic Devices

Riccardo Fratti [1,*], Niccolò Marini [1], Manfredo Atzori [2], Henning Müller [1,3,4], Cesare Tiengo [2,5] and Franco Bassetto [2,5]

1 Informatics Institute, University of Applied Sciences Western Switzerland (HES-SO Valais), 3960 Sierre, Switzerland
2 Department of Neuroscience, University of Padua, 35122 Padua, Italy
3 Medical Informatics, University of Geneva, 1205 Geneva, Switzerland
4 The Sense Innovation and Research Center, 1007 Lausanne, Switzerland
5 Clinic of Plastic Surgery, University Hospital of Padua, 35128 Padova, Italy
* Correspondence: riccardo.fratti@hevs.ch

Abstract: Advancements in neural network approaches have enhanced the effectiveness of surface Electromyography (sEMG)-based hand gesture recognition when measuring muscle activity. However, current deep learning architectures struggle to achieve good generalization and robustness, often demanding significant computational resources. The goal of this paper was to develop a robust model that can quickly adapt to new users using Transfer Learning. We propose a Multi-Scale Convolutional Neural Network (MSCNN), pre-trained with various strategies to improve inter-subject generalization. These strategies include domain adaptation with a gradient-reversal layer and self-supervision using triplet margin loss. We evaluated these approaches on several benchmark datasets, specifically the NinaPro databases. This study also compared two different Transfer Learning frameworks designed for user-dependent fine-tuning. The second Transfer Learning framework achieved a 97% F1 Score across 14 classes with an average of 1.40 epochs, suggesting potential for on-site model retraining in cases of performance degradation over time. The findings highlight the effectiveness of Transfer Learning in creating adaptive, user-specific models for sEMG-based prosthetic hands. Moreover, the study examined the impacts of rectification and window length, with a focus on real-time accessible normalizing techniques, suggesting significant improvements in usability and performance.

Keywords: surface electromyography (sEMG); hand gesture recognition; transfer learning; deep learning; domain adaptation; prosthetic hands; real-time systems; user-specific models; gradient-reversal layer; self-supervised learning; NinaPro database; triplet margin loss; fine-tuning; signal processing; adaptive systems

1. Introduction

Deep learning has achieved important improvements in surface Electromyography (sEMG) recognition systems, supporting the development of sophisticated prosthetic devices, despite the intrinsic complexity of sEMG signals. However, many challenges remain, including catastrophic forgetting, cross-user adaptation, and robustness in real-time performance [1].

Motor neurons play a crucial role in converting brain information into muscle movement [2]. The Motor Unit (MU), which consists of a motor neuron within an innervated muscle fiber, is the fundamental unit of control. The union of numerous MUs generates Motor Unit Action Potentials (MUAPs) [3] that eventually induce movement. The EMG signal acquired from the sensing devices represents the discharge properties of the Motor Units, expressing MUAP convolution in both time and space [4,5]. Typically, EMG signals have an amplitude of around ± 5000 μV and a frequency range between 6 and 500 Hz, with the most important frequency power being between 20 and 150 Hz [6]. Surface Electromyography (sEMG) is a non-invasive approach to hand gesture identification in the myoelectric control of prosthetic devices [7].

The key advantage of sEMG over other techniques, such as implanted electrodes, is its simplicity in collecting and analyzing muscle electrical data, allowing users to use their prosthetic limbs naturally through muscle activation patterns that can resemble the user's original hand movements. Despite substantial developments in this sector, sEMG-based control systems in practical applications still face obstacles [8]. In particular, unpredictable real-life situations that differ significantly from controlled laboratory conditions can bias real-world performance compared to a controlled environment [9,10]. These variations can strongly affect the shape and intensity of the signal, lowering the performance of sEMG-based control systems. Typical examples of these variations include electrode shift [11], unpredictable changes in skin electrode impedance [12], muscle fatigue, and alterations in residual limb posture [13]. As a result, achieving precise control over daily activities becomes challenging.

Many efforts have been made to investigate novel ways to improve the capabilities of myoelectric control, in order to overcome these constraints. Recent advancements have suggested the use of a larger number of sEMG electrodes placed around the residual limb, combined with sophisticated machine learning techniques, [14]. Nevertheless, increasing the number of electrodes complicates the network, increases expense, and may not always be feasible, given the arm's remaining fraction.

As prosthetic devices improve their functionality to restore the full capabilities of a missing limb, their regulation becomes increasingly challenging [15]. Therefore, new control paradigms and algorithms are being investigated to optimize grasping and manipulation tasks in diverse real-world circumstances.

Deep learning has emerged as a powerful solution in multiple areas of study, such as natural language processing, computer vision, and speech recognition [16]. Deep learning techniques have been applied to improve the precision and reliability of EMG-based gesture recognition systems, and this success has been mimicked in the field of electromyography analysis.

Atzori et al. [17] were among the first to apply Convolutional Neural Networks (CNNs) to recognize EMG patterns. Geng et al. [18] used HD-sEMG data to show the capabilities of CNNs in distinguishing between hand gestures. Their findings revealed how well deep networks can learn complex patterns in this domain. By extending this strategy with a multi-stream divide-and-conquer CNN architecture, Wei et al. [19] increased recognition accuracy over single-stream CNNs.

Hu et al. [20] investigated the incorporation of Recurrent Neural Networks (RNNs) into CNN design, and they presented a hybrid CNN-RNN network. Compared to conventional machine learning methods, this architecture showed improved generalization capabilities using unique sEMG picture representations. Progressive Neural Networks (PNNs) were used by Allard et al. [21] to transfer information from a source domain to a target domain. Based on this, they subsequently improved the PNN technique and demonstrated its efficiency with increasingly complicated gesture datasets [22].

Research has also explored the use of Transfer Learning (TL) to reduce training load and to enhance generalization. By pre-training a source model on various participant data, research has been able to enhance recognition accuracy when applied to the hand motions of a different target participant. Wang et al. [23] specifically proposed an Iterative Self-Training Domain Adaptation (STDA) method for cross-user sEMG recognition, combining discrepancy-based alignment and iterative pseudo-label updates. Islam et al. [24] developed a lightweight All-ConvNet+TL model that efficiently tackles inter-session and inter-subject variability, which successfully enhances the accuracy and speed of sEMG gesture classification, making it well suited for real-time applications. Nguyen et al. [25] took a different approach by introducing a Frequency-based Attention Neural Network (FANN) combined with Subject-Adaptive Transfer Learning.

Despite advances in TL for EMG pattern recognition, several challenges still need to be overcome. Current deep learning solutions have limits, in terms of long-term reliability and adaptability to new gestures or users [26,27]. Indeed, the user can benefit from the option to adapt and/or include new gestures into the model's repertoire of recognizable gestures. Researchers in this field have shown that deep learning may be the key to enhancing the precision and robustness of sEMG-based control. Problems such as long-term dependability, adaptation to new motions [28], and muscle activation coverage should be addressed as well [29]. Furthermore, in the case of prosthetic devices, user adaptation, energy usage, and inference time must all be considered.

The key objective of this study was to enhance model generalization among patients, to reduce individual variability, hence facilitating quick user adaptation through Transfer Learning. It also targeted key challenges in sEMG-based hand gesture recognition for prosthetics, prioritizing low-computation interfaces and efficient data splitting techniques.

To achieve these goals, different pre-processing steps, including rectification, window length, and normalization procedures, were analyzed. The study also investigated Transfer Learning with few-shot learning for new user fine-tuning. In the end, to evaluate generalization, we employed an adversarial network with gradient-reversal descent [30] and self-supervision with triplet margin loss [31].

2. Material and Methods

An extensive analysis of the methodologies adopted is provided in this section. It starts by describing the data used to assess model performance and their acquisition protocol. Two data splitting strategies, inter-subject and intra-subject, are then introduced, followed by a detailed description of the model structure, along with the training strategies employed for pre-training the model backbone. The Transfer Learning framework description is then presented. Finally, data pre-processing, normalization strategies, and parameter tuning methodologies are discussed.

2.1. Data Acquisitions

Several publicly accessible datasets were used for the model evaluation, specifically the NinaPro DB2, DB3, and DB7 databases, for which the details are listed in Table 1. The setup used Delsys Trigno electrodes for sEMG data acquisition. The subjects mimicked 40 hand gestures, each repeated six times with 3 s of rest between exercises. Visual stimuli were presented on a laptop, and relabeling was performed to account for delays. The acquisition protocols for these databases are described in further detail in [32].

A subset of 14 gestures was selected from the available gestures based on the Activities of Daily Living (ADL) shown in Figure 1. This selection was made to balance the trade-off between number of gestures and accuracy, as the study was tailored for a real prosthetic device.

Table 1. Description of NinaPro databases used. The subjects column refers to the number of patients performing the gestures, while repetitions refers to the number of iterations of the same gesture.

Name	Subjects	Channels	Classes	Repetitions	Frequency	Device
DB2	40	12	41	6	2000 Hz	Delsys Trygno
DB3	11 (A)	12	41	6	2000 Hz	Delsys Trygno
DB7	20 + 2 (A)	12	41	6	2000 Hz	Delsys Trygno

Note: In the context of this study, the letter A in *Subjects* designates amputees.

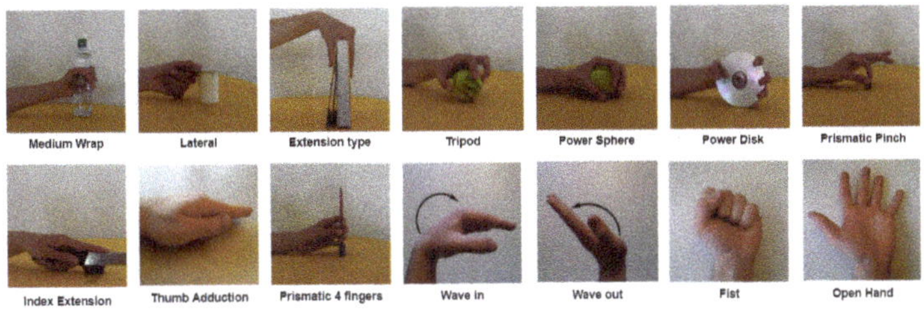

Figure 1. Selected hand gestures from Activities of Daily Living (ADL).

2.2. Data Splitting

The experimental strategy included two data splitting strategies: inter-subject and intra-subject. Inter-subject involved training on data from one group of patients and validating on another; intra-subject involved training, validating, and testing on different repetitions of the same gestures from the same pool of subjects.

For both split modalities, 15 subjects (12 healthy individuals and three amputees) were set aside to be used as test subjects for Transfer Learning. In the first method, to cover a maximum variability, the remaining patients were randomly divided, with 80% used for training and 20% for validation.

With regards to the intra-subject method, three repetitions were used for training (2, 4, 6), two for validation (1, 5), and one for testing (3). This division was introduced to account for variables such as muscular fatigue, which can induce signal fluctuations [33].

Additionally, the 15 subjects initially excluded and selected for testing Transfer Learning could also be used in both splitting methodologies to evaluate model performances over unseen subjects. This enabled a subsequent assessment of the Transfer Learning framework for user-specific fine-tuning. Both the intra-subject and inter-subject approaches applied Transfer Learning to new subjects, allowing a comparative evaluation of their effectiveness.

2.3. Data Pre-Processing and Normalization Strategies

The pre-processing involved filtering, windowing, rectification, and normalization. These strategies were selected as they did not require a significant computational effort, which is an important constraint given the expected application for real-time prosthesis use and its associated requirement for energy and efficiency and speed [34]. In order to obtain an informative bandwidth from 10 to 500 Hz, these data were filtered applying a fourth-order bandpass Butterworth filter [35]. Additionally, a notch filter at 50 Hz was applied, to remove power-line interference.

Different windows were employed, each with a 75% overlap. These windows comprised 100, 150, 200, and 250 milliseconds. Using overlapping windows reduced information loss at each window boundary, resulting in a more accurate representation of the signal's temporal features.

Normalization has long been used to reduce heterogeneity in electromyographic data and to mitigate substantial inter-subject variability [36]. Three normalization strategies (range [0 1], range [−1,1], and Z-Score) were evaluated in two ways: normalization by subject and normalization by subject and channel [37,38].

The first approach relies on the calculation of subject-specific metrics from all channels combined, which includes mean, standard deviation, minimum, and maximum. The second approach, working at the channel level, is less likely to be influenced by artifacts, spikes, and noise-related channels [39].

2.4. Model Structure

The architecture of the proposed Multi-Scale Convolutional Neural Network (MSCNN) model is shown in Figure 2. This figure provides a clear representation of the model's structure and the relationships between its various components. For a more detailed understanding of the data flow within the MSCNN, refer to Table 2, which outlines each step and operation in the process.

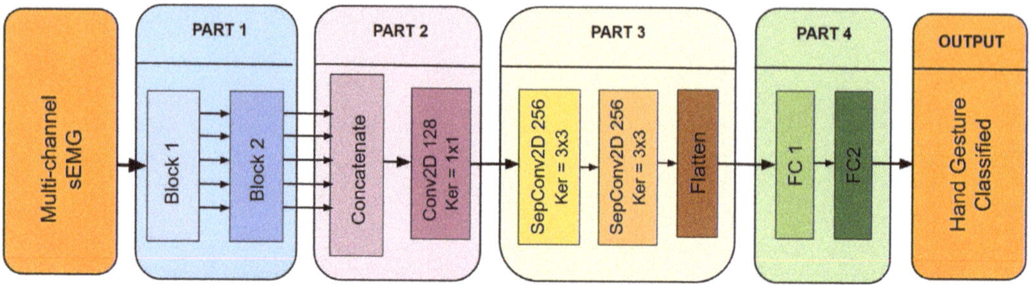

Figure 2. Architecture of the Multi-Scale CNN model.

The MSCNN model is composed of four distinct parts, each serving a unique purpose. These components are as follows:

Part 1) As seen in Figure 2, this part is divided into two distinct blocks. Block 1 aims to extract features to capture the spatio-temporal patterns in the sEMG signal. The multichannel sEMG signal is represented as an image for each sample.

A set of five filters, sized (W × 3, 2*W × 3, 3*W × 3, 4*W × 3, and 5*W × 3) are employed in Block 1, where W is selected as 1/20 of the chosen window size. The selection of these sizes was driven by the need to parameterize the model consistently across different window sizes to facilitate parameter tuning and comparison. Larger filter sizes correspond to lower-frequency related characteristics and vice versa.

Each feature map output from the parallel convolutions of Block 1 is then fed into Block 2. Block 2 has a similar structure and uses five additional convolution layers with fixed 3 × 3 filter sizes. In Block 2, separable convolutions are used to reduce the number of trainable parameters, and the number of neurons in each layer is always doubled compared to the previous ones. Furthermore, the dropout layers and Batch Normalization (BN) layers in both blocks are implemented to increase the MSCNN model's overall generalizability.

Part 2) The main objective of this part is to reduce the dimensionality of the feature maps that come from Part 1 and then fuse them. After Part 1, the number of feature maps is reduced from 320 (64 × 5) to 128 by using 1 × 1 filters in this convolutional layer. The 1 × 1 convolution prevents the use of fully connected layers, thereby decreasing the number of parameters and preventing overfitting.

Part 3) Part 3 allows the model to extract deeper features by employing other separable convolutional layers, which reduces the number of parameters and, hence, the computational load, an essential requirement for embedded applications.

Part 4) The final stage of the model involves converting the obtained features to class probabilities for classification. This is accomplished by combining two completely connected layers that process the deep features previously collected.

Notably, in *Part 1* , a customized Conv2D layer is used for the first parallel convolutions. This layer employs zero padding along the temporal axis and circular padding along the first 8 channel axes to preserve spatial relationships between electrodes.

The rationale for this approach is evident considering the acquisition setup of the databases, where the initial eight electrodes around the arm are organized in a circular arrangement. This padding accounts for preserving spatial patterns, as converting a circular architecture to a linear representation might result in loss or distortion of vital information.

Table 2. Detailed structure of the model and data flow. The number before the layer type represents the number of parallel layers of the corresponding type, e.g., 5-Conv2D represents five conv2D layers. The first dimension (kernel size) of the first set of parallel Conv2D_Circular8 layers increases linearly from a lower value to an upper value. The increment between each kernel size is W = window length/20 (window length fixed at 300 here).

Part	Layer Type	Kernel Size	Output Size	Activ	Options
1	1-Input	-	1/300 × 12	-	-
	5-Conv2D_Circular8	[15–75] × 3	32/300 × 14	ReLU	pad = 'circular'
	5-BatchNorm2D	-	32/300 × 14	-	-
	5-MaxPooling2D	15 × 1	32/20 × 14	-	-
	5-Dropout2D	-	-	20 × 14	rat e = 0.2
	5-SeparableConv2D	3 × 3	64/20 × 14	ReLU	pad = 0
	5-BatchNorm2D	-	64/20 × 14	-	-
	5-MaxPooling2D	2 × 2	64/10 × 7	-	-
	5-Dropout2D	-	64/10 × 7	-	rate = 0.2
2	1-Concatenate	-	320/10 × 7	-	-
	1-Conv2D	128/1 × 1	128/10 × 7	ReLU	-
3	1-SeparableConv2D	3 × 3	256/10 × 7	ReLU	pad = 0
	1-BatchNorm2D	-	256/10 × 7	-	-
	1-MaxPool2D	2 × 2	256/5 × 3		-
	1-Dropout2D	-	256/5 × 3	-	rate = 0.2
	1-SeparableConv2D	3 × 3	256/5 × 3	ReLU	pad = 0
	1-BatchNorm2D	-	256/5 × 3	-	-
	1-MaxPool2D	2 × 2	256/2 × 1	-	-
	1-Dropout2D	-	256/2 × 1	-	rate = 0.2
4	1-Flatten	-	512	-	-
	1-Linear	-	128	ReLU	-
	1-Linear	-	14	SoftMax	-

2.5. Model Pre-Training Strategies

This study employed three training strategies for the backbone of the model before applying Transfer Learning: standard training, pre-training with triplet margin loss, and domain adversarial network with gradient-reversal implementation.

Standard training involves a typical classification task using the softmax function and cross-entropy loss at the end of the linear layers, to predict class probabilities.

Triplet margin loss enhances domain adaptation by learning data representations that increase the similarity between positive pairs and separate negative pairs in the embedding space. The loss is defined as

$$L_{\text{triplet}} = \sum_{i}[\max(0, \text{margin} + d(a_i, p_i) - d(a_i, n_i))], \quad (1)$$

where a_i is the anchor sample, p_i is a positive sample, n_i is a negative sample, and margin specifies a minimum distance between positive and negative pairings. In our example, a positive sample is considered as a "different subject with the same label", while a negative sample is simply identified by a different label. This technique drives the model to improve discrimination between subject-specific features, allowing for effective classification in the following trials.

The gradient-reversal technique penalizes the primary model loss function depending on the performance of the domain classifier, encouraging the model to learn features that are domain-invariant (different subjects). The objective function to be reduced is expressed as follows:

$$\min \frac{1}{n} \sum_{i=1}^{n} L_c(f_i; y_i) - \lambda \cdot \frac{1}{m} \sum_{j=1}^{m} L_d(g_j; d_j), \quad (2)$$

Here, L_c is the classification loss and L_d is the domain classifier loss. The parameter λ determines the trade-off between classification and domain loss. The approach applies an exponential scaling factor to shift the model's attention from classification task to domain-invariant learning across training epochs.

As illustrated in Figure 3, the embeddings for the triplet margin loss are extracted from the output of the flatten layer in *Part 3*, which directly precedes the classification head. In the domain adversarial network with gradient reversal, an additional head, identical to *Part 4*, is integrated into the model architecture connected to the output of *Part 3*. The objective of this secondary head is to classify subjects, thereby encouraging the backbone to become invariant to subject-specific features.

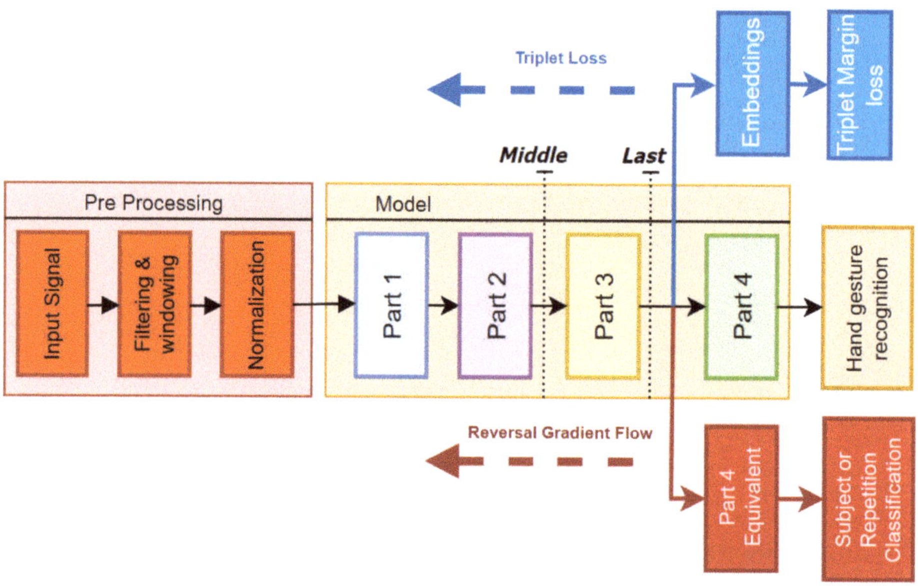

Figure 3. Diagram illustrating the architecture used during pre-training with triplet loss and gradient reversal. The two black dotted lines indicate the portions of the model retrained under two different configurations: *last* and *middle*.

2.6. Transfer Learning Framework

Two different fine-tuning approaches were tested, as depicted in Figure 3. In the first approach, referred to as *last*, only the two final linear layers from *Part 4* were retrained, for a total number of ~70,000 trainable parameters. In contrast, for the second framework, named *middle*, only the *Part 1* and *Part 2* layers were frozen, leading to ~200,000 trainable parameters.

Experiments were conducted with varying data quantities to be used in retraining (from 1 to 4 repetitions) for each subject. For each amount of data, the model was re-trained five times by varying the combination of repetitions. This allowed us to observe the model's behavior and its vulnerability to overfitting during the fine-tuning process.

Moreover, dividing data into several repetitions mimicked a real-world online scenario in which the model has no prior on the specific validation and test iterations of the same exercise beforehand. This allowed us to assess the ability to adapt to unseen data more realistically.

To minimize the impact of randomness and to obtain a more reliable comparisons between backbones, a 5-fold cross-validation was used. This means that for each subject in the test group, we trained each pre-trained model five times, each with a different number of repetitions for re-training. In each fold, a different set of repetitions was employed for validation and testing, ensuring the model performance was evaluated on unseen repetition-specific data across all folds.

This approach provided a more robust and reliable assessment of the generalizability of the models and the ability to handle variations in data.

2.7. Hyperparameter Optimization Using Grid Search

Hypothetical correlations between normalization techniques, activation function, and rectification were also investigated. Once the model was fully parameterized, three exhaustive grid searches were performed by varying the normalization methods. These searches were used to evaluate the impact and effectiveness of the hyperparameters listed in Table 3. For this experiment, an intermediate window of 300 points, corresponding to 150 ms, was chosen. This ensured that there was sufficient time (≈100–150 ms) to process, infer, and perform movements of the prosthetic hand before reaching the critical threshold of 300 ms, considered as the upper limit for the user to perceive a real-time control [40,41].

Interestingly, one search focused on the effects of the $[-1, 1]$ normalization method, particularly on the impact of forcing values into the positive range (e.g., rectification, Z-score) over information preservation. For more details, see the 'Hyperparameter Optimization via Grid search' subsection in the 'Results' section.

Table 3. This table presents the hyperparameters explored in the grid search. The keys *N_Multi-kernel, N_Post-Concatenation, and N_Separable_Conv* represent the number of neurons in the hidden convolutional layers of *Parts 1, 2, and 3*, respectively. Notably, in *Part 1*, Block 2 always employed twice as many hidden neurons as Block 1. The **bold** values are the ones selected for the tuned model.

Hyperparameter	Values Chosen
Batch size	64, 128, **256**, 512
Optimizer	SGD, **Adam**
Learning rate	$1 \times 10^{-2}, 1 \times 10^{-3}, 1 \times 10^{-4}, 1 \times 10^{-5}$
Decay rate	$0, 1 \times 10^{-2}, 1 \times 10^{-3}, 1 \times 10^{-4}, 1 \times 10^{-5}$
Activation function	ReLu, **LeakyReLU**, ELU, PReLU
Pooling type	**MaxPooling**, AveragePooling
N_Multi-kernel	16, 32, **64**, 128
N_Post-Concatenation	64, **128**, 256
N_Separable_Conv	64, 128, **256**

3. Results

This section begins with an evaluation of the effect of the window length. It then presents the results of the grid search technique used for the hyperparameter tuning. Following this, the performance of the model across the databases is showcased, before delving into the main aspect of Transfer Learning.

The *middle* Transfer Learning framework achieved accuracy of 98% ± 0.02 on a new subject after retraining for just two epochs over four repetitions within a 150 ms window, demonstrating the model's potential for real-time application. The code is available at GitHub.

3.1. Window Length Impact

This subsection compares the various window sizes for the intra-subject experiment on the combined database. Windows of size 200, 300, and 400 points, corresponding to 100, 150, and 200 ms, respectively, were tested with the fixed parameters showed in Table 2.

Furthermore, a 20 ms window (comprising 40 points) was chosen, considering the potential use of a majority voting strategy, as this has been shown to be a valid option to improve reliability [42]. The results shown in Table 4 highlight the significant performance drops associated with the smallest window size, as expected. No substantial differences were observed among the other three window sizes. Consequently, subsequent experiments were conducted using only the 200- and 300-point windows, as they are more suitable for real-time prosthesis control.

Table 4. Performance metrics for several window lengths over the merged database. Bold values indicate the best in each metric.

Window Length (ms)	Validation Loss	Accuracy	F1 Score	Best Epoch
100	0.976 ± 0.011	0.775 ± 0.012	$\mathbf{0.720 \pm 0.010}$	62 ± 4.2
150	$\mathbf{0.971 \pm 0.017}$	$\mathbf{0.778 \pm 0.015}$	0.713 ± 0.012	51 ± 3.7
200	0.988 ± 0.019	0.769 ± 0.016	0.718 ± 0.014	56 ± 3.5
20	1.340 ± 0.025	0.689 ± 0.024	0.607 ± 0.019	36 ± 4.7

3.2. Hyperparameter Optimization via Grid Search

From the very first experiments, the channel-dependent normalization mode revealed a clear benefit, since it was better at mitigating errors or abnormalities that were present in some channels. For this reason, the results presented were all performed with channel-wise normalization.

The grid search was performed employing several normalization techniques. For this evaluation, the standard training algorithm was used over the merged database with intra-subject splitting.

Table 5 summarizes the best three configurations identified using three grid searches with changing normalization techniques. For each of them, 20 models were generated, employing a fixed window length of 300 time-points, that is, with a 2000 Hz Delsys Trigno Device, 150 ms per window. For better readability, the table excludes the parameters that seemed to have minimal impact on performance. In particular, parameters such as the number of neurons in `N_Multi-kernel`, `N_Post-Concatenation`, and the maxpool type were omitted, as they were consistent across the top three configurations.

Table 5. The three best model configurations resulting from grid searching, varying the normalization strategies.

Norm Type	Rectification	Accuracy	F1 Score	N_SepConv	Activation
Z_score	Yes	0.8068	0.7514	256	PReLU
Z_score	Yes	0.7971	0.7451	256	LeakyReLU
Z_score	Yes	0.7893	0.7325	256	LeakyReLU
Range (−1; 1)	No	0.7826	0.7280	128	ELU
Range (−1; 1)	No	0.7722	0.7217	128	ELU
Range (−1; 1)	No	0.7758	0.7187	256	LeakyReLU
Range (−1; 1)	Yes	0.7399	0.6692	128	PReLU
Range (−1; 1)	Yes	0.7487	0.6749	128	PReLU
Range (−1; 1)	Yes	0.7338	0.6610	128	PReLU

Compared to other normalization methods, Z-score normalization exhibited better accuracy and faster convergence, in line with other studies [43]. However, the top three configurations when exploiting normalization within the range (−1; 1) and no rectification achieved similar performance with half the number of neurons in the *N_Separable_Conv* layers of *Part 3* of the architecture. This suggests that accounting for negative values may allow the model to learn effectively with a simpler architecture. In contrast, the last three rows show that rectifying the value and then normalizing between (-1; 1) might distort information, leading to important performance drops.

3.3. Model Performance Across Databases

In Table 6, the performance of the fine-tuned standard model with respect to the different databases for the intra-subject modality are summarized.

Table 6. Performance metrics over different databases.

Database	Window 100 ms		Window 150 ms	
	Accuracy	**F1 Score**	**Accuracy**	**F1 Score**
DB3	0.503 ± 0.012	0.454 ± 0.004	0.508 ± 0.022	0.458 ± 0.012
DB2	0.824 ± 0.010	0.770 ± 0.016	0.819 ± 0.009	0.768 ± 0.016
DB7	0.856 ± 0.005	0.797 ± 0.004	0.862 ± 0.007	0.796 ± 0.010
Merged	0.789 ± 0.011	0.741 ± 0.009	0.798 ± 0.014	0.768 ± 0.011

DB3 exhibited the lowest performance, likely due to higher variability among amputee subjects.

Moreover, there was a noticeable shift in the dynamics of the performance, especially for DB2 and DB7. Interestingly, DB7 had the best precision, despite having two amputee patients. One possible explanation for this phenomenon could be that because DB7 had only 22 subjects, compared to the 40 of DB2, the model tended to overfit on the subjects employed, which improved its overall accuracy.

Meanwhile, because DB3 was included in the composite database, its performance inevitably suffered as a result. Still, a respectable level of accuracy was obtained, most likely as a result of exploiting more data during training, which typically results in the development of more reliable models. To leverage the benefits of a more comprehensive data pool, all future experiments will be performed using the merged database.

3.4. Transfer Learning Framework Effectiveness

In this section, we examine the effects of several techniques on final-user fine-tuning within the Transfer Learning framework. In this context, final-user fine-tuning refers to the process of personalizing the model for users who will be using a myoelectric-controlled prosthesis for hand gesture recognition. The previously excluded subjects are used as new, unseen patients for applying Transfer Learning and re-training on the subject-specific data. All the training was conducted using a Tesla V100-PCIE-32GB GPU (NVIDIA Corporation, Santa Clara, CA, USA) with CUDA version 12.1. The average inference time, measured over 100 single-batched samples, was 3.16 ± 0.31 milliseconds.

Table 7 shows the *last* experiment, where only *Part 4* of the model architecture was retrained. The results show the average and standard deviation over the 15 subjects that were previously discarded from training and validation. The F1 Score is also shown, to account for the slightly imbalanced dataset. This table includes several window sizes, pre-training algorithms, and splitting modalities.

The table provides useful insights. First, the intra-subject partitioning produced the best results for regular training. This was most likely due to the model's ability to train over multiple epochs, as opposed to patient splitting, which stops early due to overfitting over training subjects.

Table 7. Average performance for the 15 previously excluded patients for TL-*last* experiments, by retraining only *Part 4* for four repetitions. Results are shown for the two pre-training data splitting approaches: inter-subject and intra-subject. Bold values indicate the best performance across different configurations.

Backbone Training	intra-Subject				Inter-Subject			
	Window 100 ms		Window 150 ms		Window 100 ms		Window 150 ms	
	Accuracy	F1 Score	Accuracy	F1 Score	Accuracy	F1 Score	Accuracy	F1 Score
Standard	**0.87 ± 0.20**	**0.86 ± 0.20**	**0.90 ± 0.18**	**0.90 ± 0.18**	0.83 ± 0.19	0.83 ± 0.19	0.86 ± 0.19	0.86 ± 0.19
Pre-triplet	0.66 ± 0.19	0.66 ± 0.20	0.68 ± 0.19	0.67 ± 0.20	0.53 ± 0.19	0.50 ± 0.20	0.66 ± 0.18	0.65 ± 0.19
Reversal	0.84 ± 0.18	0.84 ± 0.19	0.87 ± 0.17	0.87 ± 0.17	**0.85 ± 0.18**	**0.84 ± 0.18**	**0.88 ± 0.16**	**0.87 ± 0.16**
Reversal Amputee	0.84 ± 0.19	0.84 ± 0.20	0.88 ± 0.18	0.87 ± 0.18	0.84 ± 0.19	0.83 ± 0.20	0.87 ± 0.18	0.87 ± 0.19

The inter-subject splitting precluded the model from learning invariant features during standard training. In this scenario, the adversarial network outperformed all the others, showing that it had successfully learned some invariant properties. This pattern was also seen in the reverse amputee training, which ranked second in the inter-subject trials.

Self-supervision with triplet margin loss performed poorly. During pre-training, it did not use task loss, and simply forcing embeddings with triplet loss did not provide useful insights into this challenging problem.

Figure 4 shows the variability among individuals, with a focus on the stronger ones between amputees. This graph, based on the inter-subject experiment with the reversal backbone, shows the difference in accuracy values for healthy people alone versus the complete group, including amputees.

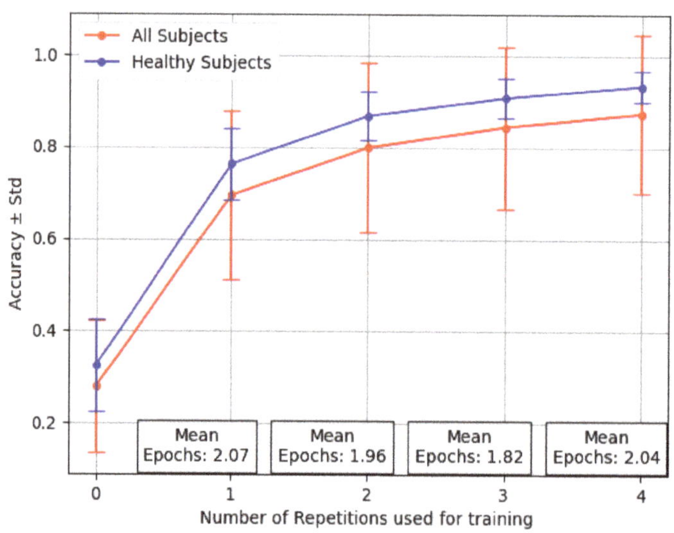

Figure 4. Performance averaged across 12 healthy patients and three amputees for the Transfer Learning *last* with varying numbers of repetitions. Where $X = 0$ (no retraining), this reflects the average results of testing over new, unseen subjects.

The increased standard deviation observed when the pool of 15 subjects included the 3 amputees is clear, mostly considering that the number of amputees was 3 out of 15. In particular, the number of epochs averaged across all the individuals was, on average, less than two, demonstrating that Transfer Learning does not require extensive and computationally expensive training sessions. However, the success of Transfer Learning relies on the consistency of the data used during pre-training within the selected subject.

Figure 5 exemplifies this challenge. It showcases the results of the TL framework for the three amputees with the reversal backbone pre-trained with inter-subject splitting.

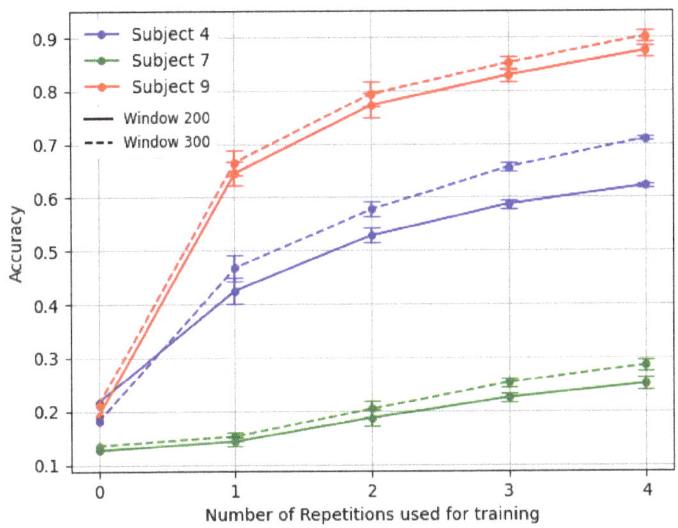

Figure 5. Comparison of the TL *last* for the three amputees. Model backbone pre-trained with reversal gradient and inter-subject splitting technique.

Table 8 shows the *middle* experiment, where both *Part 3* and *Part 4* of the model's architecture were re-trained. The results show the average and standard deviation over the 15 subjects that were previously discarded from training and validation. The F1 Score is also shown, to account for the slightly imbalanced dataset. This table includes several window sizes, pre-training algorithms, and splitting approaches.

Table 8. Average performance for the 15 previously excluded patients for the TL-*middle* experiments, by retraining both *Part 3* and *Part 4* for four repetitions. Results are shown for the two pre-training data splitting approaches: inter-subject and intra-subject. Bold values indicate the best performance across different configurations.

Backbone Training	Intra-Subject				Inter-Subject			
	Window 100 ms		Window 150 ms		Window 100 ms		Window 150 ms	
	Accuracy	F1 Score	Accuracy	F1 Score	Accuracy	F1 Score	Accuracy	F1 Score
Standard	**0.98 ± 0.02**	**0.98 ± 0.02**	**0.98 ± 0.02**	**0.98 ± 0.02**	**0.95 ± 0.04**	**0.95 ± 0.04**	0.96 ± 0.03	0.96 ± 0.03
Pre-triplet	0.94 ± 0.04	0.94 ± 0.04	0.95 ± 0.03	0.95 ± 0.03	0.93 ± 0.05	0.93 ± 0.04	0.95 ± 0.02	0.95 ± 0.02
Reversal	0.95 ± 0.05	0.95 ± 0.05	0.97 ± 0.03	0.97 ± 0.03	0.95 ± 0.05	0.95 ± 0.06	**0.97 ± 0.03**	**0.97 ± 0.03**
Reversal Amputee	0.95 ± 0.05	0.95 ± 0.05	0.97 ± 0.03	0.97 ± 0.03	0.94 ± 0.05	0.94 ± 0.05	0.96 ± 0.03	0.96 ± 0.03

In Figure 6, it is possible to observe the difference in both mean and standard deviation for accuracy and F1 Score, with respect to Figure 4. Similarly, in Figure 7 it is possible to appreciate such a consistent change in performance even for the three amputees, in particular for the problematic subject number 7.

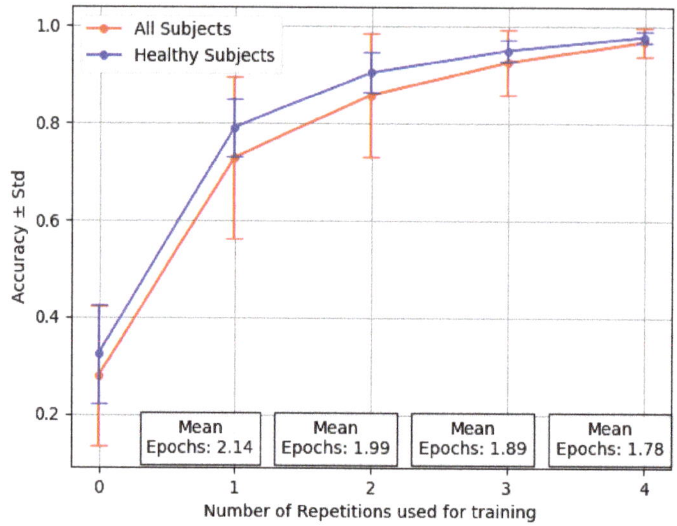

Figure 6. Performance averaged across 12 healthy patients and three amputees for the Transfer Learning *middle* with varying numbers of repetitions.

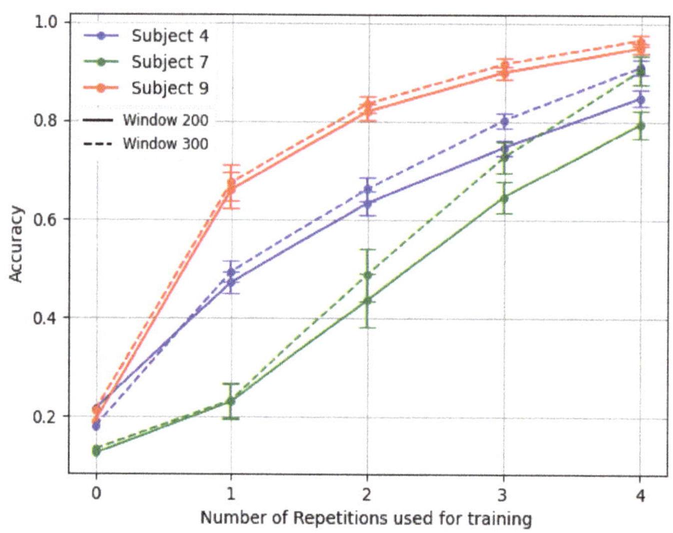

Figure 7. Comparison of the TL *middle* for the three amputees. The Model backbone was pre-trained with gradient reversal and inter-subject splitting.

Table 9 compares our results with other studies on subject-specific Transfer Learning. Direct comparisons are challenging, due to the differences in chosen classes, input data formats, model architectures, number of parameters, and Transfer Learning techniques. However, this table provides our findings alongside recent work in the field. It aims to place our results in the broader context of subject-specific Transfer Learning research.

Table 9. Performance comparison of various Transfer Learning frameworks using NinaPro databases.

Citation	Database	Classes	Input	Window	Model	Accuracy %
Zhai 2017 [44] [1]	DB2 DB3	10	Spectrogram	200 ms	2D-CNN	90.06 85.53
Wei 2019 [19] [2]	DB2	50	Features	200 ms	Multiview 2D-CNN	83.7
Soroushmojdehi 2022 [45] [3]	DB2	17	Raw sEMG	250 ms	PFCnet	82.87
Lehmler 2022 [46] [4]	DB2 DB3	17	Raw sEMG	200 ms	1D-CNN	68.52 50.72
Zabihi 2023 [47] [4]	DB2	17	Raw sEMG	200 ms	TraHGR	88.72
Qamar 2024 [48] [5]	DB2 DB3	50	Spectrogram	200 ms	DP-CNN	94.00 85.36
Fan 2023 [49] [6]	DB2+TL DB3	17	Feature	100 ms	CNN	67.5
Lin 2023 [50] [6]	DB2+TL DB3	17	Features	200 ms	LE-CNN	83.5
Ours	DB2+DB3+DB7	14	Raw sEMG	150 ms	MSCCNN-last MSCNN-middle	98.04 87.70

Overview of Transfer Learning strategies implemented: [1] Adaptive self-calibration system utilizing a consensus-based voting mechanism across four separate repetitions. [2] Personalized adaptive Batch Normalization applied to each subject's data. [3] Merging and freezing subject-specific and multi-subject features previously extracted before employing a new classifier training phase. [4] Retaining the weights of earlier layers while retraining the final layers of the model. [5] Subject-specific training from scratch without pre-existing models. [6] Initial training performed on DB2, with subsequent layer freezing and retraining of the final layers using DB3 subject-specific data.

4. Discussion

The results indicate that a more personalized model that learns subject-specific features rather than merely retraining the final linear layers is crucial for improving the adaptability and performance of sEMG-based prosthetic hands. Nonetheless, even in this situation, the number of epochs before overfitting were, on average, fewer than two in all cases, suggesting the potential for efficient and embedded fine-tuning. Given the approximately $\approx$3 ms inference time, even on a powerful GPU that is not intended for embedded devices, this performance suggests that the system is sufficiently fast for real-time applications or even for a major voting strategy.

Another consideration regards the amputees. These patients present unique challenges for gesture recognition, due to greater differences in their physiological signals, resulting from factors such as the percentage of muscle remaining after amputation along with residual sensations of the phantom limb [51]. Therefore, the creation of user-adaptive models is a crucial area of research in the field of EMG-based prosthetic devices.

For example, subjects number 7 and number 4 of DB3 demonstrated this variability particularly well, as shown in Figure 5. Subject 7 exhibited a complete absence of phantom limb sensation and zero percentage of remaining forearm, leading to the usage of 10 channels instead of 12. Meanwhile, patient number 4 had a unique combination of low phantom sensation, no prior myoelectric experience, and the highest DASH score in the dataset. In contrast, patient number 9 represented better the average clinical profile of the other amputee subjects.

Factors like phantom limb sensation intensity, remaining forearm percentage, DASH score, and prior myoelectric prosthesis experience seemed to influence the model's ability to adapt to new users.

5. Conclusions

This study demonstrates that Transfer Learning techniques improve the effectiveness of sEMG-based prosthetic hands by customizing models to individual users. Retraining with more layers produces better outcomes, emphasizing the benefit of allowing models to learn subject-specific features. The findings underline the importance of managing intra-subject heterogeneity, particularly among amputees, in order to maintain consistent performance between individuals.

Strategies such as adversarial network with gradient-reversal descent and channel-wise normalization can help in building a more robust pre-trained model. Furthermore, allowing an embedded device to extract daily parameters for normalization and re-training in loco should also be considered. Implementing these ideas in prosthetic devices could result in more adaptive, efficient, and user-specific solutions, thereby improving the functionality and usability of these technologies.

Future Work

During the re-training phase in both modalities, we used standard training, and we split the data by repetitions. However, to account for intra-subject variability, it is possible to retrain layers using gradient-reversal techniques across different repetitions or sessions of the same subject.

This approach would help to maintaining performance consistency and adapt to the user's unique patterns over time.

Furthermore, in order to overcome performance degradation due to daily variations in prosthetic devices, built-in functions should be implemented. These implementation tools could either make it easier to acquire new data for fine-tuning or provide alternative techniques to avoid such computation on the embedded device itself.

In our case, the alternative would be to collect a single sample for each exercise, which could then be used to quickly update the necessary values for normalizing each channel, based on daily circumstances. Using this single-sample strategy, the system might adjust to daily variations without incurring computational overhead, resulting in efficient and effective performance in real-world scenarios.

In the future, fuzzy similarity formulations [52] could improve sEMG pattern discrimination, whilst adaptive fuzzification techniques [53] could improve system adaptability. These advances have the potential to improve prosthesis control, particularly for smooth multi-gesture recognition.

Improving inference time is a major concern for future work, with a focus on faster computations and reduced energy consumption for prosthetic devices. Faster inference time would also enable the implementation of major voting strategies, which could significantly increase the system's reliability. Promising advances have been made in this regard by establishing quantization techniques, which may slightly reduce accuracy but dramatically improve energy efficiency and inference time in prosthetic devices [54,55].

Author Contributions: Conceptualization, R.F., M.A. and F.B.; methodology, R.F. and N.M.; validation, R.F. and N.M.; formal analysis, R.F. and N.M.; investigation, R.F.; resources, H.M. and M.A.; data curation, R.F., M.A. and H.M.; writing—original draft preparation, R.F., N.M. and M.A.; writing—review and editing, R.F., N.M., M.A., H.M. and C.T.; supervision, N.M. and M.A.; project administration. M.A. and H.M. All authors have read and agreed to the published version of the manuscript.

Funding: This research received no external funding.

Institutional Review Board Statement: Not applicable.

Informed Consent Statement: Informed consent was obtained from all subjects involved in the study.

Data Availability Statement: NinaPro data are available at: http://ninaweb.hevs.ch/.

Conflicts of Interest: The authors declare no conflict of interest.

References

1. Palermo, F.; Cognolato, M.; Gijsberts, A.; Müller, H.; Caputo, B.; Atzori, M. Repeatability of grasp recognition for robotic hand prosthesis control based on sEMG data. In Proceedings of the 2017 International Conference on Rehabilitation Robotics (ICORR), London, UK, 17–20 July 2017; pp. 1154–1159. [CrossRef]
2. Farina, D.; Holobar, A. Characterization of Human Motor Units From Surface EMG Decomposition. *Proc. IEEE* **2016**, *104*, 353–373. [CrossRef]
3. Oliveira, D.D.; Casolo, A.; Balshaw, T.; Maeo, S.; Lanza, M.; Martin, N.; Maffulli, N.; Kinfe, T.; Eskofier, B.; Folland, J.P.; et al. Neural decoding from surface high-density EMG signals: Influence of anatomy and synchronization on the number of identified motor units. *J. Neural Eng.* **2022**, *19*, 046029. [CrossRef] [PubMed]
4. Farina, D.; Jiang, N.; Rehbaum, H.; Holobar, A.; Graimann, B.; Dietl, H.; Aszmann, O. The extraction of neural information from the surface EMG for the control of upper-limb prostheses: Emerging avenues and challenges. *IEEE Trans. Neural Syst. Rehabil. Eng.* **2014**, *22*, 797–809. [CrossRef]
5. Muceli, S.; Merletti, R. Surface EMG detection in space and time: Best practices. *J. Electromyogr. Kinesiol.* **2019**, *49*, 102363.
6. Konrad, P. *The ABC of EMG: A Practical Introduction to Kinesiological Electromyography*; Noraxon Inc.: Scottsdale, AZ, USA, 2005.
7. Micera, S.; Carpaneto, J.; Raspopovic, S. Control of hand prostheses using peripheral information. *J. Neuroeng. Rehabil.* **2010**, *7*, 48–68. [CrossRef]
8. Atzori, M.; Müller, H. Control Capabilities of Myoelectric Robotic Prostheses by Hand Amputees: A Scientific Research and Market Overview. *Front. Syst. Neurosci.* **2015**, *9*, 162. [CrossRef]
9. Cifrek, V.M.T.O.M.; Sarabon, N.J.; Markovic, N.S. Surface EMG based muscle fatigue evaluation in biomechanics. *Clin. Biomech.* **2009**, *24*, 327–340. [CrossRef] [PubMed]
10. Raez, H.M.M.Y.F.; Hussaini, M.S.J.; Mohd-Yasin, P.G.M.G.A. Techniques of EMG signal analysis: Detection, processing, classification and applications. *Biol. Proced. Online* **2006**, *8*, 11–35. [CrossRef]
11. Muceli, S.; Jiang, N.; Farina, D. Extracting signals robust to electrode number and shift for online simultaneous and proportional myoelectric control by factorization algorithms. *IEEE Trans. Neural Syst. Rehabil. Eng.* **2014**, *22*, 623–633. [CrossRef]
12. Hakonen, M.; Piitulainen, H.; Visala, A. Current state of digital signal processing in myoelectric interfaces and related applications. *Biomed. Signal Process. Control* **2015**, *18*, 334–359. [CrossRef]
13. Hwang, H.J.; Hahne, J.M.; Müller, K.R. Real-time robustness evaluation of regression based myoelectric control against arm position change and donning/doffing. *PLoS ONE* **2017**, *12*, e0186318. [CrossRef] [PubMed]
14. Fougner, A.; Stavdahl, O.; Kyberd, P.J.; Losier, Y.G.; Parker, P.A. Control of upper limb prostheses: Terminology and proportional myoelectric control—A review. *J. Neuroeng. Rehabil.* **2012**, *20*, 663–677. [CrossRef] [PubMed]
15. Peerdeman, B.; Boere, D.; Witteveen, H.; in 't Veld, R.H.; Hermens, H.; Stramigioli, S.; Rietman, H.; Veltink, P.; Misra, S. Myoelectric forearm prostheses: State of the art from a user-centered perspective. *J. Prosthetics Orthot.* **2011**, *48*, 719. [CrossRef] [PubMed]
16. LeCun, Y.; Bengio, Y.; Hinton, G. Deep learning. *Nature* **2015**, *521*, 436–444. [CrossRef] [PubMed]
17. Atzori, M.; Cognolato, M.; Müller, H. Deep Learning with Convolutional Neural Networks Applied to Electromyography Data: A Resource for the Classification of Movements for Prosthetic Hands. *Front. Neurorobot.* **2016**, *10*, 9. [CrossRef]
18. Geng, W.D.; Du, Y.; Jin, W.G.; Wei, W.T.; Hu, Y.; Li, J.J. Gesture recognition by instantaneous surface EMG images. *Sci. Rep.* **2016**, *6*, 36571. [CrossRef]
19. Wei, W.T.; Wong, Y.K.; Du, Y.; Hu, Y.; Kankanhalli, M.; Geng, W.D. A multi-stream convolutional neural network for sEMG-based gesture recognition in muscle-computer interface. *Pattern Recogn. Lett.* **2019**, *119*, 131–138. [CrossRef]
20. Hu, Y.; Wong, Y.; Wei, W.; Du, Y.; Kankanhalli, M.; Geng, W.J.P. A novel attention-based hybrid CNN-RNN architecture for sEMG-based gesture recognition. *PLoS ONE* **2018**, *13*, e0206049. [CrossRef]
21. Cote-Allard, U.; Fall, C.L.; Campeau-Lecours, A.; Gosselin, C.; Laviolette, F.; Gosselin, B. Transfer Learning for sEMG Hand Gestures Recognition Using Convolutional Neural Networks. In Proceedings of the 2017 IEEE International Conference on Systems, Man, and Cybernetics (SMC), Banff, AB, Canada, 5–8 October 2017; pp. 1663–1668.
22. Cote-Allard, U.; Fall, C.L.; Drouin, A.; Campeau-Lecours, A.; Gosselin, C.; Glette, K.; Laviolette, F.; Gosselin, B. Deep learning for electromyographic hand gesture signal classification using transfer learning. *IEEE Trans. Neural Syst. Rehabil. Eng.* **2019**, *27*, 760–771. [CrossRef]
23. Wang, K.; Chen, Y.; Zhang, Y.; Yang, X.; Hu, C. Iterative Self-Training Based Domain Adaptation for Cross-User sEMG Gesture Recognition. *IEEE Trans. Neural Syst. Rehabil. Eng.* **2023**, *31*, 2974–2987. [CrossRef]
24. Islam, M.R.; Massicotte, D.; Massicotte, P.Y.; Zhu, W.P. Surface EMG-Based Inter-Session/Inter-Subject Gesture Recognition by Leveraging Lightweight All-ConvNet and Transfer Learning. *arXiv* **2024**, arXiv:2305.08014.
25. Nguyen, P.T.T.; Su, S.F.; Kuo, C.H. A Frequency-Based Attention Neural Network and Subject-Adaptive Transfer Learning for sEMG Hand Gesture Classification. *IEEE Robot. Autom. Lett.* **2024**, *9*, 7835–7842. [CrossRef]
26. ur Rehman, M.Z.; Waris, A.; Gilani, S.O.; Jochumsen, M.; Niazi, I.K.; Jamil, M.; Farina, D.; Kamavuako, E.N. Multiday EMG-Based Classification of Hand Motions with Deep Learning Techniques. *Sensors* **2018**, *18*, 2497. [CrossRef] [PubMed]
27. Ketykó, I.; Kovács, F.; Varga, K.Z. Domain Adaptation for sEMG-based Gesture Recognition with Recurrent Neural Networks. In Proceedings of the 2019 International Joint Conference on Neural Networks (IJCNN), Budapest, Hungary, 14–19 July 2019; pp. 1–7. [CrossRef]

28. Rodríguez-Tapia, B.; Soto, I.; Martínez, D.M.; Arballo, N.C. Myoelectric Interfaces and Related Applications: Current State of EMG Signal Processing–A Systematic Review. *IEEE Access* **2020**, *8*, 7792–7805. [CrossRef]
29. Chen, L.; Fu, J.; Wu, Y.; Li, H.; Zheng, B. Hand gesture recognition using compact CNN via surface electromyography signals. *Sensors* **2020**, *20*, 672. [CrossRef]
30. Ganin, Y.; Ustinova, E.; Ajakan, H.; Germain, P.; Larochelle, H.; Laviolette, F.; Lempitsky, V. Domain-Adversarial Training of Neural Networks (Version 4). *arXiv* **2015**, arXiv:1505.07818.
31. Balntas, V.; Riba, E.; Ponsa, D.; Mikolajczyk, K. Learning local feature descriptors with triplets and shallow convolutional neural networks. *Bmvc* **2016**, *1*, 3.
32. Atzori, M. Electromyography data for non-invasive naturally-controlled robotic hand prostheses. *Sci. Data* **2014**, *1*, 1–13. [CrossRef]
33. Makaram, N.; Karthick, P.A.; Swaminathan, R. Analysis of Dynamics of EMG Signal Variations in Fatiguing Contractions of Muscles Using Transition Network Approach. *IEEE Trans. Instrum. Meas.* **2021**, *70*, 4003608. [CrossRef]
34. Farrel, T.R.; Weir, R.F. The Optimal Controller Delay for Myoelectric Prostheses. *IEEE Trans. Neural Syst. Rehabil. Eng.* **2007**, *15*, 111–118. [CrossRef]
35. Merletti, R.; Cerone, G. Tutorial. Surface EMG detection, conditioning and pre-processing: Best practices. *J. Electromyogr. Kinesiol.* **2020**, *54*, 102440. [CrossRef] [PubMed]
36. Lehman, G.J.; McGill, S.M. The importance of normalization in the interpretation of surface electromyography: A proof of principle. *J. Manip. Physiol. Ther.* **1999**, *22*, 444–446. [CrossRef] [PubMed]
37. Lin, Y.; Palaniappan, R.; Wilde, P.D.; Li, L. A Normalisation Approach Improves the Performance of Inter-Subject sEMG-based Hand Gesture Recognition with a ConvNet. In Proceedings of the 2020 42nd Annual International Conference of the IEEE Engineering in Medicine & Biology Society (EMBC), Montreal, QC, Canada, 20–24 July 2020. [CrossRef]
38. Jain, R.; Garg, V.K. An Efficient Feature Extraction Technique and Novel Normalization Method to Improve EMG Signal Classification. In Proceedings of the 2022 3rd International Conference on Intelligent Engineering and Management (ICIEM), London, UK, 27–29 April 2022. [CrossRef]
39. Cram, J.; Kasman, G.; Holtz, J. Comparison of Normalization Methods for the Analysis of EMG Signals During Walking. *J. Electromyogr. Kinesiol.* **1998**, *8*, 13–19. [CrossRef]
40. Smith, L.H.; Hargrove, L.J.; Englehart, K.B.; Lock, B.A. Determining the Optimal Window Length for Pattern Recognition-Based Myoelectric Control: Balancing the Competing Effects of Classification Error and Controller Delay. *IEEE Trans. Neural Syst. Rehabil. Eng.* **2011**, *19*, 186–192. [CrossRef]
41. Jaramillo-Yánez, A.; Benalcázar, M.E.; Mena-Maldonado, E. Real-Time Hand Gesture Recognition Using Surface Electromyography and Machine Learning: A Systematic Literature Review. *Sensors* **2020**, *20*, 2467. [CrossRef]
42. Wahid, M.F.; Tafreshi, R.; Langari, R. A Multi-Window Majority Voting Strategy to Improve Hand Gesture Recognition Accuracies Using Electromyography Signal. *IEEE Trans. Neural Syst. Rehabil. Eng.* **2020**, *28*, 314–322. [CrossRef]
43. Tanaka, T.; Nambu, I.; Maruyama, Y.; Wada, Y. Sliding-Window Normalization to Improve the Performance of Machine-Learning Models for Real-Time Motion Prediction Using Electromyography. *Sensors* **2022**, *22*, 5005. [CrossRef]
44. Zhai, X.; Jelfs, B.; Chan, R.H.M.; Tin, C. Self-Recalibrating Surface EMG Pattern Recognition for Neuroprosthesis Control Based on Convolutional Neural Network. *Front. Neurosci.* **2017**, *11*, 379. [CrossRef] [PubMed]
45. Soroushmojdehi, R.; Javadzadeh, S.; Pedrocchi, A.; Gandolla, M. Transfer Learning in Hand Movement Intention Detection Based on Surface Electromyography Signals. *Front. Neurosci.* **2022**, *16*, 977328.
46. Lehmler, S.J.; ur Rehman, M.S.; Tobias, G.; Iossifidis, I. Deep transfer learning compared to subject-specific models for sEMG decoders. *J. Neural Eng.* **2022**, *19*, 056039. [CrossRef]
47. Zabihi, S.; Rahimian, E.; Asif, A.; Mohammadi, A. TraHGR: Transformer for Hand Gesture Recognition via Electromyography. *IEEE Trans. Neural Syst. Rehabil. Eng.* **2023**, *31*, 4211–4224. [CrossRef] [PubMed]
48. Qamar, H.G.M.; Qureshi, M.F.; Mushtaq, Z.; Zubariah, Z.; Rehman, M.Z.U.; Samee, N.A.; Mahmoud, N.F.; Gu, Y.H.; Al-Masni, M.A. EMG gesture signal analysis towards diagnosis of upper limb using dual-pathway convolutional neural network. *Math. Biosci. Eng.* **2024**, *21*, 5712–5734. [CrossRef] [PubMed]
49. Fan, J.; Jiang, M.; Lin, C.; Fiaidhi, J.; Ma, C.; Wu, W. Improving sEMG-based motion intention recognition for upper-limb amputees using transfer learning. *Neural Comput. Appl.* **2023**, *35*, 16101–16111. [CrossRef]
50. Lin, C.; Niu, X.; Zhang, J.; Fu, X. Improving Motion Intention Recognition for Trans-Radial Amputees Based on sEMG and Transfer Learning. *Appl. Sci.* **2023**, *13*, 11071. [CrossRef]
51. Atzori, M.; Gijsberts, A.; Castellini, C.; Caputo, B.; Hager, A.G.M.; Elsig, S.; Müller, H. Effect of clinical parameters on the control of myoelectric robotic prosthetic hands. *J. Rehabil. Res. Dev.* **2016**, *53*, 345–358. [CrossRef] [PubMed]
52. Ajiboye, A.B.; Weir, R.F. A heuristic fuzzy logic approach to EMG pattern recognition for multifunctional prosthesis control. *IEEE Trans. Neural Syst. Rehabil. Eng.* **2005**, *13*, 280–291. [CrossRef]
53. Versaci, M.; Angiulli, G.; Crucitti, P.; Carlo, D.D.; Laganà, F.; Pellicanò, D.; Palumbo, A. A Fuzzy Similarity-Based Approach to Classify Numerically Simulated and Experimentally Detected Carbon Fiber-Reinforced Polymer Plate Defects. *Sensors* **2022**, *22*, 4232. [CrossRef]

54. Benatti, S.; Montagna, F.; Kartsch, V.; Rahimi, A.; Rossi, D.; Benini, L. Online Learning and Classification of EMG-Based Gestures on a Parallel Ultra-Low Power Platform Using Hyperdimensional Computing. *IEEE Trans. Biomed. Circuits Syst.* **2019**, *13*, 516–528. [CrossRef]
55. Zanghieri, M.; Benatti, S.; Burrello, A.; Kartsch, V.; Conti, F.; Benini, L. Robust Real-Time Embedded EMG Recognition Framework Using Temporal Convolutional Networks on a Multicore IoT Processor. *IEEE Trans. Biomed. Circuits Syst.* **2020**, *14*, 244–256. [CrossRef]

sensors

MDPI

Article

Electromyographic Comparison of Traditional Fitness Machines, Outdoor Fitness Equipment Without Load Selectors, and Outdoor Fitness Equipment with Load Selectors in a Seated Chest Press Exercise in Trained Young Men

Tomás Abelleira-Lamela [1], Pablo Jorge Marcos-Pardo [2,3,*], J. Arturo Abraldes [4,*], Noelia González-Gálvez [1], Alejandro Espeso-García [1], Francisco Esparza-Ros [5] and Raquel Vaquero-Cristóbal [4]

[1] Facultad del Deporte, UCAM Universidad Católica de Murcia, 30107 Murcia, Spain; tabelleira@ucam.edu (T.A.-L.); ngonzalez@ucam.edu (N.G.-G.); aespeso@ucam.edu (A.E.-G.)
[2] SPORT Research Group (CTS-1024), CIBIS (Centro de Investigación para el Bienestar y la Inclusión Social), University of Almeria, 04120 Almeria, Spain
[3] Department of Education, Faculty of Education Sciences, University of Almeria, 04120 Almeria, Spain
[4] Research Group Movement Sciences and Sport (MS&SPORT), Department of Physical Activity and Sport Sciences, Faculty of Sport Sciences, University of Murcia, 30720 San Javier, Spain; raquel.vaquero@um.es
[5] Injury Prevention in Sport Research Group, International Chair of Kinanthropometry, UCAM Universidad Católica de Murcia, 30107 Murcia, Spain; fesparza@ucam.edu
* Correspondence: pjmarcos@ual.es (P.J.M.-P.); abraldes@um.es (J.A.A.); Tel.: +34-950-01-53-52 (P.J.M.-P.); +34-868-88-85-00 (J.A.A.)

Abstract: Outdoor fitness equipment (OFE) are strength training installations comparable to those found in indoor gyms but are located outdoors with greater accessibility. However, the scientific evidence supporting their effectiveness remains limited. The objective of this study was to analyze and compare the electromyographic (EMG) activity of upper limb muscle groups during the use of a traditional seated chest press (SCP) machine, a classic OFE SCP (OFE-SCP), and a new OFE-SCP featuring a load selector system (BIOFIT-SCP). The sample was composed of 34 active young men. EMG activity of five muscle groups was analyzed: the anterior deltoid (AD), the clavicular (CP) and sternal (SP) heads of the pectoralis major, and the lateral (LHTB) and long heads of the triceps brachii (LongHTB), under different intensities. The OFE-SCP showed significantly lower EMG activity compared with the SCP and BIOFIT-SCP in all muscles and phases ($p < 0.001$). Significantly lower EMG activity for the SP in all three phases was found on the BIOFIT-SCP compared with the SCP ($p < 0.001$), but it was significantly higher for the LHTB and AD in the BIOFIT-SCP compared with the SCP during the full and concentric phases ($p < 0.05$ to $p < 0.001$). In conclusion, training with the OFE-SCP generates less EMG activity than traditional machine training, while, in general, BIOFIT-SCP proved effective for strength training comparable to the SCP.

Keywords: outdoor gym; pectoralis major; resistance training; selectorized stack machine; surface EMG; upper limb muscle

1. Introduction

Strength training is becoming increasingly popular in the population thanks to the health benefits it brings to its practitioners [1,2]. Muscle mass gains and increased strength levels are the main objectives of most users who practice it [1,3]. Multiple training methods and tools can be used to train for strength, with guided machines being one of the most widely used [4].

However, in recent years, outdoor fitness equipment (OFE) has become popular as a more accessible alternative to guided machines in a fitness room [5,6]. These are known as sports equipment that are operated in a manner similar to traditional gym machines, although they are commonly located in public outdoor spaces such as parks [6–8]. Despite the popularity of OFE and the economic investment made for their installation and maintenance, there is some controversy regarding their effectiveness. In this line, while some studies indicate that training with OFE could improve body composition and strength production capacity in older adults [9], as well as improve physical functionality in seniors [10]; other studies indicate that training with this equipment may not be effective in adults and older adults, as a result of the reduced training intensity involved [11,12]. This is because traditional OFE are limited to working with one's own body weight [12], so they do not allow for a controlled intensity progression within training sessions. Therefore, new models are emerging that allow the user to regulate the external load through a system that can be used to increase the weight, similar to that of traditional fitness room guided machines [9,13].

In recent years, machines that allow for pushing-based exercises in a guided manner have increased in popularity and can be found in both traditional guided machine training facilities [14] and OFE [5]. This is because these types of machines have been seen as an alternative to the use of free weights and can provide the same results but with greater safety and less reliance on exercise technique control [14]. Their use is common for both aesthetic purposes and to strengthen the upper body musculature, which is essential for performing daily tasks such as lifting or pushing objects [15].

Along this line, several previous studies have analyzed the efficacy of the seated chest press (SCP) machine by performing a surface electromyographic analysis (EMG) of the activation of the upper limb musculature during its execution [4,16–18]. The choice of the SCP machine was made because it is one of the most popular guided machines for push-work in fitness [19]. These studies have shown that the SCP is an effective machine for activating the main muscle groups involved in these pushing exercises, such as the anterior deltoid, the pectoralis major, and the triceps brachii [16,17,20].

In turn, within most OFE machine lines, SCP machines (OFE-SCP) are also included. However, so far, no studies have been found that analyze the effectiveness of OFE-SCP machines in terms of muscle activity of the a priori involved musculature. Therefore, for practical purposes, it is not known whether this machine could be effective in achieving adaptations in upper limb muscles after training with it. There is, however, a previous study that compared the EMG between traditional leg machines and the OFE leg machine, finding that, in general terms, the OFE showed less activation of the muscles involved in the lower limbs, which could be because it only allows working with one's own weight [13].

Likewise, it has not been empirically proven whether the inclusion of an adjustable external weight in the OFE can improve the effectiveness of SCP machines. Only one previous study compared the EMG readings between a traditional guided leg machine and an OFE that allowed external weight adjustment, finding that at a similar intensity as a percentage of 1RM, a similar activation of the main lower limb musculature occurred, although the activation of the synergist musculature varied slightly because of the biomechanical differences between the exercises performed [13].

Therefore, the main objective of the present investigation was (a) to analyze and compare the EMG recorded in the upper limb muscle groups during the use of a SCP, OFE-SCP, and a push machine belonging to a new line of OFE that allows the external load to be regulated (BIOFIT-SCP); and (b) to determine the differences in the EMG of the upper limb muscle groups at 60 and 75% of 1RM in the SCP and BIOFIT-SCP machines. In relation to the hypothesis, it was proposed that: (a) the EMG of the upper limb musculature will be similar between the machines that allow the external load selection (SCP and BIOFIT-SCP), and activation will be greater in SCP and BIOFIT-SCP machines compared with a machine that only allows working with one's own body weight (OFE-SCP); and (b) the EMG recorded will be greater at 75% of 1RM than in 60% of 1RM in SCP and BIOFIT-SCP machines.

2. Materials and Methods

2.1. Design

This research consisted of a randomized cross-sectional study. The 1RM was calculated on the SCP (Figure 1a) and the new BIOFIT seated chest press (BIOFIT-SCP), which is an OFE that has an external weight selector (Figure 1b). The calculation of the 1RM from each of the machines was done in a different session (session one and session two). Furthermore, the EMG activity of five muscle groups (anterior deltoid (AD), clavicular (CP) and sternal (SP) heads of the pectoralis major, and the lateral (LHTB) and long head of the triceps brachii (LongHTB)) was analyzed on three different types of chest press machines, i.e., SCP, OFE-SCP, and BIOFIT-SCP in another session (session three). In the SCP and BIOFIT-SCP machines, an EMG activity analysis was performed at 60 and 75% of the 1RM, while in the OFE-SCP machine, the analysis was performed with the subject's own body weight. A period of 72 h was given between each of the evaluation sessions of each participant. The study utilized a specific research model, as illustrated in Table S1.

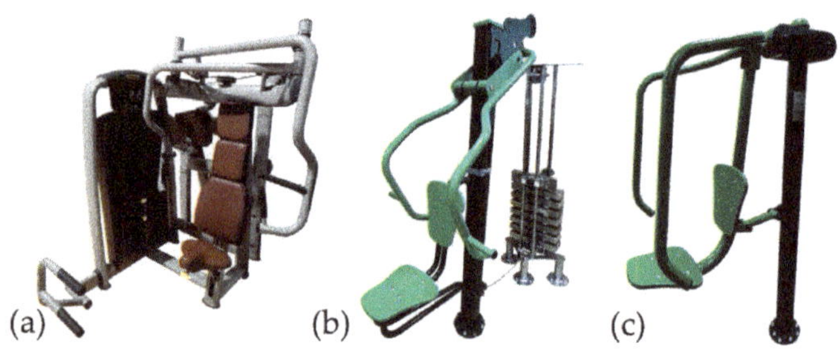

Figure 1. Fitness equipment used. (**a**) Seated chest press (SCP) (Technogym; Cesena, Italia); (**b**) outdoor fitness equipment seated chest press with selectorized system (OFE-SCP) (Entorno Urbano, Murcia, Spain); (**c**) outdoor fitness equipment seated chest press (BIOFIT-SCP) (Entorno Urbano, Murcia, Spain).

The machines selected for this research were: a machine commonly used in the fitness industry, the SCP model (Technogym; Cesena, Italy), with external load adjustment via a selectorized system [3] (Figure 1a); a machine belonging to a new line of OFE called BIOFIT-SCP (Entorno Urbano, Murcia, Spain), which allows the external load adjustment by means of a selectorized system [3], and which is patented (Patent registration for the OFE-SCP at the Spanish Patent and Trademark Office, application number: ES1296848Y; and patent registration for the external load selector, Spanish Patent and Trademark Office application number: P202231017) (Figure 1b); and a classic OFE machine line, corresponding with the same machine model, the Gemini (OFE-SCP) (Entorno Urbano, Murcia, Spain), which uses the user's body weight as the external weight [10] (Figure 1c).

This research was conducted in accordance with the principles outlined in the Declaration of Helsinki and received approval from the institutional ethics committee (Ethical Approval Reference: CE111908). Before starting the study, all volunteers were informed about the different types of tests and voluntarily signed an informed consent form to participate in the present investigation.

2.2. Participants

A methodology utilizing the standard deviation (SD) was employed to determine the sample size [21]. For this analysis, RStudio version 3.15.0 was utilized. The calculations for sample size and statistical power were conducted based on the standard deviation of the variable peak pectoralis major EMG activity during the use of a SCP selectorized plate machine (SD = 63.82 μV) demonstrated by a sample of physically active young adults (n = 15) [18]. An estimated error (d) of 32.30 μV, along with a 95% confidence interval (95%

CI), a statistical power of 95%, and a significance level of 0.05, indicated that a minimum of 34 participants was required for the study.

The study included a total of 34 male participants (mean age = 21.54 ± 1.80 years old; mean stretch stature = 176.62 ± 6.67 cm; mean body mass = 74.27 ± 9.99 kg). The inclusion criteria were: (1) being male, (2) a minimum of 2 years of experience using gym machines for training, along with currently engaging in strength training with these machines at least three times per week, and (3) being within the age range of 19 to 24 years old. The exclusion criteria were: (1) engaging in physical activity within 48 h before the initial session or between assessment sessions [19], (2) any recent injuries to the upper limbs or trunk within the previous 6 months [13], and (3) present a diagnosed muscle abnormalities or joint conditions [13].

2.3. *Measurements*

2.3.1. Questionnaire

All the participants were asked to complete an ad hoc questionnaire on their socio-demographic characteristics, their experience in strength training, the use of guided machines, and possible injuries and pathologies suffered. This questionnaire was created based on previous studies [13].

2.3.2. Anthropometric Measurements

For the measurement of the anthropometric variables stretch stature and body mass, the indications from the International Society for the Advancement of Kinanthropometry (ISAK) [22] were followed by an accredited ISAK level-2 anthropometrist. An HR001 portable stadiometer (Tanita, Arlington Heights, IL, USA) and a Tanita BC-545N scale (Tanita, Arlington Heights, IL, USA) were used. The variables were measured twice or three times if the difference between the first two was greater than 1%, with the final value being the mean or median, respectively.

2.3.3. Assessment of 1RM

The 1RM was calculated in the SCP and BIOFIT-SCP machines. The OFE-SCP was discarded for this test because it does not have any external weight control system [12]. The order of use of the SCP and BIOFIT-SCP machines to determine the 1RM in each machine was randomized. The protocol recommended by the National Strength and Conditioning Association [23] was utilized for the calculation of 1RM in both machines. The participants performed a warm-up, comprised of a set with an intensity that allowed them to complete between 5 and 10 repetitions. Following a 1 min rest period, the intensity was increased by 10 to 20%, at which point they were required to perform at least three but no more than five repetitions. Following a 2 min rest period, the intensity was increased again by 10 to 20%. At this new intensity, participants were required to complete between two and three repetitions. After this, the intensity was adjusted upward by 10 to 20% for a single repetition attempt. Depending on whether the lift was successfully completed, the load was either increased or decreased by 5 to 10%, repeating this process until the 1RM was identified. A maximum of five attempts at varying intensities were performed, with four minutes between each attempt [23]. A total of 72 h later, the protocol was repeated to determine the 1RM on the other machine, following the prior randomization.

2.3.4. Electromyographic (EMG) Analysis

Seventy-two hours after the second 1RM calculation session, the EMG signal was recorded in AD, CP, SP, LHTB, and LongHTB. The subjects performed, in a randomized order, five repetitions in the SCP machine with intensities of 60 and 75% of 1RM, five repetitions in the BIOFIT-SCP machine with intensities of 60 and 75% of 1RM, and five repetitions in the OFE-SCP machine with their own weight. This protocol followed the procedure of previous studies [13,24]. These intensities were selected for the SCP and

BIOFIT-SCP machines as they are commonly used in different guided machines training methods [13,17].

The electrodes were arranged according to the recommendations from the Surface Electromyography for the Non-invasive Assessment of Muscles (SENIAM) project [25]. The areas where they were placed were shaved and cleaned with 96% alcohol and sterile gauze to avoid possible erroneous signals and reduced skin impedance [26]. The gel electrodes used were placed at a 2 cm center-to-center distance in the longitudinal direction of the muscle fibers [25]. Following the methodology from previous studies, bilateral symmetry was assumed during exercise execution, so all electrodes were located on the right side of the body in a standardized manner [27]. The parts recorded were CP, SP, LHTB, LongHTB, and AD (Figure 2). EMG activity was recorded during exercise using the MuscleLab surface electromyography system (Ergotest Innovation AS, Stathelle, Norway) at a sampling rate of 1500 Hz.

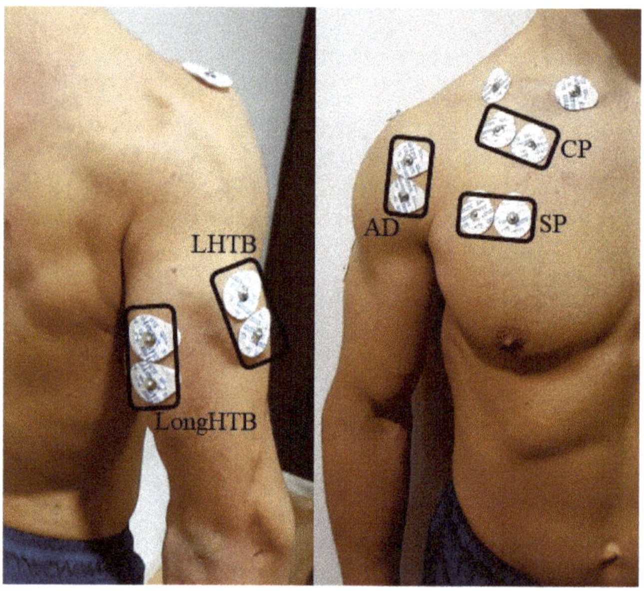

Figure 2. Electrode location on the right arm. AD = anterior deltoid; CP = clavicular pectoralis; LHTB = lateral head of the triceps brachii; LongHTB = long head of the triceps brachii; SP = sternal pectoralis.

2.4. Randomization and Blinding

The sequence of machine usage for the 1RM assessment, the order of machine use for EMG analysis, and the initial 1RM intensity for each machine during EMG analysis were all randomized following the methodology of previous studies [28]. The procedure was conducted by the principal investigator in the presence of external researchers, utilizing a computer-generated random number table. All measurements were carried out following the randomization process.

The researchers who performed the EMG measurements were not involved in the 1RM calculation tests. In the measurements of EMG, the two investigators involved in the data collection were also blinded transversely, where one controlled the exercise execution technique and provided the guidelines for the development of the protocol, while the other focused exclusively on verifying the accuracy of the EMG data recording. This procedure was carried out following previous research [13].

2.5. Procedure

Figure 3 shows the procedures carried out during the three sessions in schematic form, following previous studies [13]. In session one, the volunteers filled out the ad hoc questionnaire, followed by the anthropometric measurements. Subsequently, the essential guidelines for proper exercise execution on each of the three machines (SCP, OFE-SCP, and BIOFIT-SCP) were provided to ensure they were familiar with the required execution technique for each exercise to facilitate accurate data collection. Specifically, they were informed that exercises on the various machines should be performed using a full range of motion., from 90° elbow flexion to full extension. On the SCP machine, the seat was placed at height number 3 so that the height of the seat to the grips was the same as on the BIOFIT-SCP machine.

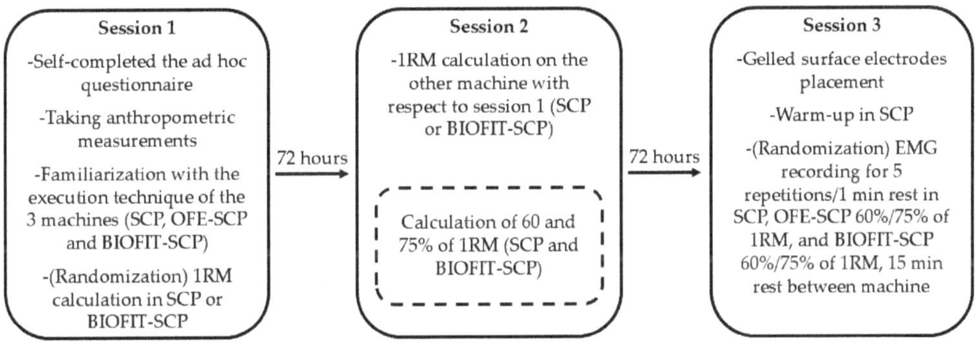

Figure 3. Flow diagram.

Subsequently, participants carried out the 1RM calculation procedure using either the SCP or BIOFIT-SCP machine, with the order of execution assigned randomly. In the second session, the procedure was repeated to assess the 1RM in the other machine. Using the data collected from this procedure, the intensities corresponding to 60% and 75% of the 1RM were subsequently calculated for each participant on both the SCP and BIOFIT-SCP machines.

In the third session, following electrode placement, participants completed a warm-up that included 12 repetitions at 30% of 1RM, 10 repetitions at 50% of 1RM, 8 repetitions at 70% of 1RM, and 2 repetitions at 90% of 1 RM on the SCP, as this was the machine on which the participants had the greatest prior experience, similar to that reported in related studies [13,29]. EMG measurements were then initiated, with the assigned machine (SCP, OFE-SCP, or BIOFIT-SCP) selected randomly. For the SCP and BIOFIT-SCP machines, the intensity level (either 60% or 75% of 1RM) was also randomized. Participants were instructed to perform five repetitions on the initially assigned machine and intensity, maintaining a consistent speed throughout [26]. A 1 min rest was taken between each repetition. After a 15 min rest, the same protocol was repeated on the same machine but using the other intensity. Once the full protocol was completed on the first machine, participants rested for 15 min before beginning the protocol on the second machine [13,26]. In the case of OFE-SCP, a single set of five repetitions was carried out using the participant's body weight, following the same procedure, since the external load could not be modified [6,10]. Rest periods were established based on previous methodologies to prevent fatigue from impacting the results [13,30].

For data analysis, all repetitions that had not been executed correctly from a technical point of view were discarded [13]. The correct execution was considered to involve the following assumptions: (a) that the repetition was performed with a constant rhythm throughout the entire range of motion, without stopping, (b) that a full range of motion from 90° elbow flexion to full extension was involved, and (c) that the buttocks were in constant contact with the seat and the feet were resting on the floor during the entire execution [31].

The data analysis was conducted using the methodology outlined in a previous study [26]. The first and last repetitions were excluded, and from the remaining three, the repetition with the highest root mean square (RMS) for the primary muscle group involved in the movement, the SP, was chosen, provided it was executed correctly. Data for all the muscles involved were recorded [13]. The data were processed using MuscleLab v10.5.67 software (MuscleLab Ergotest Innovation AS, Norway) by converting the raw amplified EMG signals into RMS values [32]. The RMS variable was measured during the entire repetition, as well as separately for its concentric and eccentric phases, to enable comparisons between machines, consistent with previous studies [33]. A MuscleLab twin-axis electrogoniometer (Ergotest Innovation AS, Stathelle, Norway) was utilized as a movement control variable to distinguish between the concentric and eccentric phases during exercise. The device was positioned on the participant's left arm, near the elbow joint, on the outer side. The distal sensor was aligned with the forearm, while the fixed sensor was aligned with the upper arm [34]. Calibration was performed using a hand-held goniometer set at 90° of elbow flexion while the participant maintained a stable, relaxed position. The point of maximum elbow extension was verified in situ, allowing differentiation of the concentric phase as the data captured from the initial 90° flexion to full elbow extension and the eccentric phase as the data collected from maximum extension back to 90° flexion. The full phase encompassed the analysis of the entire range of motion, beginning and ending at 90° flexion [13].

All measurement sessions were spaced 72 h apart to prevent any potential fatigue effects [35]. All participants were scheduled to attend each session at 8:00 a.m. [13]. All measurements were conducted in the same room, under controlled conditions with a consistent temperature of 20 °C and a stable humidity level of 60%.

2.6. Statistical Analysis

Data analysis was conducted using SPSS software (Version 25; IBM Corporation, Armonk, NY, USA). The Kolmogorov–Smirnov test and Mauchly's W-test were applied to assess data normality. Mean values and standard deviations were computed. To compare muscle activation across different machines and intensities, a repeated-measures ANOVA with Bonferroni correction was conducted. To control for Type I error, Bonferroni's correction was applied, setting statistical significance at $p = 0.005$. A threshold of $p < 0.05$ was defined for acceptable error.

3. Results

Table 1 shows the 1RM strength test results for the traditional SCP and BIOFIT-SCP machines.

Table 1. Results of the 1RM strength test.

	1RM	60% 1RM	75% 1RM
SCP (kg)	88.02 ± 20.72	57.21 ± 12.43	66.02 ± 15.54
BIOFIT-SCP (kg)	44.51 ± 15.06	26.71 ± 9.04	33.38 ± 11.30

1RM = one-repetition maximum; BIOFIT-SCP = BIOFIT seated chest press; SCP = traditional seated chest press.

Table 2 shows the mean and standard deviation of the root mean square recorded in each of the muscles analyzed when using the SCP at 60 and 75% of 1RM, the BIOFIT-SCP at 60 and 75% of 1RM, and the OFE-SCP with their own weight. The muscle group that showed a greater RMS was the AD during the concentric phase and the LHTB during the eccentric phase, except for the OFE-SCP, in which the AD muscle group showed a higher activation in both phases. With respect to the complete phase, the highest activations were observed in both the AD and LHTB, with the most notable being those recorded in the BIOFIT-SCP. Regarding the pectoralis, it was observed that CP presented a greater activation than SP in the two types of OFE (BIOFIT-SCP and OFE-SCP), while in the SCP, the greatest activation was presented by SP. In the triceps, LHTB presented the highest activation in the three machines and during the three phases analyzed compared with the LongHTB.

Table 2. Root mean square recording of the electromyographic signal in the different machines with different 1RM intensities.

	SP (M ± SD)	CP (M ± SD)	LongHTB (M ± SD)	LHTB (M ± SD)	AD (M ± SD)
Complete					
SCP 60%	324.27 ± 173.47	294.47 ± 163.67	223.78 ± 138.17	377.78 ± 167.07	395.81 ± 181.21
SCP 75%	388.32 ± 195.96	348.72 ± 174.16	298.41 ± 153.99	483.80 ± 213.02	475.87 ± 216.16
BIOFIT-SCP 60%	170.39 ± 104.46	294.61 ± 135.61	240.59 ± 76.97	477.74 ± 209.51	491.45 ± 236.06
BIOFIT-SCP 75%	242.52 ± 140.62	345.74 ± 158.54	293.61 ± 92.29	562.39 ± 231.58	565.45 ± 256.92
OFE-SCP	92.13 ± 48.52	131.16 ± 52.86	118.06 ± 43.67	233.81 ± 97.10	282.89 ± 134.02
Concentric					
SCP 60%	405.24 ± 214.63	354.11 ± 174.17	276.03 ± 151.85	449.94 ± 203.93	492.81 ± 221.87
SCP 75%	464.00 ± 231.01	400.29 ± 185.58	357.95 ± 154.14	556.01 ± 241.40	570.43 ± 258.28
BIOFIT-SCP 60%	228.02 ± 137.01	373.41 ± 168.02	314.12 ± 96.52	606.76 ± 249.69	641.30 ± 290.93
BIOFIT-SCP 75%	310.05 ± 174.62	415.49 ± 179.38	362.90 ± 110.13	675.58 ± 269.82	709.09 ± 316.85
OFE-SCP	108.66 ± 58.08	150.04 ± 62.60	141.33 ± 65.01	277.29 ± 140.53	344.46 ± 156.46
Eccentric					
SCP 60%	218.49 ± 130.02	211.61 ± 149.48	160.86 ± 136.70	293.15 ± 152.20	277.87 ± 141.80
SCP 75%	279.58 ± 145.02	271.33 ± 166.85	220.94 ± 163.04	382.85 ± 183.09	355.95 ± 167.18
BIOFIT-SCP 60%	72.34 ± 53.03	171.94 ± 90.23	146.97 ± 63.24	294.86 ± 130.77	275.66 ± 173.07
BIOFIT-SCP 75%	110.50 ± 80.07	216.06 ± 115.29	190.81 ± 81.11	377.98 ± 148.77	340.04 ± 184.52
OFE-SCP	65.07 ± 30.76	103.21 ± 43.51	87.98 ± 35.07	177.75 ± 78.50	196.61 ± 113.77

AD = anterior deltoid; BIOFIT-SCP = BIOFIT seated chest press; CP = clavicular pectoralis; LHTB = lateral head of the triceps brachii; LongHTB = long head of the triceps brachii; OFE-SCP = classic OFE; SCP = seated chest press; SP = sternal pectoralis.

Tables 3–5 and Figure 4 show the differences between machines and between different intensities in the full execution, concentric phase, and eccentric phase, respectively. At the same percentage of 1RM (60% 1RM vs. 60% 1RM, or 75% 1RM vs. 75% 1RM), the SCP presented a significantly greater activation of the SP in the three phases of execution ($p < 0.001$) and of the CP in the eccentric phase during execution at 75% of 1RM ($p < 0.05$) with respect to the RMS recorded in the BIOFIT-SCP. In the other phases and intensities, the CP did not present significant differences ($p > 0.05$). On the other hand, both LHTB ($p < 0.01$) and AD ($p < 0.05$) showed significantly higher activations in the full and concentric phases during the use of the BIOFIT-SCP with respect to the SCP at similar intensities, with no differences in the eccentric phase ($p > 0.05$). No difference in LHTB was found between SCP and BIOFIT-SCP at equal intensity ($p > 0.05$).

Table 3. Differentiation of peak electromyographic activity between different machines and intensities of 1RM during full execution.

Complete Repetition		SCP 60%	SCP 75%	BIOFIT-SCP 60%	BIOFIT-SCP 75%	OFE-SCP
Sternal pectoralis	SCP 60%	-	−64.05 ± 7.33 †	153.88 ± 19.24 †	81.75 ± 19.47 **	232.14 ± 25.54 †
	SCP 75%		-	217.93 ± 21.28 †	145.80 ± 19.93 †	296.19 ± 29.60 †
	BIOFIT-SCP 60%			-	−72.13 ± 8.12 †	78.26 ± 15.25 †
	BIOFIT-SCP 75%				-	150.39 ± 21.45 †
	OFE-SCP					-
Clavicular pectoralis	SCP 60%	-	−54.25 ± 6.35 †	−0.14 ± 15.64	−51.26 ± 16.01 *	163.31 ± 25.98 †
	SCP 75%		-	54.11 ± 17.22 *	2.98 ± 16.46	217.55 ± 27.62 †
	BIOFIT-SCP 60%			-	−51.12 ± 7.24 †	214.57 ± 25.44 †
	BIOFIT-SCP 75%				-	214.57 ± 25.44 †
	OFE-SCP					-

Table 3. *Cont.*

Complete Repetition		SCP 60%	SCP 75%	BIOFIT-SCP 60%	BIOFIT-SCP 75%	OFE-SCP
LongHTB	SCP 60%	-	−74.63 ± 8.21 †	−16.82 ± 21.09	−69.83 ± 22.03 *	105.71 ± 21.92 †
	SCP 75%		-	57.81 ± 23.17	4.80 ± 23.21	180.34 ± 23.96 †
	BIOFIT-SCP 60%			-	−53.01 ± 5.43 †	122.53 ± 10.59 †
	BIOFIT-SCP 75%				-	175.54 ± 13.57 †
	OFE-SCP					-
LHTB	SCP 60%	-	−106.02 ± 12.90 †	−99.96 ± 17.90 †	−184.62 ± 22.30 †	143.97 ± 17.26 †
	SCP 75%		-	6.06 ± 17.69	−78.60 ± 17.92 **	249.99 ± 23.80 †
	BIOFIT-SCP 60%			-	−84.66 ± 9.72 †	243.92 ± 25.18 †
	BIOFIT-SCP 75%				-	328.58 ± 28.63 †
	OFE-SCP					-
Anterior deltoid	SCP 60%	-	−80.06 ± 13.14 †	−95.64 ± 25.48 **	−169.64 ± 27.18 †	112.92 ± 22.97 †
	SCP 75%		-	−15.58 ± 25.56	−89.58 ± 25.83 *	192.98 ± 28.10 †
	BIOFIT-SCP 60%			-	−74.01 ± 8.27 †	208.56 ± 32.78 †
	BIOFIT-SCP 75%				-	282.56 ± 36.24 †
	OFE-SCP					-

BIOFIT-SCP = BIOFIT seated chest press; LHTB = lateral head of the triceps brachii; LongHTB = long head of the triceps brachii; OFE-SCP = classic OFE; SCP = seated chest press; * = p value < 0.05; ** = p value < 0.01; † = p value < 0.001.

Table 4. Differentiation of peak electromyographic activity between different machines and 1RM intensities during the concentric phase.

Concentric Phase		SCP 60%	SCP 75%	BIOFIT-SCP 60%	BIOFIT-SCP 75%	OFE-SCP
Sternal pectoralis	SCP 60%	-	−58.76 ± 8.28 †	177.22 ± 23.83 †	95.19 ± 24.31 **	296.58 ± 32.02 †
	SCP 75%		-	235.98 ± 24.57 †	153.95 ± 23.43 †	355.34 ± 35.23 †
	BIOFIT-SCP 60%			-	−82.03 ± 9.70 †	119.36 ± 20.10 †
	BIOFIT-SCP 75%				-	201.39 ± 26.60 †
	OFE-SCP					-
Clavicular pectoralis	SCP 60%	-	−46.19 ± 8.71 †	−19.30 ± 17.99	−61.38 ± 18.61 *	204.06 ± 28.57 †
	SCP 75%		-	26.88 ± 20.15	−15.20 ± 18.40	250.25 ± 30.34 †
	BIOFIT-SCP 60%			-	−42.08 ± 10.12 **	223.37 ± 28.25 †
	BIOFIT-SCP 75%				-	265.45 ± 30.12 †
	OFE-SCP					-
LongHTB	SCP 60%	-	−81.92 ± 10.20 †	−38.09 ± 20.47	−86.88 ± 22.06 **	134.70 ± 22.52 †
	SCP 75%		-	43.82 ± 20.25	−4.96 ± 20.38	216.61 ± 21.79 †
	BIOFIT-SCP 60%			-	−48.78 ± 6.19 †	172.79 ± 13.90 †
	BIOFIT-SCP 75%				-	221.57 ± 16.35 †
	OFE-SCP					-
LHTB	SCP 60%	-	−106.07 ± 15.19 †	−156.82 ± 19.44 †	−225.63 ± 23.34 †	172.65 ± 19.97 †
	SCP 75%		-	−50.75 ± 19.10	−119.56 ± 17.23 †	278.72 ± 24.40 †
	BIOFIT-SCP 60%			-	−68.82 ± 11.79 †	329.47 ± 28.76 †
	BIOFIT-SCP 75%				-	398.29 ± 31.03 †
	OFE-SCP					-
Anterior deltoid	SCP 60%	-	−77.62 ± 15.64 †	−148.49 ± 29.17 †	−216.28 ± 31.89 †	148.35 ± 29.06 †
	SCP 75%		-	−70.87 ± 29.33	−138.66 ± 30.74 **	225.96 ± 32.72 †
	BIOFIT-SCP 60%			-	−67.79 ± 11.52 †	296.84 ± 39.90 †
	BIOFIT-SCP 75%				-	364.62 ± 43.89 †
	OFE-SCP					-

BIOFIT-SCP = BIOFIT seated chest press; LHTB = lateral head of the triceps brachii; LongHTB = long head of the triceps brachii; OFE-SCP = classic OFE; SCP = seated chest press; * = p value < 0.05; ** = p value < 0.01; † = p value < 0.001.

Table 5. Differentiation of peak electromyographic activity between different machines and 1RM intensities during the eccentric phase.

Eccentric Phase		SCP 60%	SCP 75%	BIOFIT-SCP 60%	BIOFIT-SCP 75%	OFE-SCP
Sternal pectoralis	SCP 60%	-	−61.09 ± 8.07 †	146.15 ± 19.42 †	107.99 ± 18.17 †	153.41 ± 20.74 †
	SCP 75%		-	207.24 ± 22.35 †	169.08 ± 20.31 †	214.50 ± 23.91 †
	BIOFIT-SCP 60%			-	−38.16 ± 7.55 †	7.26 ± 9.42
	BIOFIT-SCP 75%				-	45.43 ± 14.51 *
	OFE-SCP					-
Clavicular pectoralis	SCP 60%	-	−59.72 ± 5.72 †	39.68 ± 16.51	−4.45 ± 15.46	108.40 ± 22.12 †
	SCP 75%		-	99.39 ± 18.78 †	55.27 ± 16.78 *	168.12 ± 25.04 †
	BIOFIT-SCP 60%			-	−44.12 ± 8.30 †	68.73 ± 11.99 †
	BIOFIT-SCP 75%				-	112.85 ± 16.31 †
	OFE-SCP					-
LongHTB	SCP 60%	-	−60.07 ± 6.92 †	13.90 ± 22.49	−29.94 ± 23.21	72.89 ± 23.55 *
	SCP 75%		-	73.97 ± 26.27	30.13 ± 26.70	132.96 ± 28.10 †
	BIOFIT-SCP 60%			-	−43.84 ± 6.50 †	58.99 ± 10.72 †
	BIOFIT-SCP 75%				-	102.83 ± 12.82 †
	OFE-SCP					-
LHTB	SCP 60%	-	−89.70 ± 12.84 †	−1.71 ± 18.26	−84.83 ± 20.58 **	115.40 ± 20.94 †
	SCP 75%		-	87.99 ± 23.55 **	4.87 ± 24.39	205.11 ± 27.57 †
	BIOFIT-SCP 60%			-	−83.12 ± 10.43 †	117.11 ± 17.33 †
	BIOFIT-SCP 75%				-	200.23 ± 19.84 †
	OFE-SCP					-
Anterior deltoid	SCP 60%	-	−78.08 ± 14.19 †	2.21 ± 24.45	−62.17 ± 23.57	81.26 ± 21.50 **
	SCP 75%		-	80.30 ± 25.32 *	15.91 ± 23.76	159.35 ± 23.67 †
	BIOFIT-SCP 60%			-	−64.39 ± 8.76 †	79.05 ± 27.79
	BIOFIT-SCP 75%				-	143.44 ± 30.00 †
	OFE-SCP					-

BIOFIT-SCP = BIOFIT seated chest press; LHTB = lateral head of the triceps brachii; LongHTB = long head of the triceps brachii; OFE-SCP = classic OFE; SCP = seated chest press; * = p value < 0.05; ** = p value < 0.01; † = p value < 0.001.

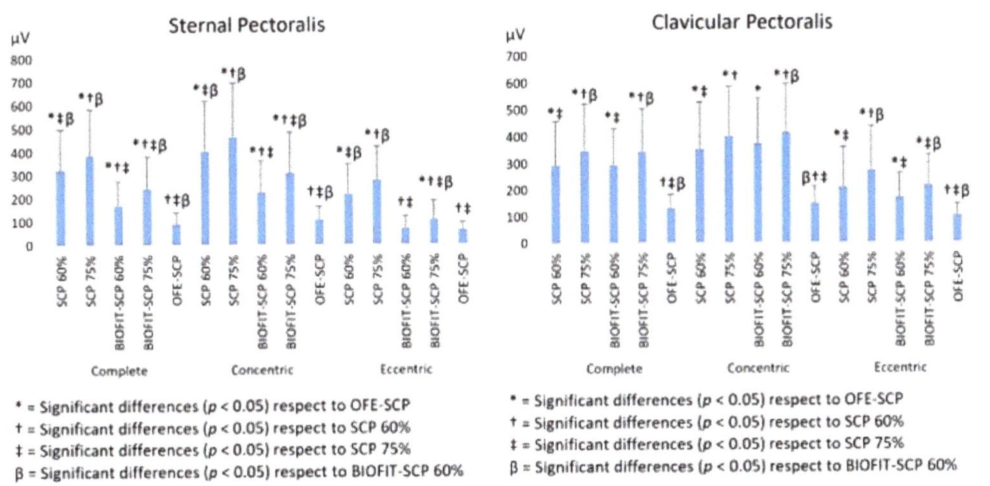

Figure 4. *Cont.*

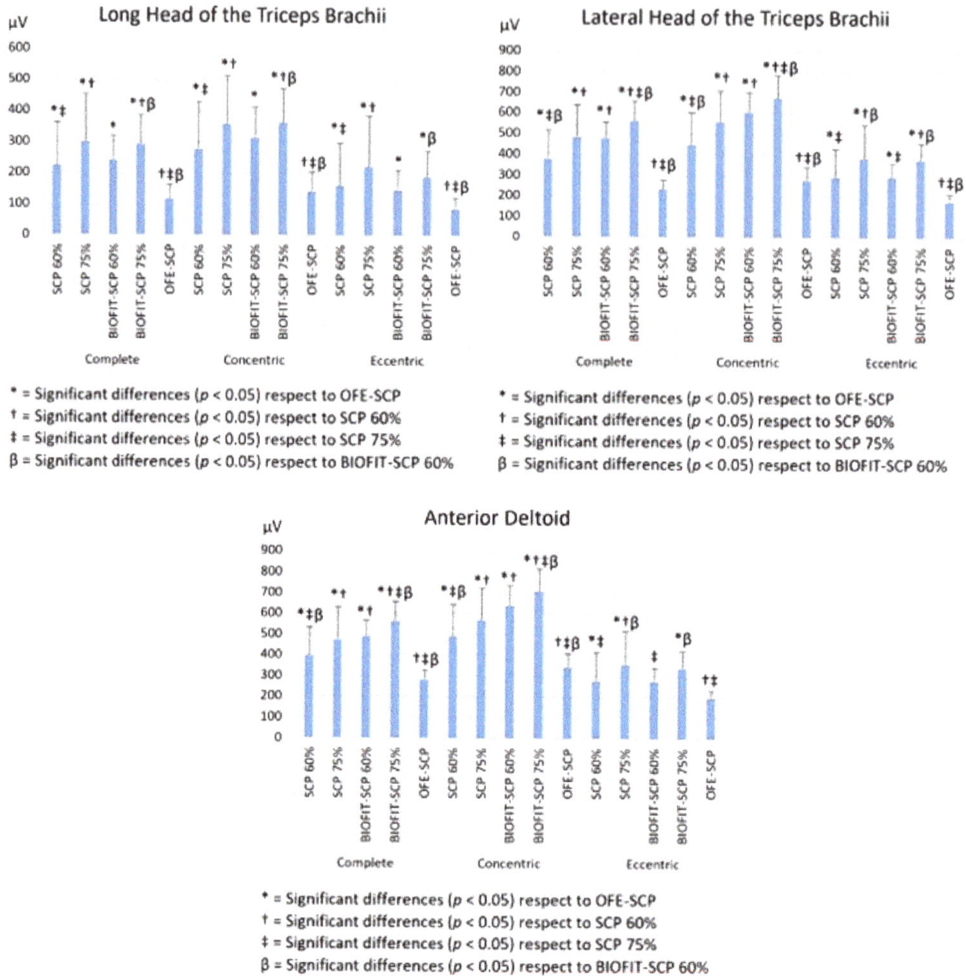

Figure 4. Differentiation of maximal electromyographic activity between different machines and 1RM intensities during the complete, concentric, and eccentric phases in different muscle groups.

When comparing OFE-SCP with SCP and OFE-SCP with BIOFIT-SCP, a significantly lower activation was found in OFE-SCP for all muscle groups in the full execution ($p < 0.001$), concentric ($p < 0.001$) and eccentric phases ($p < 0.05$), except for the SP and AD during the use of BIOFIT-SCP at 60% of 1RM ($p > 0.05$) during the eccentric phase. When comparing each machine (SCP and BIOFIT-SCP) with each of the two different intensities, significant differences were found in all muscle groups and during all the phases ($p < 0.01$).

4. Discussion

The first objective of this research was to analyze and compare the EMG activity of the muscle groups of the upper limbs produced when exercising with the SCP, OFE-SCP, and BIOFIT-SCP machines. Based on previous studies, the performance of exercises involving internal rotation with shoulder abduction and flexion, together with elbow extension, regardless of the biomechanical differences between exercises, involve the same major muscle groups, for example, the pectoral musculature, the AD, or the triceps brachii [20,36]. Other studies have analyzed EMG activity in exercises such as the pec deck, chest press, or bench press, finding that they all activate the same major muscle groups, despite their technical differences, as they are all upper limb press exercises [19,20,36]. Furthermore, in all the exercises performed on stable guided machines, the stabilizing musculature of the trunk may be less involved, as they rely less on the postural control, compared with exercises with more degrees of freedom of movement, as proven in previous studies that showed significant differences in rectus abdominis activation, being lower during the use of a plate machine compared with a cable-based machine [20]. Based on this, in the present investigation, the decision was made to perform a comparison of the EMG activity with different types of machines, of the muscle groups directly involved in the movement, and not so much of the muscles that could be involved as stabilizers.

A notable finding of this research was that in all three machines, the muscle group that showed a greater RMS was the AD during the concentric phase and the LHTB during the eccentric phase in general. These results coincide with those found in similar works, with the AD being the upper limb muscle that showed the greatest activation during the use of the standing cable press [18]. As for the activation of the LHTB, it could be due to the need for stabilization of the movement, as proven in previous works in which stronger signals were recorded in exercises with a freer movement with respect to chest press machines for this muscle group [20].

An outstanding result of the present investigation was that when comparing EMG activity between machines with the same percentage of intensity, more specifically, SCP and the BIOFIT-SCP, the SP presented a significantly greater EMG signal in all phases recorded in the SCP with respect to the BIOFIT-SCP at the same relative intensities; while the CP showed greater activation in the SCP with respect to the BIOFIT-SCP, but only in the eccentric phase at 75% of the 1RM. This may be due to the fact that the bundles of the SP and CP parts have a recruiting advantage during horizontal glenohumeral flexion and may be due to the better alignment of the muscle fibers with respect to the direction of movement [19,33] during the execution of the SCP, but that the fibers did not maintain the same linearity during the execution of the BIOFIT-SCP. Furthermore, in previous studies where a comparison of the EMG activity of each muscle between exercises and grips was performed, it was observed that a bench press with a grip at 50% of the biacromial distance compared with a grip of 150% of the biacromial distance, presented a lower activation of the SP [4]. Thus, the fibers of the CP and SP were able to collaborate to a greater extent in all phases in the SCP machine with respect to the BIOFIT-SCP machine due to the closest grip in the exercise performed on traditional machines [33]. Given these preliminary results, questions remain as to the importance of fixed or adaptive grip distance during the use of this type of machinery on EMG activation in upper limb pushing exercises.

Continuing with the comparison between the SCP and the BIOFIT-SCP, in the case of the elbow extensors, the LongHTB showed no differences between machines at equal intensity, while the LHTB showed greater activation in the BIOFIT-SCP in the full and concentric phases, both at 60% and 75%, with no differences in the eccentric phase. Studies have justified the participation of the triceps as a synergist muscle in upper limb pushing exercises [20] and thrust exercises that increased its activation by elevating the firing frequency through the inclusion of a backrest that improved the intervention of the main musculature because it reduced the dependence of the stabilizing muscle during the exercise [18]. However, in most studies, the triceps brachii was analyzed uniquely as a muscle group, but locating the electrodes in only one of the muscle bellies [4,16,18,20], with this being the first article to analyze on different machines, the activation of two triceps muscle bellies, the LHTB and the LongHTB. The differences found between both machines in the EMG activity of the triceps brachii musculature may be conditioned by the angulation of the shoulder joint because the BIOFIT-SCP involves a shoulder flexion to a greater extent, while the SCP an adduction flexion, which would condition the EMG involvement of the triceps based on previous studies [19]. Given these promising results, future research is needed to analyze the biomechanics of both machines in order to confirm the results of the present investigation.

Similarly, when comparing EMG activation between SCP and BIOFIT-SCP, the AD showed greater activation in the BIOFIT-SCP in the full and concentric phases, both at 60% and 75%, with no differences in the eccentric phase. This could also be a consequence of the differences in shoulder abduction between the two machines. It has been shown that the AD is involved, to a greater extent, in exercises where the main movement consists of shoulder flexion, compared with those where the main movement is horizontal shoulder adduction [19]. In that sense, the movement in the BIOFIT-SCP is based on a shoulder flexion and an elbow extension with a small shoulder internal rotation, while the movement in the SCP involves a shoulder internal rotation and elbow extension, together with a steady shoulder abduction. Previous results on this issue are contradictory. Thus, while in some studies, it has been found that when comparing different types of plate-loaded chest press and cable-based strength training machines, the AD did not show significant differences between machines [18]. In other studies, it was found that the AD increased its activation during a chest press exercise as the inclination of the bench changed, leading to an increase in shoulder flexion [19]. The differences between studies may be due to the dependence of the execution technique in each of the variants of the exercises because the contribution of a specific muscle will depend on factors such as the specific movement performed in the joint and the anatomical position of the muscle [19]. These promising results indicate the need for future research to analyze the biomechanics of both machines to validate the findings of this study.

Another relevant result of this study was that, when comparing the EMG activity of the BIOFIT-SCP and SCP machines to the OFE-SCP, those machines equipped with an external weight selector showed significantly higher EMG values across all muscle groups, irrespective of the intensity and phase, with only a few exceptions. However, OFE has gained significant popularity in recent years [8,10,12]. No prior studies have been identified that have examined EMG activity using this type of machine, leaving its effectiveness uncertain [37]. However, prior research has indicated that the choice of intensity influences muscle fiber recruitment, thereby affecting the recorded signal [38]. Given that the traditional OFE line only permits exercises using one's own body weight [12], the results of this study may be attributed to the fact that the participants' body weight did not correspond to an intensity near 60% or 75% of their 1RM when using the OFE-SCP. Thus, based on the results of this study, OFE-SCP may have limited effectiveness in generating significant EMG activity [37]. Thus, based on the findings of this study, OFE-SCP may have limited effectiveness in generating significant EMG activity [39], being able to provide only an adequate stimulus in populations with low fitness levels, such as older, deconditioned individuals [40].

Based on the results found in the present investigation, the first hypothesis of this research can be partially accepted, as we found higher EMG activation in SCP and BIOFIT-SCP, as compared to OFE-SCP. However, the second part of the hypothesis, which hypothesized that the EMG activation between the machines with a load selector would be similar (SCP vs. BIOFIT-SCP), has to be rejected, as differences were found in four of the five muscle groups analyzed.

The second aim of this study was to assess the differences in EMG activity of upper limb muscle groups at 60% and 75% of the 1RM using machines that permit intensity adjustment (SCP and BIOFIT-SCP). The results indicated a rise in EMG activity across all the muscle groups analyzed when exercising at 75% of the 1RM compared to 60% of the 1RM for both the full range of motion as well as during the concentric and eccentric phases in the SCP and BIOFIT-SCP. The results are consistent with previous studies, which also reported an increase in EMG activity as intensity increased [41]. This occurs because, at higher intensities, there is an increase in both the recruitment of motor units and their firing frequency [38]. These results support the acceptance of the second hypothesis, as the EMG activity recorded at 75% of 1RM was higher than at 60% of 1RM in both SCP and BIOFIT-SCP conditions.

The primary practical implication of this research is the introduction of a new type of OFE, accessible to both trainers and users, supported by scientific evidence demonstrating its effectiveness in recruiting target muscles similarly to conventional gym equipment. Furthermore, the effectiveness of the OFE was enhanced by implementing an external load selection system. This approach not only ensured adherence to the principle of progressive intensity but also improved the control of training intensity when using these machines.

The main strength of the present research is that it is one of the first studies that has analyzed EMG activity in OFE with a load selector. As studies were not found that compared similar models of indoor and outdoor machines for upper limb exercises, the present study is a pioneer in this aspect. It is also one of the first articles that has analyzed the EMG activation generated by classic OFE, which works with the user's own weight and whose shortcomings have been widely discussed. In addition, this work opens up new avenues of research by suggesting biomechanical and electromyographic analyses of the new patented systems and validities, offering opportunities to optimize their design and functionality in various contexts. In addition, it is one of the few works that includes both the eccentric and concentric phases, as well as the separate analysis of the sternal and clavicular parts of the pectoralis major muscle. In addition to the above, it is the only study that analyzed two of the triceps muscle bellies in guided machines.

Although this study offers a novel approach compared to prior research on the effectiveness of OFE, it is not free from limitations. Firstly, this study did not include an analysis of EMG results for antagonist muscles like the biceps brachii, latissimus dorsi, or other areas such as the lateral deltoid. The location of the electrodes in the back area makes measurement difficult because of the backrest of the machines. On the other hand, future work is encouraged to include an analysis of a larger number of muscle groups whenever possible. A second limitation was that the three machines presented slight variations in execution biomechanics due to the specific mechanics of each device, making direct comparisons between the exercises imprecise. However, it is important to note that studies with an electromyographic component often aim to highlight these biomechanical variations, providing valuable information on how different equipment or variations within the same exercise uniquely affect muscle activation. These comparisons are essential to understanding the nuances of muscle responses. The third limitation of this study was the exclusion of the maximum voluntary contraction test as a variable in the EMG analysis. This work compared the machines using the absolute values recorded. Including this variable could have provided insights into the percentage of activation of each muscle group relative to its maximum during each of the exercises assessed. In addition, it was not possible to calculate the 1RM in the OFE-SCP due to the impossibility of adjusting the load on the machine when working with one's own body weight and mobilizing it with levers. As a final limitation, although the study incorporated an electrogoniometer for elbow joint monitoring during electromyographic signal recording, it did not incorporate a goniometric analysis of the elbow and shoulder joints, which might have provided information on joint angles and their relationship to muscle activation patterns. In addition, it could have provided findings by linking specific joint mechanics to differences in muscle recruitment across machines, such as the difference in angular velocity between shoulder and elbow and whether this changes the data recorded in the interfering muscle groups.

5. Conclusions

It was observed that the EMG activity presented by the current OFE designed to strengthen the upper limb musculature (OFE-SCP) is significantly lower than that of the SCP, a machine commonly used in strength training in fitness centers for the pectoralis, with higher activation in triceps. In contrast, classic OFE, which only allows self-loading work, was found to be less effective than seated chest press machines with external load selectors. Machines with external weight selectors (SCP and BIOFIT-SCP) consistently showed higher EMG activations across all muscle groups compared with self-loading OFE, suggesting that these may be less effective in generating sufficient muscle activation for hypertrophy and strength adaptations, except for populations with lower fitness levels. Thus, the study supports the idea that OFE with external load selectors represent a valuable tool for strength training, as they show EMG activity levels closer to those of conventional gym equipment. This innovation could improve the applicability of OFE in progressive training programs by allowing controlled intensity adjustments. Additionally, in general terms, a higher intensity implied a greater activation of the involved musculature. This study is a pioneer in the analysis of EMG activation in OFE with external load selection, as it included a separate analysis of the pectoralis major parts and two triceps muscle heads. Future studies should address the limitations of this research, proposing as future lines of research the inclusion of the analysis of antagonist muscles, the goniometric analysis of elbow and shoulder joint angles, as well as taking into account the differentiation of biomechanics of execution between machines, and the maximal voluntary contraction test, in order to know what percentage of activation is generated in each muscle group and to refine the understanding of muscle activation patterns. In addition, it will be necessary to expand the sample to be analyzed, taking into account the female population as well as the older population in future research in order to achieve a greater transfer to the population.

In summary, OFE equipped with external load selectors has the potential to bridge the gap between traditional self-weight OFE and gym-based strength training equipment, offering a controlled approach to muscle activation and adaptation for outdoor fitness applications.

6. Patents

This study presents data on two patents registered by the Spanish Patent and Trademark Office: a new machine designed for the outdoor fitness equipment line (code: ES2975897) and an innovative external load selection system of the exercise for this machine (code: ES2975886).

Supplementary Materials: The following supporting information can be downloaded at: https://www.mdpi.com/article/10.3390/s24237740/s1, Table S1. Research model.

Author Contributions: Conceptualization, T.A.-L., P.J.M.-P., N.G.-G., A.E.-G. and R.V.-C.; methodology, T.A.-L., P.J.M.-P., N.G.-G., A.E.-G. and R.V.-C.; formal analysis, T.A.-L. and N.G.-G.; investigation, T.A.-L. and A.E.-G.; resources, F.E.-R. and R.V.-C.; data curation, T.A.-L., N.G.-G. and A.E.-G.; writing—original draft preparation, T.A.-L. and R.V.-C.; writing—review and editing, T.A.-L., P.J.M.-P., J.A.A., N.G.-G., A.E.-G., F.E.-R. and R.V.-C.; visualization, J.A.A.; supervision, R.V.-C.; project administration, P.J.M.-P. and N.G.-G.; funding acquisition, P.J.M.-P. All authors have read and agreed to the published version of the manuscript.

Funding: This project was funded by the Ministry of Science, Innovation and Universities of the Government of Spain in the call Retos-Colaboración 2017, under the title RTC-2017-6145-1, 2017. T.A.-L.'s participation in the present research is the result of a 2020–2021 Research Staff Training grant awarded by the UCAM Universidad Católica de Murcia.

Institutional Review Board Statement: This study was conducted in accordance with the Declaration of Helsinki and approved by the Institutional Ethics Committee of the Universidad Católica de Murcia (Ethical Application Ref: CE111908; 29 November 2019).

Informed Consent Statement: Informed consent was obtained from all subjects involved in the study.

Data Availability Statement: The data presented in this study are available on request from the corresponding author due to privacy.

Acknowledgments: We would like to thank the volunteers for their participation in the study. In addition, we would also like to thank the company Entorno Urbano S.L.U., Copele S.L.U., the Centro Tecnológico del Metal (CTM) of the Region of Murcia for providing us with the outdoor fitness equipment and managing transport and assembly of the equipment. T.A.-L. would like to thank the UCAM Universidad Católica de Murcia for its support through the FPI grant for the development of this research toward his doctoral thesis. This paper is part of the doctoral thesis of T.A.-L.

Conflicts of Interest: The authors declare no conflicts of interest.

References

1. Armstrong, R.; Baltzopoulos, V.; Langan-Evans, C.; Clark, D.; Jarvis, J.; Stewart, C.; O'Brien, T. An Investigation of Movement Dynamics and Muscle Activity during Traditional and Accentuated-Eccentric Squatting. *PLoS ONE* **2022**, *17*, e0276096. [CrossRef] [PubMed]
2. Martín-Fuentes, I.; Oliva-Lozano, J.M.; Muyor, J.M. Evaluation of the Lower Limb Muscles' Electromyographic Activity during the Leg Press Exercise and Its Variants: A Systematic Review. *Int. J. Environ. Res. Public Health* **2020**, *17*, 4626. [CrossRef] [PubMed]
3. Escamilla, R.F.; Fleisig, G.S.; Zheng, N.; Lander, J.E.; Barrentine, S.W.; Andrews, J.R.; Bergemann, B.W.; Moorman, C.T. Effects of Technique Variations on Knee Biomechanics during the Squat and Leg Press. *Med. Sci. Sports Exerc.* **2001**, *33*, 1552–1566. [CrossRef]
4. Muyor, J.; Rodríguez-Ridao, D.; Oliva-Lozano, J.M. Comparison of Muscle Activity between the Horizontal Bench Press and the Seated Chest Press Exercises Using Several Grips. *J. Hum. Kinet.* **2023**, *87*, 23–34. [CrossRef]
5. Chow, H.; Ho, C.H. Does the Use of Outdoor Fitness Equipment by Older Adults Qualify as Moderate to Vigorous Physical Activity? *PLoS ONE* **2018**, *13*, e0196507. [CrossRef] [PubMed]
6. Jansson, A.K.; Lubans, D.R.; Smith, J.J.; Duncan, M.J.; Haslam, R.; Plotnikoff, R.C. A Systematic Review of Outdoor Gym Use: Current Evidence and Future Directions. *J. Sci. Med. Sport* **2019**, *22*, 1335–1343. [CrossRef] [PubMed]
7. Abelleira-Lamela, T.; Vaquero-Cristóbal, R.; González-Gálvez, N.; Esparza-Ros, F.; Espeso-García, A.; Marcos-Pardo, P.J. Sagittal Spine Disposition and Pelvic Tilt during Outdoor Fitness Equipment Use and Their Associations with Kinanthropometry Proportions in Middle-Aged and Older Adults. *PeerJ* **2021**, *9*, e12657. [CrossRef]
8. Chow, H.; Wu, D.-R. Outdoor Fitness Equipment Usage Behaviors in Natural Settings. *Int. J. Environ. Res. Public Health* **2019**, *16*, 391. [CrossRef]
9. Marcos-Pardo, P.J.; Espeso-García, A.; Vaquero-Cristóbal, R.; Abelleira-Lamela, T.; González-Gálvez, N. The Effect of Resistance Training with Outdoor Fitness Equipment on the Body Composition, Physical Fitness, and Physical Health of Middle-Aged and Older Adults: A Randomized Controlled Trial. *Healthcare* **2024**, *12*, 726. [CrossRef]
10. Chow, H.; Chang, K.T.; Fang, I.Y. Evaluation of the Effectiveness of Outdoor Fitness Equipment Intervention in Achieving Fitness Goals for Seniors. *Int. J. Environ. Res. Public Health* **2021**, *18*, 12508. [CrossRef]
11. Chow, H.-W.; Mowen, A.; Wu, G. Who Is Using Outdoor Fitness Equipment and How? The Case of Xihu Park. *Int. J. Environ. Res. Public Health* **2017**, *14*, 448. [CrossRef] [PubMed]
12. Liu, Y.-C.; Yang, W.-W.; Fang, I.-Y.; Pan, H.L.-L.; Chen, W.-H.; Liu, C. Training Program with Outdoor Fitness Equipment in Parks Offers No Substantial Benefits for Functional Fitness in Active Seniors: A Randomized Controlled Trial. *J. Aging Phys. Act.* **2020**, *28*, 828–835. [CrossRef] [PubMed]
13. Abelleira-Lamela, T.; Marcos-Pardo, P.J.; Abraldes, J.A.; González-Gálvez, N.; Espeso-García, A.; Esparza-Ros, F.; Vaquero-Cristóbal, R. Comparative Electromyographic Analysis in Leg Press of Traditional Fitness Equipment, Traditional Outdoor Fitness Equipment, and a New Model of Outdoor Fitness Equipment in Trained Young Men. *Appl. Sci.* **2024**, *14*, 7390. [CrossRef]
14. Haugen, M.E.; Vårvik, F.T.; Larsen, S.; Haugen, A.S.; van den Tillaar, R.; Bjørnsen, T. Effect of Free-Weight vs. Machine-Based Strength Training on Maximal Strength, Hypertrophy and Jump Performance—A Systematic Review and Meta-Analysis. *BMC Sports Sci. Med. Rehabil.* **2023**, *15*, 103. [CrossRef]
15. Hik, F.; Ackland, D.C. The Moment Arms of the Muscles Spanning the Glenohumeral Joint: A Systematic Review. *J. Anat.* **2019**, *234*, 1–15. [CrossRef]
16. Coratella, G.; Tornatore, G.; Longo, S.; Esposito, F.; Cè, E. Specific Prime Movers' Excitation during Free-weight Bench Press Variations and Chest Press Machine in Competitive Bodybuilders. *Eur. J. Sport Sci.* **2019**, *20*, 571–579. [CrossRef]
17. López-Vivancos, A.; González-Gálvez, N.; Orquín-Castrillón, F.J.; Vale, R.G.d.S.; Marcos-Pardo, P.J. Electromyographic Activity of the Pectoralis Major Muscle during Traditional Bench Press and Other Variants of Pectoral Exercises: A Systematic Review and Meta-Analysis. *Appl. Sci.* **2023**, *13*, 5203. [CrossRef]
18. Signorile, J.F.; Rendos, N.K.; Heredia Vargas, H.H.; Alipio, T.C.; Regis, R.C.; Eltoukhy, M.M.; Nargund, R.S.; Romero, M.A. Differences in Muscle Activation and Kinematics Between Cable-Based and Selectorized Weight Training. *J. Strength Cond. Res.* **2017**, *31*, 313–322. [CrossRef]
19. Trebs, A.A.; Brandenburg, J.P.; Pitney, W.A. An Electromyography Analysis of 3 Muscles Surrounding the Shoulder Joint during the Performance of a Chest Press Exercise at Several Angles. *J. Strength Cond. Res.* **2010**, *24*, 1925–1930. [CrossRef]
20. Cacchio, A.; Don, R.; Ranavolo, A.; Guerra, E.; McCaw, S.T.; Procaccianti, R.; Camerota, F.; Frascarelli, M.; Santilli, V. Effects of 8-Week Strength Training with Two Models of Chest Press Machines on Muscular Activity Pattern and Strength. *J. Electromyogr. Kinesiol.* **2008**, *18*, 618–627. [CrossRef]
21. Bhalerao, S.; Kadam, P. Sample Size Calculation. *Int. J. Ayurveda Res.* **2010**, *1*, 55. [CrossRef] [PubMed]
22. Esparza-Ros, F.; Vaquero-Cristóbal, R.; Marfell-Jones, M.J. *International Standards for Anthropometric Assessment*; International Society for Advancement in Kinanthropometry: Potchefstroom, South Africa, 2019.
23. Miller, T.A. *NSCA's Guide to Tests and Assessments*; Human Kinetics: Champaign, IL, USA, 2012.
24. Duffey, M.J.; Challis, J.H. Fatigue Effects on Bar Kinematic during the Bench Press. *J. Strength Cond. Res.* **2007**, *21*, 556–560. [PubMed]
25. Stegeman, D.F.; Hermens, H.J. Standards for Surface Electromyography: The European Project "Surface EMG for Non-Invasive Assessment of Muscles (SENIAM)". *Enschede Roessingh Res. Dev.* **2007**, *10*, 8–12.

26. Da Silva, E.M.; Brentano, M.A.; Cadore, E.L.; De Almeida, A.P.V.; Kruel, L.F.M. Analysis of Muscle Activation During Different Leg Press Exercises at Submaximum Effort Levels. *J. Strength Cond. Res.* **2008**, *22*, 1059–1065. [CrossRef] [PubMed]
27. Snarr, R.L.; Esco, M.R. Electromyographic Comparison of Traditional and Suspension Push-Ups. *J. Hum. Kinet.* **2013**, *39*, 75–83. [CrossRef]
28. Schoenfeld, B.J.; Contreras, B.; Willardson, J.M.; Fontana, F.; Tiryaki-Sonmez, G. Muscle Activation during Low- versus High-Load Resistance Training in Well-Trained Men. *Eur. J. Appl. Physiol.* **2014**, *114*, 2491–2497. [CrossRef]
29. Andersen, V.; Fimland, M.S.; Mo, D.A.; Iversen, V.M.; Vederhus, T.; Rockland Hellebø, L.R.; Nordaune, K.I.; Saeterbakken, A.H. Electromyographic Comparison of Barbell Deadlift, Hex Bar Deadlift, and Hip Thrust Exercises: A Cross-over Study. *J. Strength Cond. Res.* **2018**, *32*, 587–593. [CrossRef]
30. Gonzalez, A.M.; Ghigiarelli, J.J.; Sell, K.M.; Shone, E.W.; Kelly, C.F.; Mangine, G.T. Muscle Activation during Resistance Exercise at 70% and 90% 1-Repetition Maximum in Resistance-Trained Men. *Muscle Nerve* **2017**, *56*, 505–509. [CrossRef]
31. Mausehund, L.; Werkhausen, A.; Bartsch, J.; Krosshaug, T. Understanding Bench Press Biomechanics—The Necessity of Measuring Lateral Barbell Forces. *J. Strength Cond. Res.* **2022**, *36*, 2685–2695. [CrossRef]
32. Van den Tillaar, R.; Andersen, V.; Saeterbakken, A.H. Comparison of Muscle Activation and Kinematics during Free-Weight Back Squats with Different Loads. *PLoS ONE* **2019**, *14*, e0217044. [CrossRef]
33. Albarello, J.C.d.S.; Cabral, H.V.; Leitão, B.F.M.; Halmenschlager, G.H.; Lulic-Kuryllo, T.; da Matta, T.T. Non-Uniform Excitation of Pectoralis Major Induced by Changes in Bench Press Inclination Leads to Uneven Variations in the Cross-Sectional Area Measured by Panoramic Ultrasonography. *J. Electromyogr. Kinesiol.* **2022**, *67*, 102722. [CrossRef] [PubMed]
34. Akima, H.; Maeda, H.; Koike, T.; Ishida, K. Effect of Elbow Joint Angles on Electromyographic Activity versus Force Relationships of Synergistic Muscles of the Triceps Brachii. *PLoS ONE* **2021**, *16*, e0252644. [CrossRef] [PubMed]
35. Picerno, P.; Iannetta, D.; Comotto, S.; Donati, M.; Pecoraro, F.; Zok, M.; Tollis, G.; Figura, M.; Varalda, C.; Di Muzio, D.; et al. 1RM Prediction: A Novel Methodology Based on the Force–Velocity and Load–Velocity Relationships. *Eur. J. Appl. Physiol.* **2016**, *116*, 2035–2043. [CrossRef] [PubMed]
36. Stastny, P.; Gołaś, A.; Blazek, D.; Maszczyk, A.; Wilk, M.; Pietraszewski, P.; Petr, M.; Uhlir, P.; Zajac, A. A Systematic Review of Surface Electromyography Analyses of the Bench Press Movement Task. *PLoS ONE* **2017**, *12*, e0171632. [CrossRef]
37. Mehr, K. Surface Electromyography in Orthodontics—A Literature Review. *Med. Sci. Monit.* **2013**, *19*, 416–423. [CrossRef]
38. Cormie, P.; McGuigan, M.R.; Newton, R.U. Developing Maximal Neuromuscular Power Part 1-Biological Basis of Maximal Power Production. *Sports Med.* **2011**, *41*, 17–38. [CrossRef]
39. Schwanbeck, S.R.; Cornish, S.M.; Barss, T.; Chilibeck, P.D. Effects of Training with Free Weights Versus Machines on Muscle Mass, Strength, Free Testosterone, and Free Cortisol Levels. *J. Strength. Cond. Res.* **2020**, *34*, 1851–1859. [CrossRef]
40. Kim, D.-I.; Lee, D.H.; Hong, S.; Jo, S.; Won, Y.; Jeon, J.Y. Six Weeks of Combined Aerobic and Resistance Exercise Using Outdoor Exercise Machines Improves Fitness, Insulin Resistance, and Chemerin in the Korean Elderly: A Pilot Randomized Controlled Trial. *Arch. Gerontol. Geriatr.* **2018**, *75*, 59–64. [CrossRef]
41. Snyder, B.J.; Fry, W.R. Effect of Verbal Instruction on Muscle Activity Duringthe Bench Press Exercise. *J. Strength. Cond. Res.* **2012**, *26*, 2394–2400. [CrossRef]

sensors

MDPI

Article

Time-Normalization Approach for fNIRS Data During Tasks with High Variability in Duration

Anna Falivene [1,*,†], Charlotte Johnson [2,3,†], Katrijn Klingels [3], Pieter Meyns [3], Evi Verbecque [3], Ann Hallemans [2], Emilia Biffi [1], Caterina Piazza [1] and Alessandro Crippa [1]

1 Scientific Institute IRCCS E. Medea, 23842 Bosisio Parini, Italy; emilia.biffi@lanostrafamiglia.it (E.B.); caterina.piazza@lanostrafamiglia.it (C.P.); alessandro.crippa@lanostrafamiglia.it (A.C.)
2 Research Group MOVANT, Department of Rehabilitation Sciences and Physiotherapy (REVAKI), University of Antwerp, 2610 Wilrijk, Belgium; charlotte.johnson@uantwerpen.be (C.J.); ann.hallemans@uantwerpen.be (A.H.)
3 Research Centre (REVAL), Faculty of Rehabilitation Sciences and Physiotherapy, Hasselt University, 3590 Diepenbeek, Belgium; katrijn.klingels@uhasselt.be (K.K.); pieter.meyns@uhasselt.be (P.M.); evi.verbecque@uhasselt.be (E.V.)
* Correspondence: anna.falivene@lanostrafamiglia.it
† These authors contributed equally to this work.

Abstract: Functional near-infrared spectroscopy (fNIRS) is particularly suitable for measuring brain activity during motor tasks, due to its portability and good motion tolerance. In such cases, the trials' duration may vary depending on the experimental conditions or the participant's response, therefore a comparison of hemodynamic responses across repetitions cannot be properly performed. In this work, we present a MATLAB (R2023a) function (*TaskNorm.m*) developed for time-normalizing fNIRS data recorded during trials with different durations. It is based on a spline interpolation method that rescales the time -axis to the percentage of the trial with a fixed number of samples. This allows us to successively average across repetitions to obtain the mean hemodynamic responses and complete the standard data processing. The algorithm was tested on eight subjects (four with developmental coordination disorder, age: 9.78 ± 0.30 and four typically developing children, age: 9.02 ± 0.30) performing three different tasks. The results show that the *TaskNorm* function works as expected, allowing both a comparison and averaging of the data across multiple repetitions. The performance of the function is independent of the task or the pre-processing pipeline applied. The proposed function is publicly available and importable into the HomER3 package (v1.72.0), representing a further step in the ongoing standardization process of fNIRS data analysis.

Keywords: data time-normalization; functional near-infrared spectroscopy; spline interpolation; self-paced tasks; MATLAB

1. Introduction

In the last decades, the use of functional near-infrared spectroscopy (fNIRS) has seen an important increase as a tool for investigating functional brain activity with a wide range of applications in the field of neuroscience [1,2]. Functional NIRS is an optical neuroimaging technique that monitors hemodynamic changes within the brain through optical sensors placed on the surface of the scalp. In particular, light at different wavelengths (typically between 650 and 850 nm) is emitted by sources onto the surface of the head. Modifications in the optical absorption are then recorded by detectors to measure changes in blood oxygenation in terms of cortical oxyhemoglobin (ΔHbO_2), deoxyhemoglobin (ΔHbR), and total hemoglobin (ΔHbT) concentrations. During evoked activity in the cortex of the brain, the increase in the blood flow in the active region alters the concentrations of the oxygenated and deoxygenated hemoglobin in the brain, resulting in changes in the absorption of light detected by the optodes [3]. Therefore, an increase in HbO_2 and HbT and a corresponding decrease in HbR are expected in the activated areas [4,5].

Thus, fNIRS allows us to estimate brain activity indirectly based on hemodynamic changes in the brain, providing a valid alternative to functional magnetic resonance imaging (fMRI), which is considered the gold standard methodology for the assessment of cortical and subcortical activity [6]. Nonetheless, fMRI suffers from some important limitations related to its costs, sensitivity to movement artefacts, and the restricted range of motion in the scanner that affects both the tolerability of the technique and the types of tasks that can be performed [7].

Although each available neuroimaging technique (e.g., fMRI, electroencephalography-EEG, magnetoencephalography, and positron emission tomography) has advantages and disadvantages, when compared to others, fNIRS is low-cost, non-invasive, and portable, with a relatively good spatial/temporal resolution, and a good tolerance to motion [8]. Additionally, fNIRS is particularly suitable to monitor brain activity in clinical and pediatric populations, which traditionally experience more difficulties in undergoing fMRI [9]. Consequently, fNIRS has been classically exploited to assess task-related cortical activation during both cognitive (e.g., picture identification [10]; emotion recognition [11,12]; and memory tasks [13]), and motor activities [1] (e.g., walking [14,15]; balance control tasks [16,17] involving walking over obstacles [18,19], and stepping tasks [20]), as well as activities involving dual tasks (typically walking with a concomitant cognitive and/or motor task [21,22]).

Experimental fNIRS protocols commonly include a baseline period and tasks with fixed durations, but in some cases, the trials' duration may vary depending either on the experimental conditions or the participant's response. This circumstance occurs when subjects are asked to perform a specific number of repetitions of a given task as fast as possible or when subjects walk a given distance at their preferred speed (i.e., self-paced task). This is particularly true when a motor activity is involved [23], but it can also occur when the brain activity is recorded during real-life conditions, such as working activities or navigating through the environment [24,25], or even for brain–computer interfaces [26].

This variability may lead to statistical errors related to the aggregation of data that are not directly comparable [27]. Moreover, most of the commonly used methods for analysis are based on the event-locked or task-locked averaging of specific portions of the continuous fNIRS signal. This is inaccurate in the case of periods of interest with different time durations. In such cases, a time-normalization procedure is needed to match the duration of different segments and concurrently preserve the main features while reducing variability [27], so that results can be comparable between different trials and later across participants.

Temporal normalization of the signal has been previously implemented for EEG data using linear interpolation. In particular, in the case of motor activities, such as walking, the single-trial signals are time-warped so that the different epochs are aligned and analyzed as a percentage of the gait cycle [28–30].

As for fNIRS applications, the dynamic time warping (DTW) technique can be exploited in the case of variable latencies in task-induced activities to account for the temporal variation in the alignment between two signals that need to be averaged as an enhanced alternative to the most common method of point-by-point averaging, as in [31]. Furthermore, the DTW algorithm can also be applicable with time series of different lengths, with the goal of finding an optimal alignment between the two given sequences through local compressions and extensions of the temporal axes, with a minimal overall cost [32]. Specifically, the time-axis of one signal is warped so that the maximum coincidence is attained with the other [33]. However, in some cases, the over-stretching or over-compression [34] may not fully preserve the signal shape, leading to a loss of information.

To date, to the best of our knowledge, a time-scale normalization of the fNIRS signal that could address this issue is yet to be developed.

The aim of the present work was therefore to develop a MATLAB function (*TaskNorm.m*) that time-normalizes the fNIRS concentration data, recorded during trials with different durations, as a percentage of the trial by exploiting a spline interpolation method. This step, when applied at the end of the pre-processing pipeline, would allow to obtain fNIRS data of the same length and thus to compare the hemodynamic response function (HRF) of trials of the same task, where there is no a priori defined stimulus/task duration.

The developed function is publicly available and is importable into the HomER3 [9] package, which is one of the most used open-source MATLAB (The Mathworks, Inc., Natick, MA, USA) toolboxes for fNIRS data visualization and analysis. This facilitates the analysis of data recorded during trials with different durations by scientists with low programming skills and contributes to the ongoing standardization process of pipelines for fNIRS data analysis.

In order to test whether, when applying the developed algorithm, fNIRS concentrations were aligned and therefore comparable with each other, we specifically examined experimental fNIRS data that were collected during three different balance control tasks in children with developmental coordination disorder (DCD) and their typically developing (TD) peers. DCD is a neurodevelopmental condition characterized by a delay in acquiring motor skills and poorer execution of coordinated movement, with a significant impact on daily life [35]. DCD occurs in 5–6% of school-aged children, is typically present since early childhood, and cannot be explained by other conditions, such as neurological disorders, and intellectual or visual disabilities [35,36]. Children with DCD struggle with balance in up to 87% of the cases, although the manifestations of these difficulties are very heterogeneous [37]. The brain control mechanisms underlying these alterations are also poorly understood [37,38]. To this end, real-time brain imaging fNIRS recordings during movement can shed light on how the brain interacts with motor skills and balance control [39–41]. However, these signals are affected by variability due to the nature of the task being performed, so a further standardization step is required, which can be implemented using the function we proposed here.

In the following sections, detailed descriptions of the algorithm implemented in the *TaskNorm* function (Section 2.1) of the fNIRS data acquisition protocol performed on eight children during balance control tasks and of the signal pre-processing pipeline are provided (Section 2.2). The results concerning the performance and applicability of the function are reported in Section 3.

2. Materials and Methods

2.1. Algorithm Description

We implemented a MATLAB (R2023a) function *TaskNorm.m* that performs a time-normalization of the concentration data for each repetition of the task, whose duration is converted into a percentage (0–100%) of the trial duration.

The main steps of the *TaskNorm* function, and its integration in the processing pipeline, are reported as a flowchart in Figure 1.

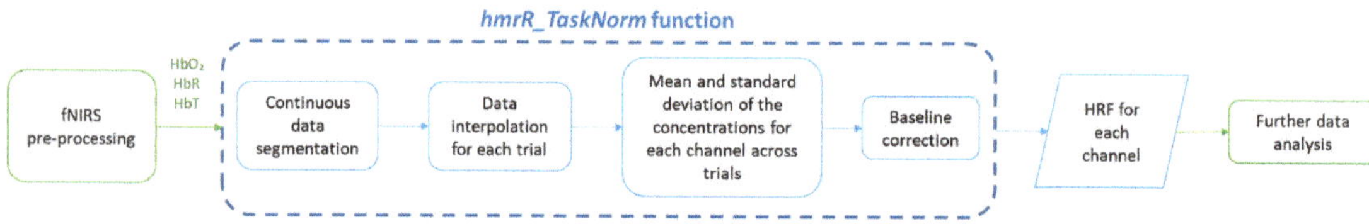

Figure 1. The flowchart of the processing pipeline. The HbO_2, HbR, and HbT data are given as an input to the *TaskNorm* function, of which each step is highlighted. The output of the function is the time-normalized signals and the HRF for each channel. Abbreviations: fNIRS = functional near-infrared spectroscopy; HbO_2 = oxyhemoglobin concentration; HbR = deoxyhemoglobin concentration; HbT = total hemoglobin concentration; and HRF = hemodynamic response function.

The main inputs of the function are the pre-processed concentration signal recorded during the selected experimental task, along with the triggers that delimit each repetition and its duration (stored directly in the raw .snirf files input in the pre-processing step); and the durations (expressed in seconds) of the baseline and the post-task periods, which are defined by the users in the two *deltat* parameters. The first *deltat* value accounts for the duration of the baseline period before the task onset, while the second value accounts for the duration of the post-task period (i.e., after the task offset). The users can choose the appropriate values of these parameters according to both their research question and experimental design. It is also possible not to consider the period after the end of the task (setting the post-task duration to 0 s), making the baseline period the common reference to measure the changes in the hemodynamic activity related to the task.

Once the *deltat* parameters are set, the concentration signal is segmented into several epochs corresponding to the task repetitions, which include the baseline, the effective task, and the post-task periods.

Afterwards, the time-normalization is performed by applying a spline interpolation to each data segment relative to a single task repetition in order to estimate the signal value in a desired and fixed number of points.

We developed two different versions of the time-normalization algorithm. In the first version, the interpolation is performed separately for the baseline, for the task, and for the post-task periods. This allows us to make the three phases distinguishable and to consider the exact onset/end of each task's repetition. The entire signal is then reassembled for further analysis. Specifically, the baseline and post-task periods are interpolated using a number of samples computed as in Equation (1):

$$\#sample_{i,j} = C \times \Delta t_{i,j} \quad (1)$$

where $\Delta t_{i,j}$ is the *deltat* parameter referring to the baseline period (i) or post-task period (j), and C is a constant value set arbitrarily to have an adequate number of points while at the same time avoiding overfitting.

The number of samples for the task period is calculated as in Equation (2):

$$\#sample = C \times m \quad (2)$$

where m is the mean duration of all the repetitions and C is a constant value set with the same criterion as above.

In the second version of the function, the whole task repetition is interpolated without distinguishing the phases within it, computing the number of samples as in Equation (2), with m being the duration of the three phases considered together. The final percentages related to the onset/end of the task were imposed by taking into account the corresponding values in each repetition, the average of which was selected.

The effect of varying the value of C, from $C = 10$ to $C = 100$, was tested by comparing the Euclidean distance between the position of the maximum and minimum peaks of the raw data with the position of the ones after the spline interpolation and the eventual reassembling phase for each repetition. This was used as an error measure of the function. The error becomes stable around $C = 40$, whereas a value of $C = 90$ or 100 would heavily increase the number of samples needed for the interpolation. For these reasons, we chose $C = 50$ as we think this could be an adequate trade-off. However, the function can be easily modified and this constant value can be set based on the experimental data of the specific research (e.g., sampling frequency of the signal).

Regardless of the selected version of the function, both the mean value across repetitions and the standard deviation of HbO_2, HbR, and HbT concentrations of each channel are then computed to obtain an average HRF for each channel. Next, a baseline correction is applied to the resulting HRF: the mean value of the HRF during only the baseline period is subtracted from the entire HRF to normalize the response, obtaining on average a value of 0 μM for the HbO_2, HbR, and HbT concentrations during the baseline period.

Furthermore, the developed function also allows the baseline-corrected mean HRF and its standard deviation to be plotted in MATLAB, for both HbO_2 and HbR concentrations for each channel, and the corresponding concentrations for each repetition (this can be done by setting the input parameter *show* equal to 1; see Supplementary Figure S1).

The function was made compatible with the HomER3 toolbox [9]. Therefore, it is possible to import the function in the HomER3 processing stream, run the stream all at once, and visualize the hemodynamic response function (HRF) outcome directly in the Main GUI, as well as to export the outcome for further analysis outside the HomER toolbox.

2.2. Testing of the Algorithm

2.2.1. Participants

The *TaskNorm* function was tested using the data acquired from eight children (age range 8–10 years old), i.e., four children with DCD (mean age: 9.78 ± 0.30, 3 males) and four typically developing children (mean age: 9.02 ± 0.30, 2 males). Prior to inclusion, all children and parents were informed about the methodology and duration of this study. Parents, on behalf of their children, signed official informed consent to participate in the study. This study was conducted according to the guidelines of the Declaration of Helsinki and approved by the Committee for Medical Ethics UZA-UAntwerp (B300201941833).

2.2.2. Experimental Protocol

All participants performed six different balance tasks in a standardized order, while undergoing fNIRS and electromyography (EMG). For the purpose of this study, EMG data were only exploited to identify the start and the ending of the proposed task by examining muscle activity. The experimental paradigm is presented in Figure 2. The fNIRS and EMG data-recording procedure lasted for approximately 30 min. The proposed tasks were chosen from a physiotherapeutic balance test: the Balance Evaluation Systems Test for Children (Kids-BESTest) [42]. Among the six tasks (described in Table 1), three were selected to test the algorithm. Specifically, we first excluded tasks with only one repetition, as the algorithm requires a minimum of two repetitions to be applied. Among the remaining tasks, we selected one task per section, as presented in the Kids-BESTest: for the 'Stability in gait' section, we selected the walking task, which is the most common and analyzed task in the literature; for the 'Anticipatory postural adjustment' section, the alternate stair touching task was chosen since it is a more comprehensive task with respect to the others; and for 'Stability limits', the leaning task was kept.

Table 1. Brief descriptions of the tasks proposed in the experimental protocol. Bold text indicates the tasks selected for the testing of the algorithm.

Kids-BEST Test Section	Task		Baseline/Rest Position		Repetitions	
	Name	Explanation	Description	Intra-Trial Mean Duration (SD) [s]	N	Mean Duration (SD) [s]
Anticipatory postural adjustment	**Alternate stair touching**	The child taps with their feet on a stool in front of them alternately with the left and right foot as fast and as controlled as possible.	Standing on two feet	12.70 (4.11)	5 times	7.81 (1.46)
Stability limits	**Leaning left and right while seated**	The child leans as far and as stable as possible sidewards while seated, without falling and keeping their feet on the ground. Arms are crossed at the chest.	Sitting	12.59 (12.16)	10 times	12.87 (2.25)
Stability in gait	**Walking**	The child walks 6 m over level ground as fluently as possible.	Standing on two feet	11.3 (6.91)	6 times	6.60 (1.30)
Anticipatory postural adjustment	Standing on one leg	The child stands on one leg for as long as possible. This exercise is then repeated with the other leg.	Standing on two feet	60.56 (22)	6 times	18.88 (11.21)
Reactive postural responses	In place response—backward	The therapist holds the child, who resists. When the therapist suddenly releases the child, he/she should keep balancewithout taking a step.	Standing on two feet	-	1 time	0.1 (0)
Reactive postural responses	Compensatory stepping correction—backward	The child leans beyond theirbackward limits against the therapist's hands. When the therapist suddenly releases the child, he/she should be able to avoid falling, perhaps even taking a step.	Standing on two feet	-	1 time	0.1 (0)

The fNIRS raw data collected during the three selected tasks are publicly available at https://doi.org/10.5281/zenodo.10124956, Zenodo.

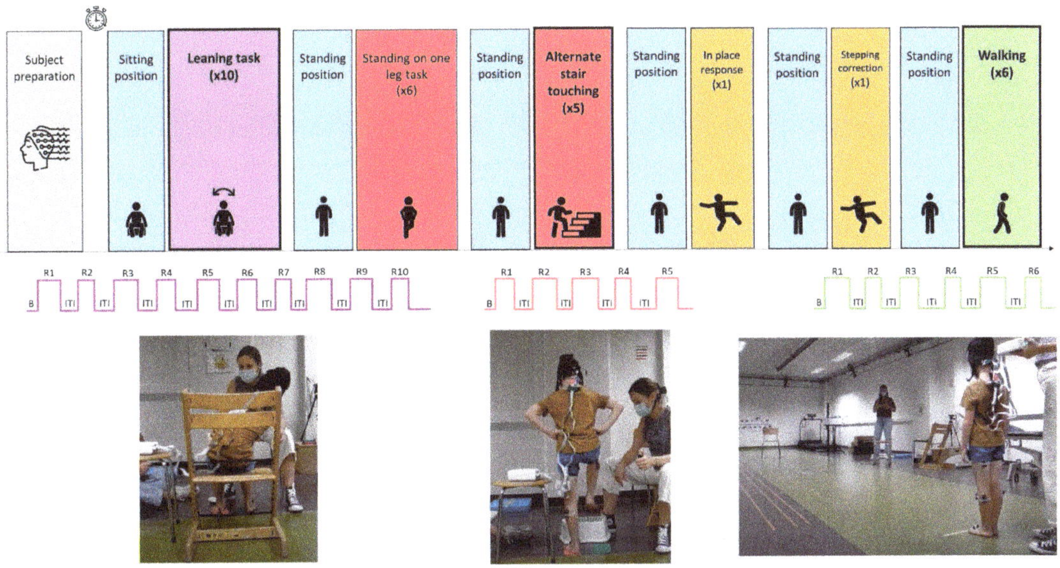

Figure 2. Complete experimental paradigm. For the tasks used for the testing procedure, the number of repetitions and inter-trial intervals and an example photo of a subject performing the task are provided. Abbreviations: B = Initial Baseline; R_i = Repetition *i*; and ITI = Inter-trial interval.

2.2.3. Signal Acquisition and Pre-Processing

The NIRSport2 device (NIRx Medical Technologies, Berlin, Germany) was used for recording fNIRS data with a sampling rate of 10.27 Hz, with two continuous wavelengths (760 nm and 850 nm), 8 sources, and 8 detectors (dual-tip). The probe configuration was generated using the fNIRS optodes locator decider (fOLD) open source toolbox (Zimeo Morais GA). Specifically, the most specific channels were chosen to represent each region of interest (ROI) (i.e., supplementary motor area—SMA/premotor cortex—PMC and inferior/superior parietal lobule—IPL/SPL brain areas), using an 8-8 optode bundle placed according to the 10-10 international system (Supplementary Figure S2). More precisely, sources were placed at FC3, FCz, FC4, C1, C2, Cp3, CPz, and CP4 positions; whereas detectors were located at FC1, FC2, C3, Cz, C4, CP1, and CP2 channel positions.

Data were acquired using Aurora fNIRS 2021.9 Acquisition Software for Windows (NIRx Medical Technologies, Berlin, Germany). Simultaneously, EMG was recorded using Delsys Trigno™ (Delsys Inc., Natick, USA). As an event marker, to identify start and stop of each trial, an analog input adapter connected to the EMG device (Delsys Inc., Natick, USA) was used.

The fNIRS signal pre-processing was performed using the HomER3 package (v1.72.0) [9] according to the workflow and parameters described in other studies [43–45]. Specifically, the raw light intensity was initially converted into optical density with the *hmrR_Intensity2OD* function.

A minimum of 0 and a maximum of 3 channels (Mean = 0.42 SD = 0.88), whose signals were too weak, too strong, or had a high standard deviation with respect to the thresholds defined in [44], were pruned with the *hmrR_PruneChannels* function (setting *dRange*(1) = 5×10^{-4}; *dRange*(2) = 1; *SNRtresh* = 10; and *SDrange* = [0, 45]) and then discarded from further analysis.

Motion artefacts correction was performed using a combination of a 5-s moving average filter and a discrete wavelet transform was applied to every channel data series exploiting the *hmrR_MovingAverage* and *hmrR_MotionCorrectWavelet* functions (setting the *interquartile range* to 0.1, as in [46]), and as performed in [43]. The remaining motion artefacts were then detected using the *hmrR_MotionArtifactByChannel* function (setting *tMotion* = 1; *tMask* = 1; *STDEVthresh* = 50.0; and *AMPthresh* = 0.60).

Moreover, a band-pass filter (*hpf* = 0.01; *lpf* = 0.2, cut-off frequencies expressed in Hz) was used to remove instrumental and physiological noises with the *hmrR_BandpassFilt* function.

The optical density data were converted to hemoglobin concentrations (HbO_2, HbR, and HbT) applying the Modified Beer–Lambert Law (MBLL), which is implemented in the *hmrR_OD2Conc* function (*ppf* = [1.8, 1.8]).

Finally, the *hmrR_TaskNorm* function (*deltat* = [2, 3]) was used to obtain the average HRF for each analyzed task. We set the duration of the baseline to 2 s, which can be considered an adequate period of time to obtain an accurate average of the baseline signal, while the post-task duration was imposed at 3. The outcome of the *hmrR_TaskNorm* function can be used for further analyses (e.g., compare HRF of different channels or groups of subjects, as done in Section 3.2).

2.2.4. Statistical Analysis

Both versions of the algorithm were applied to the three tasks to test their working principle. A quantitative validation of the function performance in preserving the shape of the raw signal was performed by comparing with non-parametric paired tests the amplitude and the time of occurrence of maximum and minimum peaks between the two signals.

In addition, for each subject and repetition, the onset and offset values were stored (see Supplementary Materials) and the difference between the outcomes of the two versions of the function was calculated to compare the error made by the second one.

Shapiro–Wilk normality test was performed to verify the distribution of the resulting data. We then investigated the differences between the performances of the two algorithms in terms of accuracy in identifying onset and offset percentages with respect to the entire task duration by means of a Wilcoxon signed-rank test, according to the normality test outcome. The One Sample t-test was applied to the error data to statistically compare the mean of the observed data to the assumed zero mean. Furthermore, the two codes were compared in terms of smoothness of signals with a Wilcoxon signed-rank test. The standard deviation of the first derivative of each task repetition was computed as an index of signal smoothness.

For all statistical tests, significance level was set at $p < 0.05$.

3. Results

3.1. Results of the TaskNorm.m Function

The example provided in Figure 3 shows how the time-scale normalization with the spline interpolation preserved the original shape of the data, while aligning the time scale. Statistical tests on pre-post interpolation differences at the time of occurrence of the characteristic points yielded no statistically significant differences ($p > 0.05$), whereas significant differences are present when considering the amplitude of such points ($p < 0.01$). However, the mean difference between the raw and interpolated signals of the amplitude of the maximum peaks and minimum peaks were $1{,}61 \times 10^{-4}$ µM and $1{,}62 \times 10^{-4}$ µM respectively, which we assume to be acceptable.

Time-normalization, thus, allowed for an effective comparison, and a subsequent averaging, across different repetitions of the tasks and consequently between subjects.

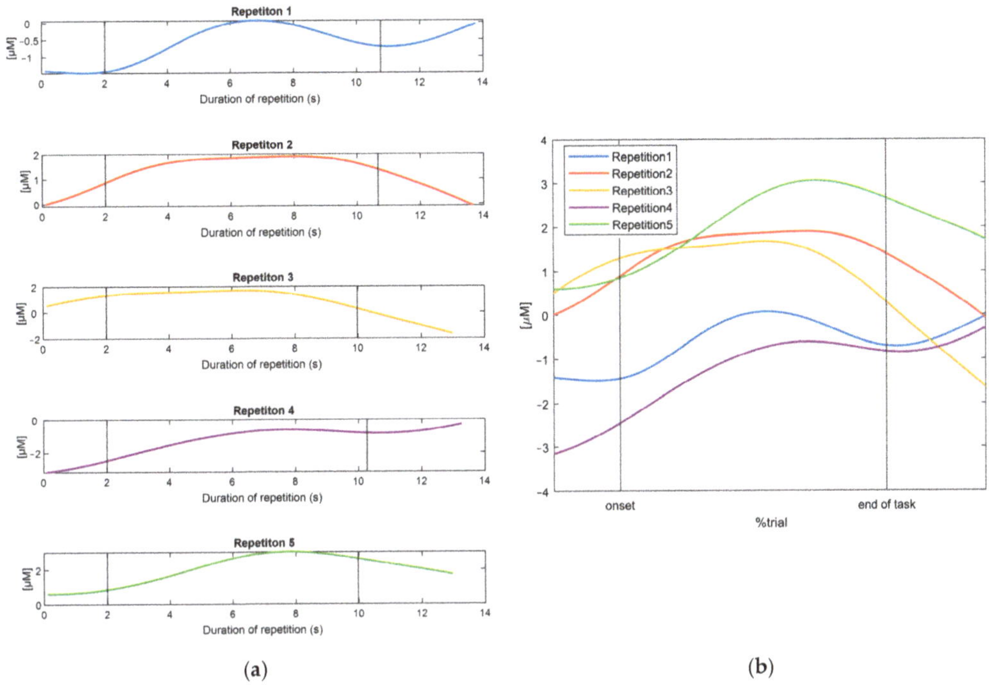

Figure 3. (**a**) The concentration data for each repetition before interpolation and (**b**) the concentration data following the time-normalization procedure. The figures are referring to the HbO_2 data recorded by one channel for one subject performing one task, as an example.

Table 2 displays the percentage of onsets and offsets in relation to task duration, along with the corresponding standard deviation, for each subject and task as obtained by the two different methods. For the sake of simplicity, the table shows only the mean values over the repetitions of each task, while the statistical tests described in Section 2.2.4 were performed on the data presented in Supplementary Table S1. In the first version, the values correspond exactly to the beginning and end of the task (this is why the resulting standard deviations are zero), since the three phases were kept distinguishable, while in the second case, the values obtained are only an estimation. The statistical analysis yielded a significant difference between the two functions only in the identification of the offsets ($p < 0.01$) and the related error values ($p < 0.001$).

As for the smoothness of the signals, the results of the performed test showed statistically significant differences between the two versions ($p < 0.01$) for each channel and for both the HbO_2 and HbR concentration signals.

Both versions of the *TaskNorm.m* are publicly available on Zenodo (https://doi.org/10.5281/zenodo.10043158). Furthermore, the first version of the function has been merged into the HomER3 master branch on GitHub, so it is now available as part of the HomER3 toolbox at https://github.com/BUNPC/Homer3/tree/master/FuncRegistry/UserFunctions (accessed on 21 December 2023). The aim of the present work is thus to describe, for the first time, to a possible user of the function, its rationale, the working principle, and its applicability in real data analysis, describing also the possible differences between the two versions.

Table 2. Difference between first and second version in identifying onset and offset. Mean values (and standard deviation) with respect to %trial are reported for each subject and task.

			First Version		Second Version	
Task	**ID Code**	**Mean Duration (SD)**	**%Mean Onset (SD)**	**%Mean Offset (SD)**	**%Mean Onset (SD)**	**%Mean Offset (SD)**
Walking	DCD_1	8.38 (1.26)	15.15 (0)	76.97 (0)	14.95 (1.26)	76.89 (2.02)
Walking	DCD_2	5.16 (0.94)	19.72 (0)	70.02 (0)	19.64 (1.78)	69.54 (2.79)
Walking	DCD_3	6.76 (0.47)	16.42 (0)	75.04 (0)	17.03 (0.69)	73.79 (1.07)
Walking	DCD_4	6.27 (0.66)	17.92 (0)	72.76 (0)	17.85 (1.04)	72.53 (1.67)
Walking	TD_1	6.25 (0.5)	17.92 (0)	72.76 (0)	17.67 (0.75)	72.56 (1.16)
Walking	TD_2	5.34 (0.37)	19.72 (0)	70.02 (0)	19.24 (0.7)	70.14 (1.09)
Walking	TD_3	6.31 (0.42)	17.92 (0)	72.76 (0)	17.67 (0.69)	72.86 (1.03)
Walking	TD_4	8.32 (0.35)	15.15 (0)	76.97 (0)	14.95 (0.34)	76.89 (0.55)
Alternate stair touching	DCD_1	10.21 (0.96)	13.12 (0)	80.05 (0)	13.08 (0.77)	79.71 (1.24)
Alternate stair touching	DCD_2	6.89 (1.07)	16.42 (0)	75.04 (0)	16.69 (1.39)	74.12 (2.2)
Alternate stair touching	DCD_3	7.63 (0.93)	15.15 (0)	76.97 (0)	15.87 (1.2)	75.5 (1.86)
Alternate stair touching	DCD_4	9.09 (0.99)	14.06 (0)	78.62 (0)	14.16 (0.93)	78.11 (1.5)
Alternate stair touching	TD_1	6.89 (1.17)	16.42 (0)	75.04 (0)	16.86 (1.61)	73.96 (2.61)
Alternate stair touching	TD_2	6.24 (0.28)	17.92 (0)	72.76 (0)	17.85 (0.49)	72.5 (0.79)
Alternate stair touching	TD_3	7.23 (0.28)	16.42 (0)	75.04 (0)	16.36 (0.3)	74.79 (0.52)
Alternate stair touching	TD_4	8.3 (0.37)	15.15 (0)	76.97 (0)	14.95 (0.43)	76.89 (0.66)
Leaning	DCD_1	12.87 (1.22)	10.93 (0)	83.39 (0)	11.23 (0.8)	82.65 (1.23)
Leaning	DCD_2	11.12 (1.43)	12.3 (0)	81.3 (0)	12.39 (1.2)	80.85 (1.9)
Leaning	DCD_3	13.17 (1.44)	10.93 (0)	83.39 (0)	11.01 (0.85)	82.98 (1.32)
Leaning	DCD_4	11.45 (1.05)	12.3 (0)	81.3 (0)	12.14 (0.74)	81.23 (1.15)
Leaning	TD_1	13.28 (1.23)	10.93 (0)	83.39 (0)	10.9 (0.69)	83.09 (1.08)
Leaning	TD_2	12.65 (1.21)	10.93 (0)	83.39 (0)	11.35 (0.78)	82.54 (1.22)
Leaning	TD_3	11.37 (0.79)	12.3 (0)	81.3 (0)	12.14 (0.6)	81.23 (0.89)
Leaning	TD_4	17.09 (2.45)	8.93 (0)	86.42 (0)	9.1 (0.94)	85.99 (1.48)

Figures 4 and 5 show two examples of the output of the function, when setting the *show* parameter to 1, related to a single channel for one subject performing the walking task. Specifically, the effect of the *deltat* parameters is highlighted: Figure 4 displays the hemodynamic response with a 2-s baseline and a 3-s post-task period (*deltat* = [2, 3]), whereas Figure 5 demonstrates the possibility of adjusting the baseline duration to normalize the signal and not considering the post-task period (*deltat* = [1, 0]). In both cases, the mean baseline-corrected HRF and its standard deviation are reported for both the HbO_2 and HbR signals, as well as the concentration data for each repetition. The present function provides the same output for each channel.

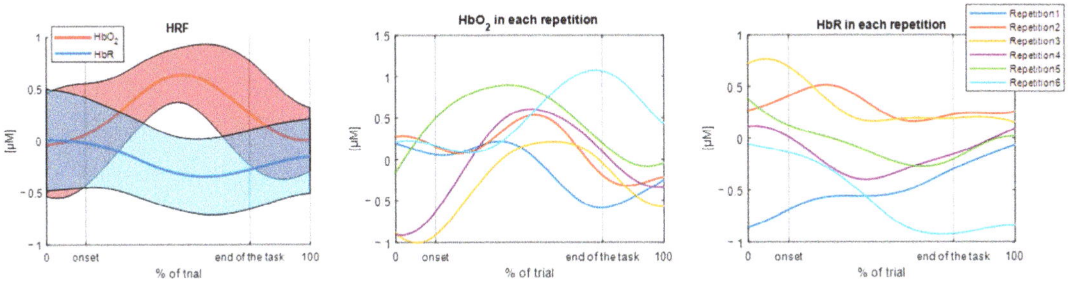

Figure 4. Example of the output of the function when setting *deltat* = [2, 3]. The output is related to a single channel during the walking task, for a single subject. Mean hemodynamic responses (and standard deviation area), time-normalized HbO_2 concentrations for each repetition, and time-normalized HbR concentrations for each repetition are shown. Abbreviations: HRF = hemodynamic response function; HbO_2 = oxyhemoglobin concentration; and HbR = deoxyhemoglobin concentration.

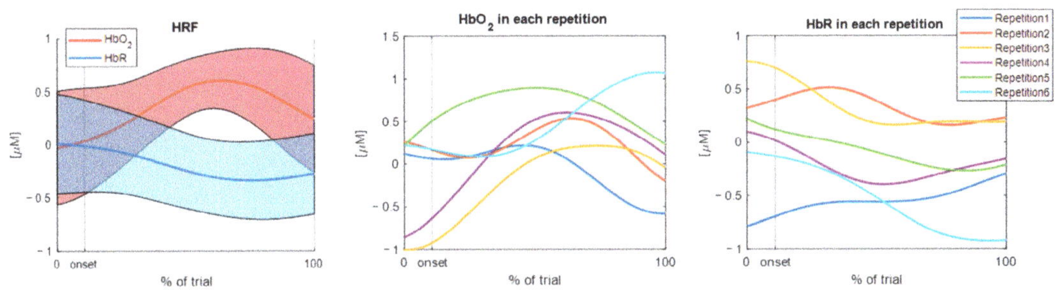

Figure 5. Example of the output of the function when setting *deltat* = [1, 0]. The output is related to a single channel during the walking task, for a single subject. In this case, the post-task period has not been considered. Mean hemodynamic responses (and standard deviation area), time-normalized HbO_2 concentrations for each repetition, and time-normalized HbR concentrations for each repetition are shown. Abbreviations: HRF = hemodynamic response function; HbO_2 = oxyhemoglobin concentration; and HbR = deoxyhemoglobin concentration.

Once the function is imported into the HomER processing stream, the HRF outcome is displayed in the Homer3 MainGUI (see Supplementary Figure S3) and may be further analyzed using the other available tools in the HomER package [9].

3.2. Results of the Testing

Once the HRF had been extracted for each subject and task, it was possible to compute the mean HRF for the two groups for the selected channel. Both versions of the algorithm were applied to the three tasks to test their working principle. For the sake of simplicity and in order to show the results derived from the use of both versions, the outcome HRF during the alternate stair touching task and the leaning task obtained with the first version of the function and the walking task HRF derived with the second version are presented in the current work.

Figure 6 depicts the HRF outcomes for the different tasks, referring to HbO_2 concentrations, averaged across DCD and TD children and across the defined channel ROIs. As for the baseline period, a reduced standard deviation can be noticed across all tasks and all the ROIs due to the baseline-correction procedure that bounds that portion of the signal in a range around 0 μm. During the effective task execution period, a task-induced increase can be generally observed in the mean HRF, relative to the HbO_2 concentration. On the other hand, a general decreasing trend in the hemodynamic responses can be observed in the post-task periods that can be considered a resting phase with lower brain activation.

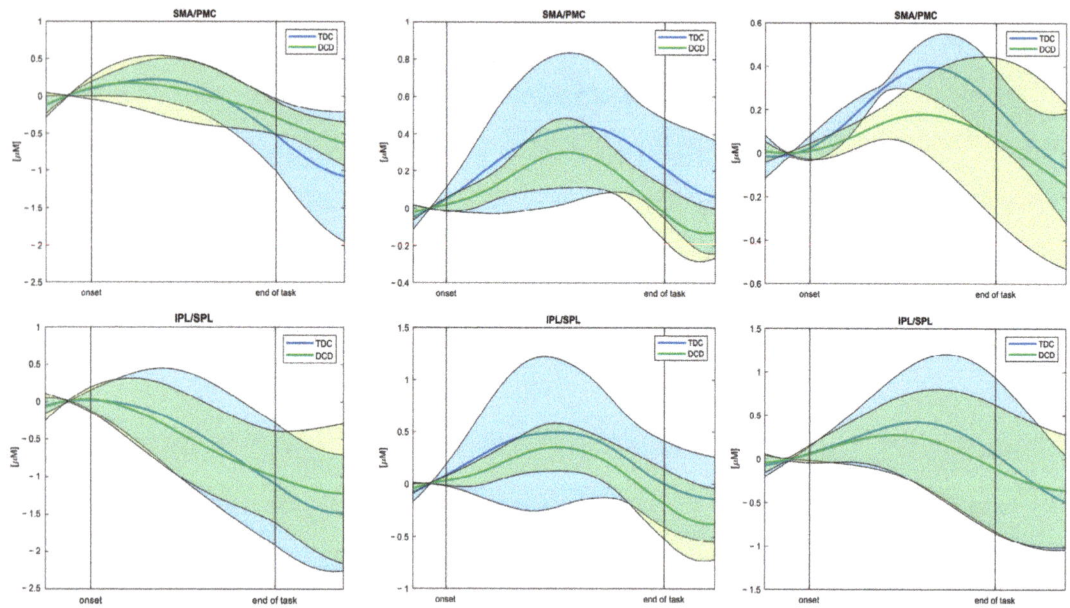

Figure 6. Example of the hemodynamic response (mean and standard deviation area) related to HbO_2 concentrations averaged across subject and channel clusters (up: SMA/PMC regions of interest, down: IPL/SPL regions of interest), divided into the three tasks: **First column**) Alternate stair touching task; **Second column**) Leaning task; and **Third column**) Walking task. An increasing trend during task execution and a subsequent decrease towards and during the post-task period can be observed. Abbreviations: SMA = supplementary motor area; PMC = premotor cortex; IPL = inferior parietal lobule; and SPL = superior parietal lobule.

The same averaging procedure was performed for the HRF related to the HbR concentrations (Figure 7). A general opposing trend can be found in the HbR concentrations with respect to the HbO_2 concentrations during the different tasks. In particular, during the execution of the task, a decrease in the HbR concentration and a subsequent increase towards and during the post-task period can be observed. HbR concentrations during the baseline periods can be interpreted in the same way as for the HbO_2.

Neither the group nor the task analyzed had an impact on the algorithm's outcomes.

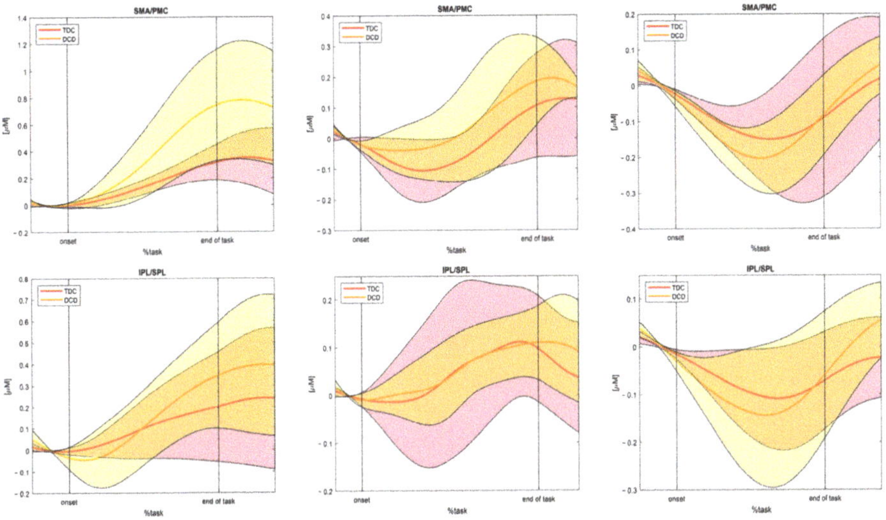

Figure 7. Example of the hemodynamic response (mean and standard deviation area) related to HbR concentrations averaged across subject and channel clusters (up: SMA/PMC regions of interest, down: IPL/SPL regions of interest), divided into the three tasks: **First column**) Alternate stair touching task; **Second column**) Leaning task; and **Third column**) Walking task. A decreasing trend during task execution and a subsequent increase towards and during the post-task period can be observed, opposite to the HbO_2 behaviour. Abbreviations: SMA = supplementary motor area; PMC = premotor cortex; IPL = inferior parietal lobule; and SPL = superior parietal lobule.

4. Discussion

The present study introduces a novel function for the time-normalization of hemodynamic responses recorded during trials in which there is no a priori defined task duration. This is particularly true for experimental designs that include motor tasks that can be performed at self-paced conditions. Indeed, in such cases it is not possible to compare data recorded during either different repetitions of the same task or between different subjects. For this reason, the function interpolates the processed data with a fixed number of samples, which are then expressed as a function of the percentage of the trial duration. This allows us to successively apply the standard procedure of analyzing fNIRS data (i.e., averaging across trials to obtain the mean hemodynamic responses) without the need of advanced programming skills.

The *TaskNorm* function operates with the same fundamental concept as the HomER3 function *hmrR_BlockAvg* [9], computing the mean baseline-corrected HRF of the concentrations of each repetition of the task. However, the original *hmrR_BlockAvg* function is not applicable in the experimental conditions as outlined above, since the inputs of this function are the baseline period before the task onset and a fixed interval period after the task onset across all trials. The variability in the duration of each repetition is therefore not considered, leading to incorrect averaging and comparisons. Hence, there is a need to develop a new function to address this issue.

Other techniques, such as the DTW algorithm, are exploited in the field of fNIRS analysis [31] to align one signal to a second time series, which can also be recorded during trials of different durations, through local compressions and extensions of the temporal axes. However, the stretching of the signals may lead to a modification of the original shape, which would not be preserved. Furthermore, DTW-based averaging can be realized by aligning the data to a reference signal (which we do not have) or the average must be obtained sequentially in a pairwise manner, with a risk of error propagation. In these cases, the temporal alignment is performed with respect to the signal of reference, which would result in an incorrect temporal representation of the final HRF. For these reasons, a direct quantitative comparison of the performance of the DTW with our function is not possible.

The present work aimed to address the aforementioned issues by implementing the *TaskNorm* function, which was designed with the primary purpose of normalizing the data to ensure a proper comparison. Two versions of the *TaskNorm* function were developed with different modalities of data interpolation. In the first version, the three phases of the task (i.e., baseline, task duration, and post-task duration) are separately interpolated, and then the signal is reassembled, while in the second version, the function interpolates the signal related to the entire task. The first version allowed the onset and the end of the task to be accurately identified as a percentage of the trial, while statistical analyses show that the second version made a significant error in identifying the offset. Thus, this first version can be particularly useful when the experimental design requires a higher temporal resolution, as in the case of rapid event-related designs or when fNIRS is combined with EEG recordings. The first version of the algorithm could also be preferred when exploring plausible between-subject differences specifically related to either the "preparation/planning phase" of the response or the post-task recovery, or when interested in analyzing individual intra-trial differences. Conversely, the signals obtained with the latter version have a significantly lower smoothness index, thus resulting in a smoother signal. This version seems therefore more appropriate for research focusing on the overall hemodynamic response throughout the entire trial.

Both versions of the function were tested on fNIRS data recorded in eight participants (four children with DCD and four TD children) during three different tasks (alternate stair touching, leaning, and walking). Participants were asked to perform each task for a fixed number of times at their own pace, so that no fixed trial duration was imposed.

The present results show that the algorithm correctly allows comparisons and averaging of data from each channel across multiple repetitions of the same task, even if they have different durations, because of the time-normalization performed.

Thereafter, hemodynamic responses for each channel were averaged across subjects and then across SMA/PMC and IPL/SPL brain ROIs for each task. The resulting HRFs are consistent with the previous literature [17,47], as a general increase in cortical activity that is attributable to task performance was expected, whereas an opposing trend characterizes the HbR responses. Nevertheless, the small sample size represents a limitation of this study, thus it is important to acknowledge that these responses can be analyzed only qualitatively, as they are the result of a comparison on a limited number of subjects. Indeed, the purpose of this study is to test the applicability of the implemented algorithm to obtain a mean HRF value across repetitions for each subject (which allows the subsequent average computation across subjects), rather than study the brain activation elicited by the different tasks or the differences between the two groups.

The final outcome of the function is independent either from the task or from the pre-processing pipeline applied to the data, since the only parameters needed are the continuous fNIRS signal of the task considered, and the durations of the baseline and the post-task periods (both parameters are set by the users). We are aware that because of our selection of the *deltat* parameters, an overlap between the HRF from one repetition and one from a successive repetition may occur. However, our purpose was to present the entire functionality of the *TaskNorm* function and statistically evaluate all the differences between the two versions (i.e., trial phase segmentation, onset/offset identification). For this reason, we decided to perform the analysis while also considering the post-task period. However, the user can select the appropriate *deltat* values for the experimental protocol in order to obtain proper results. This ensures the generalizability and flexibility of the function for different experimental designs, including motor or cognitive task analysis.

The present function introduced a time-normalization step in an already existing and widely used procedure (i.e., block-average analysis). This was done to enable this analysis to be performed correctly even under conditions where it could not be (e.g., varying tasks duration). Future developments envisage the possible integration of this step into other analysis techniques that do not take into account different trial durations.

Finally, future prospects will focus on possible applications of the algorithm to a larger number of subjects but also to different populations, such as healthy or neurologically impaired individuals, as well as pediatric and adult ages. Participants may perform self-paced motor/balance tasks but also cognitive or in real-life conditions tasks, such as working activities. Different brain regions can be investigated (such as the frontal, parietal, or temporal areas) also enabling quantitative assessments of group-related or task-related differences.

5. Conclusions

The present work responds to the need for a method suitable for fNIRS analysis to compare data across different trials and subjects in studies where task trials have variable durations. To this end, we developed the *TaskNorm.m* function, which allows rescaling of the time-axis in each trial to a common reference duration set to a percentage of the trial duration, with a spline interpolation method. The novelty of the current study lies mainly in the application of the fNIRS data analysis, since time-normalization techniques for fNIRS data in such conditions are yet to be implemented, rather than in the methodological aspects, as it exploits an already existing methodology.

The function is structured so that it is readable and importable in the HomER processing stream by the user, but is also available as part of the HomER3 package [9], which is a toolbox developed and distributed to facilitate the processing of the fNIRS signal, which still lacks a well-defined and standardized procedure due to the relative novelty of this imaging technique [8].

Moreover, the function is already available on the Zenodo platform, and both versions can be used independently, without being imported into the HomER package. This makes them accessible to those who do not use this package or are merely interested in the time-normalization procedure. Additionally, the main aspects of the function can be easily converted to other programming languages (e.g., C++ or Python), so that is widely applicable to different settings and experimental research.

Therefore, we feel that the function proposed here could be a further step in contributing to the ongoing process of developing a standardized signal analysis pipeline for fNIRS data.

Supplementary Materials: The following supporting information can be downloaded at https://www.mdpi.com/article/10.3390/s25061768/s1, Figure S1: Screenshot of the Processing Stream GUI with the parameters set for the presented analysis; Figure S2: Clustering of channels in SMA/PMC (green area) and IPL/SPL (yellow area) regions of interest. Abbreviations: SMA = supplementary motor area; PMC = premotor cortex; IPL = inferior parietal lobule; and SPL = superior parietal lobule; Figure S3: Homer3 MainGUI. Example of hemodynamic response (HRF) outcome (oxyhemoglobin-HbO_2 and deoxyhemoglobin–HbR concentrations). Table S1: Difference between first and second version in identifying onset and offset.

Author Contributions: Conceptualization, A.C., A.H., C.P., E.V., K.K. and P.M.; software, A.F.; formal analysis, A.F. and C.J.; investigation, C.J.; data curation, A.F. and C.J.; writing—original draft preparation, A.F. and C.J.; writing—review and editing, A.C., A.H., A.F., C.P., C.J., E.B., E.V., K.K. and P.M.; visualization, A.F.; supervision, A.C., A.H., E.B., E.V., K.K. and P.M.; project administration,

A.C. and C.P.; funding acquisition, A.H., E.B., E.V., K.K. and P.M. All authors have read and agreed to the published version of the manuscript.

Funding: This research was partially funded by the Italian Ministry of Health (Ricerca Corrente 2023–2024 and 2024–2025 to E. Biffi) and by the Research Foundation-Flanders (FWO) (grant number: 43498, year: 2020).

Institutional Review Board Statement: This study was conducted in accordance with the Declaration of Helsinki and approved by the Committee for Medical Ethics UZA-UAntwerp (B300201941833).

Informed Consent Statement: Informed consent was obtained from all subjects involved in this study.

Data Availability Statement: Publicly available datasets were analyzed in this study. This data can be found here: https://doi.org/10.5281/zenodo.10124956, Zenodo. Both versions of *TaskNorm.m* are publicly available on Zenodo at https://doi.org/10.5281/zenodo.10043158. The first version of the function is available as part of the HomER3 toolbox at https://github.com/BUNPC/Homer3/tree/master/FuncRegistry/UserFunctions (accessed on 21 December 2023).

Acknowledgments: Special thanks go to all the children and their parents who took part in the study.

Conflicts of Interest: The authors declare no conflicts of interest.

Abbreviations

The following abbreviations are used in this manuscript:

fNIRS	functional near-infrared spectroscopy
HbO_2	oxyhemoglobin
HbR	deoxyhemoglobin
HT	total hemoglobin
fMRI	functional magnetic resonance imaging
EEG	electroencephalography
DTW	dynamic time warping
DCD	developmental coordination disorder
TD	typically developing
HRF	hemodynamic response function
EMG	electromyography
fOLD	fNIRS optodes locator decider
SMA	supplementary motor rea
PMC	premotor cortex
IPL	inferior parietal lobule
SPL	superior parietal lobule
ROI	regions of interest

References

1. Menant, J.C.; Maidan, I.; Alcock, L.; Al-Yahya, E.; Cerasa, A.; Clark, D.J.; de Bruin, E.; Fraser, S.; Gramigna, V.; Hamacher, D.; et al. A consensus guide to using functional near-infrared spectroscopy in posture and gait research. *Gait Posture* **2020**, *82*, 254–265. [CrossRef] [PubMed]
2. Yücel, M.A.; Lühmann, A.V.; Scholkmann, F.; Gervain, J.; Dan, I.; Ayaz, H.; Boas, D.; Cooper, R.J.; Culver, J.; Elwell, C.E.; et al. Best practices for fNIRS publications. *Neurophotonics* **2021**, *8*, 012101. [CrossRef]
3. Karim, H.; Schmidt, B.; Dart, D.; Beluk, N.; Huppert, T. Functional near-infrared spectroscopy (fNIRS) of brain function during active balancing using a video game system. *Gait Posture* **2012**, *35*, 367–372. [CrossRef]
4. Hoshi, Y.; Kobayashi, N.; Tamura, M. Interpretation of near-infrared spectroscopy signals: A study with a newly developed perfused rat brain model. *J. Appl. Physiol.* **2001**, *90*, 1657–1662. [CrossRef] [PubMed]
5. Kamran, M.A.; Mannan, M.M.N.; Jeong, M.Y. Cortical signal analysis and advances in functional near-infrared spectroscopy signal: A review. *Front. Hum. Neurosci.* **2016**, *10*, 261. [CrossRef]
6. Pereira, J.; Direito, B.; Lührs, M.; Castelo-Branco, M.; Sousa, T. Multimodal assessment of the spatial correspondence between fNIRS and fMRI hemodynamic responses in motor tasks. *Sci. Rep.* **2023**, *13*, 2244. [CrossRef] [PubMed]

7. Logothetis, N.K. What we can do and what we cannot do with fMRI. *Nature* **2008**, *453*, 869–878. [CrossRef]
8. Pfeifer, M.D.; Scholkmann, F.; Labruyère, R. Signal processing in functional near-infrared spectroscopy (fNIRS): Methodological differences lead to different statistical results. *Front. Hum. Neurosci.* **2018**, *11*, 641. [CrossRef]
9. Huppert, T.J.; Diamond, S.G.; Franceschini, M.A.; Boas, D.A. HomER: A review of time-series analysis methods for near-infrared spectroscopy of the brain. *Appl. Opt.* **2009**, *48*, D280–D298. [CrossRef]
10. Ren, Y.; Cui, G.; Zhang, X.; Feng, K.; Yu, C.; Liu, P. The promising fNIRS: Uncovering the function of prefrontal working memory networks based on multi-cognitive tasks. *Front. Psychiatry* **2022**, *13*, 985076. [CrossRef]
11. Mauri, M.; Grazioli, S.; Crippa, A.; Bacchetta, A.; Pozzoli, U.; Bertella, S.; Gatti, E.; Maggioni, E.; Rosi, E.; Diwadkar, V.; et al. Hemodynamic and behavioral peculiarities in response to emotional stimuli in children with attention deficit hyperactivity disorder: An fNIRS study. *J. Affect. Disord.* **2020**, *277*, 671–680. [CrossRef] [PubMed]
12. Grazioli, S.; Crippa, A.; Mauri, M.; Piazza, C.; Bacchetta, A.; Salandi, A.; Trabattoni, S.; Agostoni, C.; Molteni, M.; Nobile, M. Association between fatty acids profile and cerebral blood flow: An exploratory fNIRS study on children with and without ADHD. *Nutrients* **2019**, *11*, 2414. [CrossRef] [PubMed]
13. Wijeakumar, S.; Huppert, T.J.; Magnotta, V.A.; Buss, A.T.; Spencer, J.P. Validating an image-based fNIRS approach with fMRI and a working memory task. *Neuroimage* **2017**, *147*, 204–218. [CrossRef]
14. Lu, C.-F.; Liu, Y.-C.; Yang, Y.-R.; Wu, Y.-T.; Wang, R.-Y. Maintaining gait performance by cortical activation during dual-task interference: A functional near-infrared spectroscopy study. *PLoS ONE* **2015**, *10*, e0129390. [CrossRef] [PubMed]
15. Miyai, I.; Tanabe, H.C.; Sase, I.; Eda, H.; Oda, I.; Konishi, I.; Tsunazawa, Y.; Suzuki, T.; Yanagida, T.; Kubota, K. Cortical mapping of gait in humans: A near-infrared spectroscopic topography study. *Neuroimage* **2001**, *14*, 1186–1192. [CrossRef]
16. Lee, B.C.; Choi, J.; Martin, B.J. Roles of the prefrontal cortex in learning to time the onset of pre-existing motor programs. *PLoS ONE* **2020**, *15*, e0241562. [CrossRef]
17. Mihara, M.; Miyai, I.; Hatakenaka, M.; Kubota, K.; Sakoda, S. Role of the prefrontal cortex in human balance control. *Neuroimage* **2008**, *43*, 329–336. [CrossRef]
18. Hawkins, K.A.; Fox, E.J.; Daly, J.J.; Rose, D.K.; Christou, E.A.; McGuirk, T.E.; Otzel, D.M.; Butera, K.A.; Chatterjee, S.A.; Clark, D.J. Prefrontal over-activation during walking in people with mobility deficits: Interpretation and functional implications. *Hum. Mov. Sci.* **2018**, *59*, 46–55. [CrossRef]
19. Lin, M.I.B.; Lin, K.H. Walking while performing working memory tasks changes the prefrontal cortex hemodynamic activations and gait kinematics. *Front. Behav. Neurosci.* **2016**, *10*, 92. [CrossRef]
20. de Lima-Pardini, A.C.; Zimeo Morais, G.A.; Balardin, J.B.; Coelho, D.B.; Azzi, N.M.; Teixeira, L.A.; Sato, J.R. Measuring cortical motor hemodynamics during assisted stepping–An fNIRS feasibility study of using a walker. *Gait Posture* **2017**, *56*, 112–118. [CrossRef]
21. Mirelman, A.; Maidan, I.; Bernad-Elazari, H.; Nieuwhof, F.; Reelick, M.; Giladi, N.; Hausdorff, J.M. Increased frontal brain activation during walking while dual tasking: An fNIRS study in healthy young adults. *J. Neuroeng. Rehabil.* **2014**, *11*, 85. [CrossRef]
22. Clark, D.J.; Rose, D.K.; Ring, S.A.; Porges, E.C. Utilization of central nervous system resources for preparation and performance of complex walking tasks in older adults. *Front. Aging Neurosci.* **2014**, *6*, 217. [CrossRef]
23. McKendrick, R.; Mehta, R.; Ayaz, H.; Scheldrup, M.; Parasuraman, R. Prefrontal Hemodynamics of Physical Activity and Environmental Complexity during Cognitive Work. *Hum. Factors* **2017**, *59*, 147–162. [CrossRef]
24. Pinti, P.; Aichelburg, C.; Lind, F.; Power, S.; Swingler, E.; Merla, A.; Hamilton, A.; Gilber, S.; Burgess, P.; Tachtsidis, I. Using fiberless, wearable fnirs to monitor brain activity in real-world cognitive tasks. *J. Vis. Exp.* **2015**, *2015*, e53336. [CrossRef]
25. Balardin, J.B.; Zimeo Morais, G.A.; Furucho, R.A.; Trambaiolli, L.; Vanzella, P.; Biazoli, C.; Sato, J.R. Imaging brain function with functional near-infrared spectroscopy in unconstrained environments. *Front. Hum. Neurosci.* **2017**, *11*, 258. [CrossRef]
26. Liu, Z.; Shore, J.; Wang, M.; Yuan, F.; Buss, A.; Zhao, X. A systematic review on hybrid EEG/fNIRS in brain-computer interface. *Biomed. Signal Process. Control* **2021**, *68*, 102595. [CrossRef]
27. Llopis-Albert, C.; Toro, W.R.V.; Farhat, N.; Zamora-Ortiz, P.; Del Pozo, Á.F.P. A new method for time normalization based on the continuous phase: Application to neck kinematics. *Mathematics* **2021**, *9*, 3138. [CrossRef]
28. Wagner, J.; Solis-Escalante, T.; Scherer, R.; Neuper, C.; Müller-Putz, G. It's how you get there: Walking down a virtual alley activates premotor and parietal areas. *Front. Hum. Neurosci.* **2014**, *8*, 93. [CrossRef]
29. Gwin, J.T.; Gramann, K.; Makeig, S.; Ferris, D.P. Electrocortical activity is coupled to gait cycle phase during treadmill walking. *Neuroimage* **2011**, *54*, 1289–1296. [CrossRef]
30. Wagner, J.; Solis-Escalante, T.; Grieshofer, P.; Neuper, C.; Müller-Putz, G.; Scherer, R. Level of participation in robotic-assisted treadmill walking modulates midline sensorimotor EEG rhythms in able-bodied subjects. *Neuroimage* **2012**, *63*, 1203–1211. [CrossRef]
31. Zhu, L.; Najafizadeh, L. Dynamic time warping-based averaging framework for functional near-infrared spectroscopy brain imaging studies. *J. Biomed. Opt.* **2017**, *22*, 066011. [CrossRef]

32. Müller, M. Dynamic Time Warping. In *Information Retrieval for Music and Motion*; Springer: Berlin/Heidelberg, Germany, 2007; pp. 69–84, ISBN 978-3-540-74048-3.
33. Sakoe, H.; Chiba, S. Dynamic Programming Algorithm Optimization for Spoken Word Recognition. *IEEE Trans. Acoust.* **1978**, *26*, 43–49. [CrossRef]
34. Li, H.; Liu, J.; Yang, Z.; Liu, R.W.; Wu, K.; Wan, Y. Adaptively constrained dynamic time warping for time series classification and clustering. *Inf. Sci.* **2020**, *534*, 97–116. [CrossRef]
35. Blank, R.; Barnett, A.L.; Cairney, J.; Green, D.; Kirby, A.; Polatajko, H.; Rosenblum, S.; Smits-Engelsman, B.; Sugden, D.; Wilson, P.; et al. International clinical practice recommendations on the definition, diagnosis, assessment, intervention, and psychosocial aspects of developmental coordination disorder. *Dev. Med. Child Neurol.* **2019**, *61*, 242–285. [CrossRef]
36. American Psychiatric Association. *Diagnostic and Statistical Manual of Mental Disorders: DSM-5TM*, 5th ed.; American Psychiatric Publishing, Inc.: Arlington, VA, USA, 2013. [CrossRef]
37. Verbecque, E.; Johnson, C.; Rameckers, E.; Thijs, A.; van der Veer, I.; Meyns, P.; Smits-Engelsman, B.; Klingels, K. Balance control in individuals with developmental coordination disorder: A systematic review and meta-analysis. *Gait Posture* **2021**, *83*, 268–279. [CrossRef]
38. Subara-Zukic, E.; Cole, M.H.; McGuckian, T.B.; Steenbergen, B.; Green, D.; Smits-Engelsman, B.C.M.; Lust, J.M.; Abdollahipour, R.; Domellöf, E.; Deconinck, F.J.A.; et al. Behavioral and Neuroimaging Research on Developmental Coordination Disorder (DCD): A Combined Systematic Review and Meta-Analysis of Recent Findings. *Front. Psychol.* **2022**, *13*, 809455. [CrossRef]
39. Al-Yahya, E.; Esser, P.; Weedon, B.D.; Joshi, S.; Liu, Y.-C.; Springett, D.N.; Salvan, P.; Meaney, A.; Collett, J.; Inacio, M.; et al. Motor learning in developmental coordination disorder: Behavioral and neuroimaging study. *Front. Neurosci.* **2023**, *17*, 1187790. [CrossRef]
40. Joshi, S.; Weedon, B.D.; Esser, P.; Liu, Y.C.; Springett, D.N.; Meaney, A.; Inacio, M.; Delextrat, A.; Kemp, S.; Ward, T.; et al. Neuroergonomic assessment of developmental coordination disorder. *Sci. Rep.* **2022**, *12*, 10239. [CrossRef]
41. Caçola, P.; Getchell, N.; Srinivasan, D.; Alexandrakis, G.; Liu, H. Cortical activity in fine-motor tasks in children with Developmental Coordination Disorder: A preliminary fNIRS study. *Int. J. Dev. Neurosci.* **2018**, *65*, 83–90. [CrossRef] [PubMed]
42. Dewar, R.; Claus, A.P.; Tucker, K.; Ware, R.; Johnston, L.M. Reproducibility of the Balance Evaluation Systems Test (BESTest) and the Mini-BESTest in school-aged children. *Gait Posture* **2017**, *55*, 68–74. [CrossRef] [PubMed]
43. Piazza, C.; Bacchetta, A.; Crippa, A.; Mauri, M.; Grazioli, S.; Reni, G.; Nobile, M.; Bianchi, A.M. *Preprocessing Pipeline for fNIRS Data in Children BT-XV Mediterranean Conference on Medical and Biological Engineering and Computing–MEDICON 2019*; Henriques, J., Neves, N., de Carvalho, P., Eds.; Springer International Publishing: Cham, Switzerland, 2020; pp. 235–244.
44. Whiteman, A.C.; Santosa, H.; Chen, D.F.; Perlman, S.; Huppert, T. Investigation of the sensitivity of functional near-infrared spectroscopy brain imaging to anatomical variations in 5- to 11-year-old children. *Neurophotonics* **2017**, *5*, 011009. [CrossRef]
45. Hu, X.-S.; Arredondo, M.M.; Gomba, M.; Confer, N.; DaSilva, A.F.; Johnson, T.D.; Shalinsky, M.; Kovelman, I. Comparison of motion correction techniques applied to functional near-infrared spectroscopy data from children. *J. Biomed. Opt.* **2015**, *20*, 126003. [CrossRef]
46. Molavi, B.; Dumont, G.A. Wavelet-based motion artifact removal for functional near-infrared spectroscopy. *Physiol. Meas.* **2012**, *33*, 259–270. [CrossRef]
47. Lim, S.B.; Louie, D.R.; Peters, S.; Liu-Ambrose, T.; Boyd, L.A.; Eng, J.J. Brain activity during real-time walking and with walking interventions after stroke: A systematic review. *J. Neuroeng. Rehabil.* **2021**, *18*, 8. [CrossRef]

sensors

MDPI

Article

Physiological Sensor Modality Sensitivity Test for Pain Intensity Classification in Quantitative Sensory Testing

Wenchao Zhu and Yingzi Lin *

Intelligent Human Machine Systems Laboratory, Department of Mechanical and Industrial Engineering, Northeastern University, Boston, MA 02155, USA
* Correspondence: yi.lin@northeastern.edu

Abstract: Chronic pain is prevalent and disproportionately impacts adults with a lower quality of life. Although subjective self-reporting is the "gold standard" for pain assessment, tools are needed to objectively monitor and account for inter-individual differences. This study introduced a novel framework to objectively classify pain intensity levels using physiological signals during Quantitative Sensory Testing sessions. Twenty-four participants participated in the study wearing physiological sensors (blood volume pulse (BVP), galvanic skin response (GSR), electromyography (EMG), respiration rate (RR), skin temperature (ST), and pupillometry). This study employed two analysis plans. Plan 1 utilized a grid search methodology with a 10-fold cross-validation framework to optimize time windows (1–5 s) and machine learning hyperparameters for pain classification tasks. The optimal time windows were identified as 3 s for the pressure session, 2 s for the pinprick session, and 1 s for the cuff session. Analysis Plan 2 implemented a leave-one-out design to evaluate the individual contribution of each sensor modality. By systematically excluding one sensor's features at a time, the performance of these sensor sets was compared to the full model using Wilcoxon signed-rank tests. BVP emerged as a critical sensor, significantly influencing performance in both pinprick and cuff sessions. Conversely, GSR, RR, and pupillometry demonstrated stimulus-specific sensitivity, significantly contributing to the cuff session but with limited influence in other sessions. EMG and ST showed minimal impact across all sessions, suggesting they are non-critical and suitable for reducing sensor redundancy. These findings advance the design of sensor configurations for personalized pain management. Future research will focus on refining sensor integration and addressing stimulus-specific physiological responses.

Keywords: machine learning; pain intensity; physiological signals; quantitative sensory testing; sensor modality; time window

1. Introduction

Pain is a complex and subjective experience, remaining one of the most significant clinical challenges, with 51.6 million U.S. adults (20.9%) experiencing chronic pain and 17.1 million (6.9%) suffering from high-impact chronic pain during 2021 [1,2]. The symptom of chronic pain causes the greatest source of disability for human beings, leading to substantial issues and affecting the quality of life for individuals and society [3,4].

Pain can be understood as a conscious interpretation of sensory stimuli that triggers nociceptive afferents, accompanied by the mental projection of these stimuli onto specific body regions. Pain assessment involves approximating an individual's subjective self-report, which serves as their ground truth [5,6]. The traditional pain assessment is performed through a survey based on participants' subjective perception of their pain, such as the numeric rating scale (NRS), the visual analogue scale (VAS), and the verbal rating scale (VRS). However, self-reported assessments are prone to bias from anxiety, memories, pain intensity, and physical activities [6–8]. The consequences of such inaccuracies can lead to under-treatment and over-treatment of pain that is either ineffective or detrimental to patient safety [9,10].

Quantitative Sensory Testing (QST) was developed as an objective, standardized way to evaluate pain sensitivity and pain perception using calibrated mechanical or thermal stimuli to measure sensory thresholds and tolerances [11]. This technique can aid in diagnosing conditions such as neuropathic pain and chronic low back pain by detecting abnormalities in QST sessions. For example, one QST session, the cuff inflation test, has shown that cLBP patients require lower cuff pressure to evoke moderate pain compared with healthy controls, and they also rate mechanical probes as more painful [12]. In addition, psychosocial factors—including emotional states and pain catastrophizing—further influence nociceptive processing and contribute to variations in pain perception [13,14].

Physiological sensors have developed as an objective measurement of human states and characteristics [15–17], such as blood volume pulse (BVP), electroencephalogram (EEG), galvanic skin response (GSR), respiration rate (RR), electromyography (EMG), skin temperature (ST), and pupillometry. For example, increased skin conductance level in GSR was detected when external noxious stimuli (e.g., pressure, thermal, cold pain) were presented [18]. Heart rate and heart rate variability, which can be derived from BVP signals, is associated with a stress response. During different pain stimuli, a decreased BVP or an increase in heart rate have been observed [19,20]. ST can be measured in the palm and the back of the hand, and decreased ST has been reported during and after painful stimuli [20,21]. EEG studies have addressed the correlations between noxious stimuli and different EEG frequency bands. For example, decreases in the alpha band have been observed as the common indicator [22,23].

Research has demonstrated the potential of using sensors in classifying pain intensity levels [24–26]. For example, Guo et al. estimated three levels of cold pain using facial expression by comparing three neural network models, and the personalized spatial–temporal framework using a convolutional long short-term memory model achieved the highest performance [27]. Another study measured the pain level via features generated from the pupillometry data using a genetic algorithm with an artificial neural network classifier, and the best performance was obtained with an accuracy of 81% [28]. EEG studies have demonstrated statistical differences in central and occipital regions, and was able to classify pain and no-pain states using multi-layer CNN frameworks [29,30]. Multimodal physiological classification with decision-level fusion and feature-level fusion proved promising in pain level detection and classification [21,31,32].

Combining multimodal physiological sensors with QST represents a frontier in pain research, enhancing the objectivity and sensitivity of pain assessments. BVP signals have been used to classify the pressure session, achieving 96.6% accuracy in a binary (threshold vs. tolerance) task [33]. GSR has assessed conditioned pain modulation between patient and healthy groups, revealing significant differences in the dominant hand ($p = 0.003$) [34]. EMG signals were evaluated under varying cuff pressures [35]. Despite these advances, few studies have examined comprehensive pain-level classification across all QST sessions. Additionally, the sensitivity of each sensor to these classifications remains largely unexplored.

Time window selection influences the interpretation of physiological signals in the task of pain assessment. One study demonstrated two distinct labeling approaches: fixed time windows and percentage-based timestamps. The fixed time window method segments data in a consistent and fixed manner to provide a straightforward approach for analyzing responses [36,37]. In contrast, percentage-based timestamping aligns labels to the individual's pain threshold and tolerance tailored to personal variations in pain perception [25]. Importantly, the chosen segmentation method directly impacts the number of samples generated for analysis and influences the dataset size available for machine learning models.

While multimodal physiological signals can aid in QST pain assessment tasks, two critical gaps remain in the current literature. First, existing studies have primarily focused on isolated noxious stimuli (e.g., cold pain, pressure, or cuff pressure) but rarely compared them across different noxious stimuli in a holistic way. Secondly, existing studies either combine all sensor data into the analysis model or exclusively analyze one of the modalities. The relative contributions of individual sensors remain unclear. Furthermore, the time window of signal segmentation under different tasks needs to be cautiously selected. Our study aims to advance the understanding of physiological sensor contributions to pain assessment and the development of individualized pain biomarkers, by addressing two research questions:

- The first question is to quantify the sensitivity of different time windows and machine learning classification model selection for pain level classification.
- The second question is to evaluate how excluding individual physiological sensors affects the model performance.

2. Materials and Methods

2.1. Participants

The study was conducted from January to May 2022 and approved by the Brigham and Women's Hospital Institutional Review Board (IRB), Boston, MA, USA (protocol code 2019P002781, 18 November 2019). Healthy participants and chronic low back pain patients who have had cLBP for at least three months with an average intensity of more than three out of ten on pain scales were recruited. All participants were neurologically intact and had no history of myocardial infarction, substantial motor or sensory deficits, or no evidence of cognitive impairment.

2.2. Apparatus

The study used sensors to monitor physiological responses during QST sessions. Pupillometry data were tracked using Tobii Pro Glasses 2 (Tobii, Danderyd, Sweden). Other sensors (FlexComp Infiniti, Thought Technology, Montreal, QC, Canada) that measured the participant's physiological responses included a BVP (SA9308M, Thought Technology) sensor for heart rate tracking through the middle finger of the non-dominant hand, a chest-mounted respiration sensor (SA9311M, Thought Technology), an EMG sensor (T9306M, Thought Technology) for muscle activity on the non-dominant forearm, an ST sensor (SA9310M, Thought Technology) on the back of the non-dominant hand, and a GSR sensor (SA9309M, Thought Technology) for electrical activity between the index and ring fingers on the non-dominant hand. A computer system was used to collect and store data (Dell Latitude E6230, Dell, Round Rock, TX, USA).

2.3. Experimental Procedures

Participants were familiarized with the QST equipment and completed the Brief Pain Inventory questionnaire [38]. The physiological data collection for each participant took approximately 80–120 min. As shown in Figure 1, the process involved the following:

(1) Participants were seated comfortably in a reclining chair.
(2) A research assistant helped participants wear all sensors, including pupillometry, BVP, GSR, EMG, ST, and RR. The setup took around 20 min.

(3) A one-minute baseline was recorded, during which the participant stayed in a natural resting condition.
(4) Data collection occurred over 30 min for one round of QST, during which participants followed instructions from the research assistant, reported pain intensities, and were asked to minimize unnecessary movement.
(5) Another one-minute baseline was recorded.
(6) Participants then performed physical maneuvers spanning about 3–5 min, with sensors disconnected.
(7) Participants then repeated steps 3 to 5 for a second round of QST collection.
(8) The sensors were removed, and participants were debriefed and compensated.

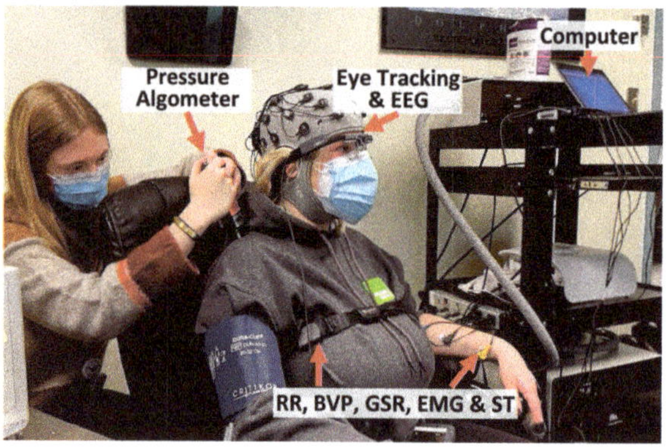

Figure 1. Experiment apparatus. In this pressure pain experiment, a digital pressure algometer was applied on the participant's trapezius. Physiological signals (RR, BVP, GSR, EMG, ST, and ET) were collected in the meantime.

Due to COVID-19 safety measures, all research staff and study participants were required to wear a face covering/mask to cover the nose and mouth, and only four people were present in the testing room at one time due to the COVID-19 period.

2.4. Quantitative Sensory Testing

QST has four sessions: pressure pain threshold and tolerance, temporal summation of mechanical pinprick pain, temporal summation of cuff pain, and conditioned pain modulation. The Temporal Summation of Pain tests the ability of the central nervous system to amplify the incoming pain over time when applying an increasing pain. It can be demonstrated in various pain modalities, including mechanical pinprick and cuff pain.

(1) Pressure pain threshold and tolerance were assessed using a digital pressure algometer. The testing sites were located on the dorsal surface of the forearm and over the trapezius muscle in the upper back and neck region. The researcher increased the pressure pain gradually via a flat round transducer on a small skin area (probe area 0.785 cm^2) at a steady speed of ~1 lb./s (0.45 kg/s). The pressure value was first recorded when the participant reported the onset of pain as a pressure pain threshold and was terminated when the participant reached their maximum pain tolerance. Four trials were performed, including the left forearm, the right forearm, the left trapezius, and the right trapezius.
(2) Mechanical pinprick pain was assessed by applying 10 calibrated force pinprick stimuli to the skin at a fixed frequency (1 Hz). Participants were asked to rate their pain intensity after the 1st, 5th, and 10th stimuli. The procedure was first applied on the left index finger and then repeated on the right index finger.

(3) Cuff pain was assessed by inflating a blood pressure cuff on the left leg to a threshold pressure level (5 out of 10 on a scale) and maintained for a fixed duration (2 min). Participants were asked to rate their pain levels every 30 s.

(4) Conditioned pain modulation was assessed by applying a noxious thermal stimulus and an increasing pressure pain simultaneously. Participants were first asked to submerge their dominant hand into the cold-water bath set at 6 degrees Celsius. Meanwhile, increasing pressure was applied to the non-dominant trapezius muscle, as described in the pressure pain steps. The participants then reported their onset of pain and their maximum pain tolerance. The post-pain rating was registered 15 s after the cessation of pressure pain.

2.5. Data Preprocessing

The overall research diagram is presented in Figure 2. First, physiological (BVP, GSR, EMG, ST, RR, and pupillometry) data were synchronized by resampling them to 50 Hz. The left-eye and right-eye pupillometry data were interpolated to fill in any missing gaps [20]. The BVP signal was filtered via a fifth-order Butterworth band-pass with [0.5, 12] Hz as cut-off frequencies. The GSR was filtered via a fifth-order 1 Hz low-pass Butterworth filter. The RR was filtered via a fifth-order Butterworth band-pass with [0.1, 1] Hz as cut-off frequencies. In addition, eight time-series HRV data were generated from BVP signals, including PPG rate, meanNN, SDNN, RMSSD, SDSD, HF, SD1, and SD2 using the NeuroKit2 package in Python 3.7.9 [28]. For extracting heart rate variability signals, a 15 s sliding time window with 50 Hz was selected.

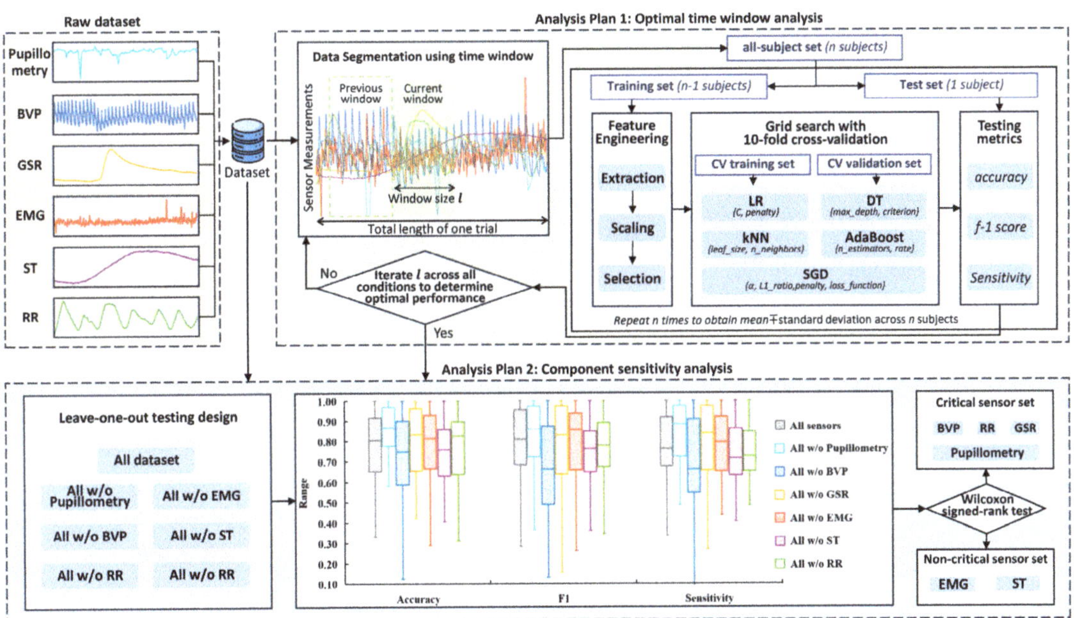

Figure 2. Diagram of the study. After collecting raw datasets from multimodal physiological sensors (BVP: blood volume pulse, GSR: galvanic skin response, EMG: electromyography, ST: skin temperature, RR: respiration rate), the dataset underwent two analysis plans: (1) perform the optimal time window analysis to select the optimal time window and hyperparameters via grid search. Time windows included 1, 2, 3, 4, and 5 s; (2) undergo component sensitivity analysis to investigate the performance across 7 distinct leave-one-out sets.

2.6. Feature Extraction and Selection

Features were extracted from all physiological sensors. GSR signals were separated into phasic and tonic signals. Statistical features were then generated from all physiological sensors, such as mean, median, range, variance, standard deviation, skewness, and kurtosis [39]. Five additional features were generated from EMGs, including mean absolute value, root mean square, variance, zero crossings, waveform length, and slope signal changes [39].

Principal Component Analysis was utilized for feature selection by setting a 90% information variance threshold to determine the cumulative features to be used.

2.7. Analysis Plan

To solve the research questions, a two-phase analysis plan was employed.

Analysis Plan 1 (optimal time window analysis): this phase focused on determining the optimal time window for signal segmentation and evaluating the performance of various machine learning models across all QST sessions. The time window candidates included 1 s, 2 s, 3 s, 4 s, and 5 s. A grid search methodology was employed to explore the relationship between time window lengths and classification performance, using a 10-fold cross-validation framework. The output from Plan 1 was to identify the combination of time window and model hyperparameters that achieved the highest accuracy, F1 score, and sensitivity for each QST session.

Analysis Plan 2 (component sensitivity analysis): this phase investigated the individual contribution of each sensor modality to classification performance. Using a leave-one-out (LOO) iteration strategy, one sensor's features were excluded at a time, and the model was retrained and tested using the remaining sensors. The performance of each LOO model was compared to the full model using statistical analysis (i.e., Wilcoxon signed-rank test).

The Wilcoxon signed-rank test is a nonparametric statistical test to compare two samples. This is a useful alternative to the paired t-test when the data do not follow a normal distribution. The differences between paired observations were computed and their absolute values were ranked. The test statistic was derived by summing the ranks of the positive and negative differences. Sensors with significant differences in Wilcoxon signed-rank test statistics (i.e., p value is below 0.05) were identified as critical sensor candidates. Sensors with minimal impact were identified as non-critical sensor candidates. The optimal time window and the optimal machine learning model were predetermined from Analysis Plan 1's results.

The classification task was to predict pain intensity states from physiological features and compare them with subjective ratings as the ground truth. In the pressure pain test, a three-class classification task was used to differentiate among three pain states: baseline (no pain), threshold (threshold of pressure), and tolerance (tolerance of pressure). This classification was applied to data combining pressure and conditioned pain modulation sessions, as both sessions employed identical labels for pressure threshold and tolerance. The classification task in both the pinprick session and cuff session was classifying pain intensity states based on numerical rating scales (0–10). Participants' self-reports were categorized into three levels: 1–3 (Mild Pain), 4–6 (Moderate Pain), and 7–10 (Severe Pain). This classification was conducted separately for stimuli applied to the left and right hands. Finally, the third task aimed to classify the pain intensity levels during a 2-minute temporal summation of cuff sensations. Similar to the pinprick classification task, the pain levels were categorized as Mild, Moderate, and Severe.

A grid search approach involving five classification models and their respective hyperparameters (detailed in Table 1) was used. These models included logistic regression (LOG), decision tree (DT), k Nearest Neighbors (KNN), Stochastic Gradient Descent (SGD), and AdaBoost (ADA). The Synthetic Minority Oversampling Technique (SMOTE) was applied to the training dataset to balance the minority classes [40]. Principal Component Analysis (PCA) was selected to reduce dimensionality with an 80% threshold.

Table 1. Grid search hyperparameters of classifiers.

Logistic Regression	C Penalty	$10^{-3}, 10^{-2}, 10^{-1}, 1, 10, 10^2, 10^3$ L1, L2
Decision Tree	Criterion Max Depth	gini, entropy 4, 5, 6, 7, 8, 9, 10, 11, 12, 15, 20, 30, 40, 50, 70, 90, 120
K Nearest Neighbors	Algorithm Leaf size N neighbors	Ball tree, kd tree, brute Range from 1 to 50 step 3 10, 13, 16, 19, 22, 25, 28
Stochastic Gradient Descent	Alpha L1 ratio Penalty Loss function	$10^{-2}, 10^{-3}, 10^{-4}$ 0.05, 0.06, 0.07, 0.08, 0.09, 0.1, 0.12, 0.13, 0.14, 0.15, 0.2 L1, L2 hinge, log, modified Huber, squared hinge
AdaBoost	Base estimator Max depth Min sample N estimators Learning rate	Decision tree 2, 5, 8, 11 5, 10 10, 50, 100, 250 0.01, 0.1

3. Experimental Results

Detailed demographic information is presented in Table 2. A total of 25 participants were initially screened, and 1 participant was excluded due to schedule conflicts. Twenty-four participants were successfully recruited. It included 17 healthy adult participants (11 females, mean age 28.8 years old) and 7 cLBP patients (5 females, mean age 44.4 years old). With an average of 14 years of pain duration, cLBP patients reported higher pain intensity and interference (Brief Pain Inventory; intensity, 5.0 ± 1.4; interference, 3.7 ± 2.5) than healthy participants (intensity, 0.3 ± 0.4; interference, 0.1 ± 0.2).

Table 2. Demographic information.

Mean ± SD or %	cLBP Patient	Healthy Group
Number of participants	7	17
Age, y	44.4 ± 14.5	28.8 ± 13.1
Female sex	5	11
Pain duration, y	14.0 ± 15.5	0
Pain intensity	5.0 ± 1.4	0.3 ± 0.4
Pain interference	3.7 ± 2.5	0.1 ± 0.2

Table 3 includes the sample size and time lengths of various session durations: baseline, the time from session start to pressure threshold, the time from pressure threshold to tolerance, pinprick, and cuff. The time variance among pressure threshold and pressure tolerance sessions is high, with standard deviations (STD) of 3.74 and 5.34, respectively, compared to their mean values of 6.86 and 13.99, respectively. The time lengths of pinprick and cuff sessions were more stable regarding the variance, with STDs of 1.44 and 3.39, respectively, compared to their mean values of 6.99 and 29.95, respectively.

Table 3. Time length statistics of QST sessions.

QST Session	Sample Size	Mean ± STD (s)
Baseline	32	59.80 ± 6.93
Pressure–Threshold	160	6.86 ± 3.74
Pressure–Tolerance	160	13.99 ± 5.34
Pinprick	128	6.99 ± 1.44
Cuff	128	29.95 ± 3.39

3.1. Analysis Plan 1—Optimal Time Window Analysis

Figure 3 shows three accuracy curve plots of three sessions based on different algorithms and five segmented time windows. For each session, the time window selected was about three factors: accuracy of classification, time variance, and size of the datapoints.

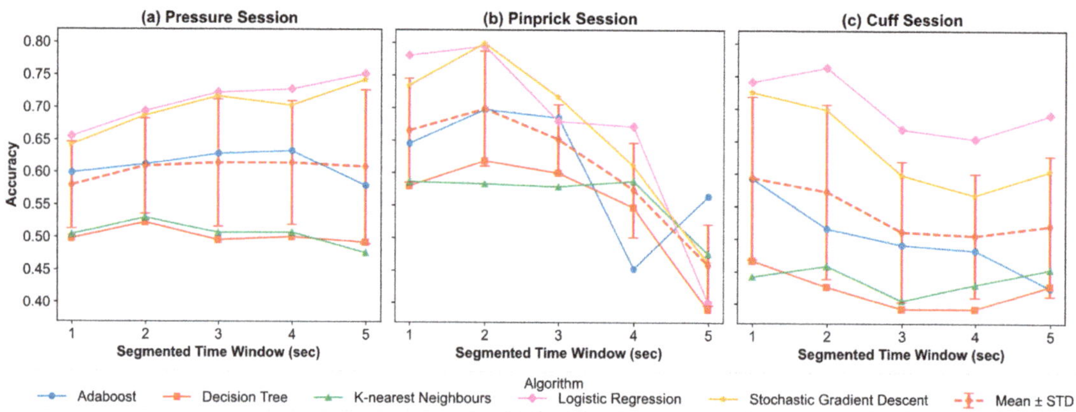

Figure 3. Accuracy curve of all algorithms and the mean accuracy of five algorithms under five segmented time windows (1, 2, 3, 4, 5 s) among pressure session (**a**), pinprick session (**b**), and cuff session (**c**). The red line in each figure shows the mean and standard deviation (STD) of all algorithms under different segmented time windows.

For the pressure session, the highest average performance was achieved when the time window was set as 3 s (accuracy = 61.4%, f-1 = 54.3%), followed by 4 s (accuracy = 61.4%, f-1 = 53.5%). However, the highest performance was achieved in the logistic regression classifier of the 5 s time window (accuracy = 75.2%, f-1 = 67.2%). The size of the datapoint generated from a 5 s time window is 558, smaller than that generated from a 3 s time window, which is 1112 datapoints. Considering all factors, 3 s was selected as the optimal time window for the pressure session.

For the pinprick session, the highest performance was achieved in the SGD classifier using a 2 s time window (accuracy = 79.9%, f-1 = 64.8%). The highest average accuracy was achieved with the 2 s time window (69.9%), followed by 1 s (66.5%), and 3 s (65.2%). The optimal time window for the pinprick session was 2 s.

For the cuff session, the highest performance was achieved when the time window was set as 2 s in the LOG classifier (accuracy = 76.4%), followed by the LOG classifier (accuracy = 74.1%) and SGD classifier (accuracy = 72.6%) under a 1 s time window. The highest average performance was achieved in the 1 s time window (accuracy = 59.3%, f-1 = 41.2%), followed by the 2 s (accuracy = 57.3%, f-1 = 41.2%). Considering the dataset size, 4478 datapoints were segmented using the 1 s time window before SMOTE, and 2047 datapoints were generated with the 2 s time window. Therefore, a 1 s time window was chosen for the cuff session.

3.2. Analysis Plan 2—Component Sensitivity Analysis

This plan investigated the optimal sensor set for all QST sessions. First, a baseline model that included features from all sensors was established. This model employed PCA and performed the classification model based on the best model from the previous section. The LOG model was identified as the optimal for machine learning for all three sessions because this model achieved the highest classification accuracy performance. Second, six additional sensor set plans were compared by iteratively removing one sensor at a time from the following set: pupillometry, BVP, EMG, GSR, RR, and ST.

For the pinprick session, the baseline model demonstrated an accuracy of 79.8% (f-1 = 62.9%). Six additional sets were evaluated, as shown in Table 4. The differences in accuracy compared to the baseline model are depicted in Figure 4a. The removal of BVP was found to significantly enhance overall performance, with increases of 6.2% in accuracy and 10.9% in f-1 ($p < 0.05$, Wilcoxon signed-rank test). Compared to the baseline, EMG and ST sensors had a negligible impact on the accuracy, with absolute accuracy differences of less than 1%. Removing GSR and RR sensors resulted in non-significant ($p > 0.05$, Wilcoxon signed-rank test) accuracy improvements of 2.3% and 2.2%, respectively. Excluding pupillometry decreased non-significant accuracy by 1.4%.

Table 4. Performance of pinprick, cuff, and pressure sessions.

	Pinprick		Cuff		Pressure	
Sensor Set	**Accuracy %**	**F-1 Score %**	**Accuracy %**	**F-1 Score %**	**Accuracy %**	**F-1 Score %**
All sensors	79.8	62.9	76.5	60.8	72.3	66.4
All w/o BVP	86 ↑	73.8 ↑	70.4 ↓	47.5 ↓	72.3	66.4
All w/o EMG	80.7	65.4	74.6	58.6	72.3	66.4
All w/o GSR	82.1	67.7	79.0 ↑	60.7 ↑	72.3	66.4
All w/o RR	80.6	63.8	74.2 ↓	53.9 ↓	72.4	66.7
All w/o ST	80.7	67.4	71.0	51.1	72.3	66.4
All w/o pupillometry	78.4	59.9	81.8 ↑	62.3 ↑	72.3	66.4

↑↓ indicates a statistically significant increase or decrease in performance (Wilcoxon signed-rank, $p < 0.05$).

For the cuff session, the baseline model achieved accuracies of 76.5% (f-1 = 60.8%). The performance from the LOO analysis is shown in Table 4 and illustrated in Figure 4b. For the LOG classifier, both the BVP and RR were observed to significantly decrease in accuracy by 6.1% ($p < 0.05$) and 2.3% ($p < 0.05$), respectively. Removing EMG and ST led to 1.9% and 5.5% decreases in accuracy, respectively, while these changes were not statistically significant. Excluding GSR and pupillometry showed an increase in performance, with accuracy improvements of 2.5% ($p < 0.05$) and 5.3% ($p < 0.05$), respectively.

In contrast, the pressure session showed minimal variation in performance across different sensor sets in Table 4. The performance of the LOG model was stable and not affected by the removal of individual sensors. The removal of BVP, EMG, GSR, ST, and pupillometry decreased the performance, with less than 2% for the accuracy and f-1 score.

In summary, it was found that removing BVP significantly improved accuracy for the pinprick session, whereas removing EMG, GSR, RR, ST, and pupillometry did not significantly impact the classification performance. Regarding the cuff session, the elimination of BVP, GSR, RR, and pupillometry significantly impacted accuracy, while EMG and ST showed non-significant performance, indicating they were non-critical sensor candidates. None of the singular sensors significantly impacted the performance of the pressure sessions.

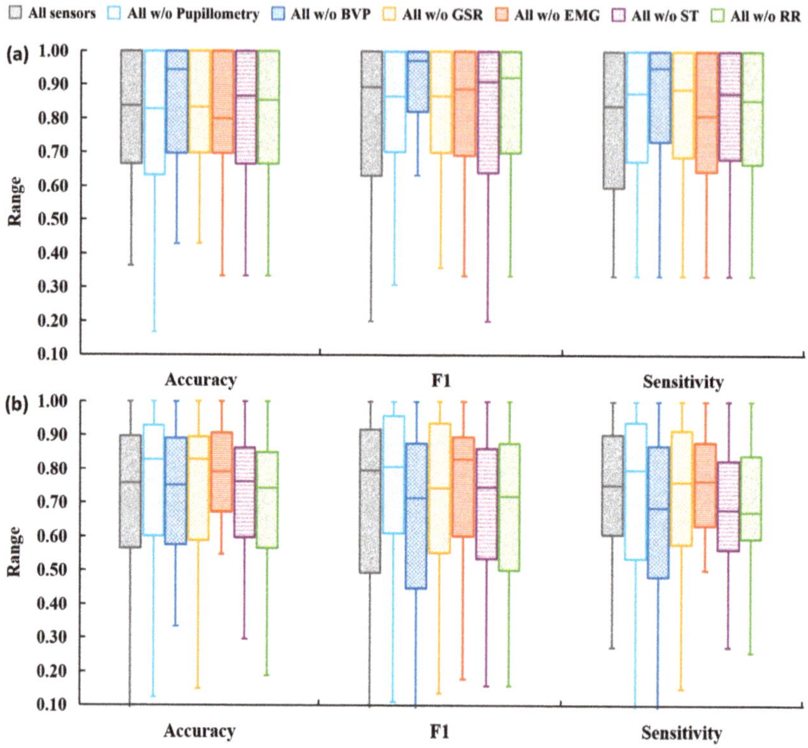

Figure 4. Performance metrics of the pinprick session and cuff session are presented in grouped boxplots (**a**,**b**). Each bar shows the mean and standard deviation.

4. Discussion

This study represents a novel approach to objectively assess and classify pain intensity levels utilizing physiological sensors across pressure, pinprick, and cuff sessions. Our methodology involved classifying traditional subjective ratings, such as baseline, threshold, and tolerance in the pressure session, and categorizing pain intensity levels into mild, moderate, and severe pain in the pinprick and cuff sessions based on physiological features and multiple classification models to achieve optimal performance. Two critical analyses were explored: (1) determining the optimal segmented time window (ranging from 1 to 5 s); (2) identifying the individual contributions of each singular sensor by implementing an LOO iteration strategy and classifying critical and non-critical sensor candidates.

Existing literature on pain intensity level classification using physiological signals typically employed fixed time windows, such as 1 s [41], 4 s [42], and 10 s [23]. Our study contributes to this field by investigating five different time windows under three QST sessions. The results highlighted uniformity among baseline, pinprick, and cuff sessions but a significant variance across the pressure sessions. Such variability presented a challenge in determining the segmented time window. Our study carefully weighed the trade-offs among the factors like accuracy, F1 score, number of datapoints, and distributions of time lengths. For instance, a 2 s time window was chosen for the pinprick session due to its superior performance in classification models. In contrast, the cuff session's time window was selected based on average performance and dataset size. However, the pressure session did not show significant performance differences between the 3, 4, and 5 s time windows. This highlights the need for discussing confounding factors such as psychosocial factors [13,43].

In analyzing optimal sensor sets for QST sessions, our baseline all-sensor models were compared against six other sets, each excluding one sensor iteratively. Our results reveal that removing BVP improved accuracy in pinprick sessions but decreased accuracy in cuff sessions. Removing EMG and ST sensors had negligible impact on pinprick session outcomes. In the cuff session, removing BVP and RR had negative impacts, whereas eliminating GSR and pupillometry significantly improved performance. Other literature analyzed the relationship between cuff sessions with singular physiological signals, such as EMG [35] and BVP [44], but very few published studies have examined the cuff pain intensity level classification via physiological signals. The pressure session presented uniformly consistent results, with an average accuracy variation of less than 2% among seven sensor sets using the LOG models. The reasons for this uniformity are not fully understood and warrant further investigation. Potential reasons include confounding factors such as the selection of time windows, the number of data points, and the participant population. The performance is consistent with one other study, which found that the highest performance of a three-class pain level classification in pressure pain session was achieved with 69% accuracy, 83.3% sensitivity, and 75% specificity [33].

Our study contributed to the field by indicating sensors (i.e., EMG and ST) that contributed minimally to classification performance and, as a result, implying a solution for cases when reducing sensor redundancy if necessary. Our study also highlighted the sensor that significantly impacted performance in both pinprick and cuff sessions (i.e., BVP), and sensors that are critical but stimuli-sensitive (i.e., GSR and RR). These stimuli-sensitive sensors should be further analyzed in sensor configuration tests. Pressure sessions demonstrated uniform performance, indicating that the pressure pain classification may rely on generalized physiological responses rather than specific sensor inputs.

The limitations of our study are multifaceted. First, the study's approach of consolidating a limited and unbalanced sample of healthy participants (N = 17) and chronic patients (N = 7) into a single group was necessary for generalizability. At the same time, it limits the sensitivity of our findings between these distinct groups. Second, the limited specificity performance of non-critical sensors (i.e., EMG and ST in cuff sessions; EMG, GSR, RR, ST in pinprick sessions) does not directly mean that they can be excluded in all cases. Psychological, environmental, and physical activity factors might lead to limited-specificity performance. Alternative solutions can be achieved in multiple ways. One solution can be to integrate multimodal deep learning models like long short-term memory models and transfer learning [45,46]. Exploring different sensor fusion methods such as feature-fusion, decision-level fusion can also be an alternative way [47]. A new field, network physiology, can be integrated into the pain assessment problem [48]. Instead of evaluating sensors in a deterministic role, this area can treat sensors as probabilistic models to analyze the connectivity between each modality [49,50]. This holistic approach acknowledges the dynamic connections between different physiological modalities, potentially resolving inconsistencies where a sensor may be effective in one context but not in another [49,51].

Exploring individual variations in response to different stimuli is a promising area to understand pain sensitivity. The current practice of pain sensitivity is assessed by patient's self-report, which cannot exclude the presence of inter- and intra-subject variability in characteristics such as psychological factors [13]. Stimulus-specific physiological responses represent a novel and critical area of exploration, linking specific physiological modalities to distinct noxious stimuli. This consideration is particularly important; when considering enhancing the portability and practicality of sensor configurations for chronic pain patients, sensor selection should be tailored to the specific type of pain being assessed. Future research can replicate this study following the described QST procedures, analysis plans, and pseudocode algorithms in the Supplementary Materials. This framework can be easily extended to explore the relationship between physiological responses to other dimensions of pain assessment, such as types of stimuli and pain locations, beyond pain intensity. The ability to identify the types and locations of pain will benefit patients who have difficulties in self-reporting [52].

In terms of broader impacts, this sensor sensitivity study paves the way for enhancing the portability and feasibility of pain assessment, especially in at-home settings. This study suggests that sensors having minimal impact on performance can be excluded from wearable pain assessment devices. This result can be used to simplify pain assessment device design by only including sensors such as GSR, BVP, and RR. That is to say, there are plenty of digital health technologies for remote data acquisition [53]. For example, the Empatica watch monitors BVP, GSR, and ST (Empatica, Empatica Inc., Cambridge, MA, USA); the Google Fitbit series collects different sets of sensors among PPG, oxygen saturation, GSR, and ST (Google, Santa Clara, CA, USA). By selecting sensor modalities sensitive to specific noxious stimuli, researchers can balance feasibility with model performance and enhance the practicality of remote pain assessment [26].

5. Conclusions

This study presented a novel framework for pain assessment using physiological sensors during QST sessions, integrating two complementary analysis plans. Analysis Plan 1 identified optimal time windows for signal segmentation, with 1–5 s windows yielding varied results across pinprick, cuff, and pressure sessions. The findings highlight the importance of tailoring time segmentation to specific stimuli to maximize classification performance. Analysis Plan 2 evaluated sensor contributions using leave-one-out iterations. BVP, GSR, RR, and pupillometry were identified as stimulus-specific critical sensor candidates, although only BVP showed significant performance across stimuli. In contrast, EMG and ST were found to be non-critical, showing minimal impact on performance across all sessions. Future research should explore stimulus-specific physiological responses further to optimize sensor configurations for different pain types. Incorporating advanced multi-sensor fusion techniques and individualization methodology can support the development of personalized, efficient, and practical wearable systems for chronic pain assessment and management.

Supplementary Materials: The following supporting information can be downloaded at: https://www.mdpi.com/article/10.3390/s25072086/s1, File S1: Supplementary Material: Pseudocode Algorithms for Physiological Sensor Modality Sensitivity Test for Pain Intensity Classification in Quantitative Sensory Testing.

Author Contributions: Conceptualization, Y.L.; methodology, W.Z. and Y.L.; software, W.Z.; validation, W.Z.; formal analysis, W.Z.; investigation, W.Z. and Y.L.; resources, Y.L.; data curation, W.Z.; writing—original draft, W.Z.; writing—review and editing, W.Z. and Y.L.; visualization, W.Z.; supervision, Y.L.; project administration, Y.L.; funding acquisition, Y.L. All authors have read and agreed to the published version of the manuscript.

Funding: This research has been financially supported by a National Science Foundation (NSF) project entitled "Novel Computational Methods for Continuous Objective Multimodal Pain Assessment Sensing System (COMPASS)" under Award #1838796.

Institutional Review Board Statement: The study was approved by the Institutional Review Board of Brigham and Women's Hospital (protocol code 2019P002781, 11/18/2019).

Informed Consent Statement: Informed consent was obtained from all subjects involved in the study. All subjects and/or their legal guardian(s) consented to the publication of identifying information/images in an online open-access publication.

Data Availability Statement: Datasets in the study are available from the corresponding author upon reasonable request.

Conflicts of Interest: The authors declare no conflicts of interest.

References

1. Rikard, S.M.; Strahan, A.E.; Schmit, K.M.; Guy, G.P. Chronic Pain Among Adults—United States, 2019–2021. *MMWR Morb Mortal Wkly Rep.* **2023**, *72*, 379–385. [CrossRef] [PubMed]
2. Raja, S.N.; Carr, D.B.; Cohen, M.; Finnerup, N.B.; Flor, H.; Gibson, S.; Keefe, F.J.; Mogil, J.S.; Ringkamp, M.; Sluka, K.A.; et al. The revised International Association for the Study of Pain definition of pain: Concepts, challenges, and compromises. *Pain* **2020**, *161*, 1976–1982. [CrossRef] [PubMed]
3. Lee, S.H.; Liang, H.W. Discriminative Changes in Sitting and Standing Postural Steadiness in Patients with Chronic Low Back Pain. *IEEE Trans. Neural Syst. Rehabil. Eng.* **2023**, *31*, 3752–3759. [CrossRef] [PubMed]
4. Slaboda, J.C.; Boston, J.R.; Rudy, T.E.; Lieber, S.J.; Rasetshwane, D.M. The use of splines to calculate jerk for a lifting task involving chronic lower back pain patients. *IEEE Trans. Neural Syst. Rehabil. Eng.* **2005**, *13*, 406–414. [CrossRef]
5. Davis, K.D.; Aghaeepour, N.; Ahn, A.H.; Angst, M.S.; Borsook, D.; Brenton, A.; Burczynski, M.E.; Crean, C.; Edwards, R.; Gaudilliere, B.; et al. Discovery and validation of biomarkers to aid the development of safe and effective pain therapeutics: Challenges and opportunities. *Nat. Rev. Neurol.* **2020**, *16*, 381–400. [CrossRef]
6. Berger, S.E.; Vachon-Presseau, É.; Abdullah, T.B.; Baria, A.T.; Schnitzer, T.J.; Apkarian, A.V. Hippocampal morphology mediates biased memories of chronic pain. *Neuroimage* **2018**, *166*, 86–98. [CrossRef]
7. Naugle, K.M.; Ohlman, T.; Naugle, K.E.; Riley, Z.A.; Keith, N.R. Physical activity behavior predicts endogenous pain modulation in older adults. *Pain* **2017**, *158*, 383–390. [CrossRef]
8. Nijs, J.; Girbés, E.L.; Lundberg, M.; Malfliet, A.; Sterling, M. Exercise therapy for chronic musculoskeletal pain: Innovation by altering pain memories. *Man. Ther.* **2015**, *20*, 216–220. [CrossRef]
9. Deyo, R.A.; Mirza, S.K.; Turner, J.A.; Martin, B.I. Overtreating chronic back pain: Time to back off? *J. Am. Board Fam. Med.* **2009**, *22*, 62–68. [CrossRef]
10. Von Korff, M.R. Health Care for chronic pain: Overuse, underuse, and treatment needs: Commentary on: Chronic pain and health services utilization—Is there overuse of diagnostic tests and inequalities in nonpharmacologic methods utilization? *Med. Care* **2013**, *51*, 857–858. [CrossRef]
11. Edwards, R.R.; Sarlani, E.; Wesselmann, U.; Fillingim, R.B. Quantitative assessment of experimental pain perception: Multiple domains of clinical relevance. *Pain* **2005**, *114*, 315–319. [CrossRef] [PubMed]
12. Meints, S.M.; Mawla, I.; Napadow, V.; Kong, J.; Gerber, J.; Chan, S.-T.; Wasan, A.D.; Kaptchuk, T.J.; McDonnell, C.; Carriere, J.; et al. The relationship between catastrophizing and altered pain sensitivity in patients with chronic low-back pain. *Pain* **2019**, *160*, 833–843. [CrossRef] [PubMed]
13. Fillingim, R.B.; Bruehl, S.; Dworkin, R.H.; Dworkin, S.F.; Loeser, J.D.; Turk, D.C.; Widerstrom-Noga, E.; Arnold, L.; Bennett, R.; Edwards, R.R.; et al. The ACTTION-American Pain Society Pain Taxonomy (AAPT): An evidence-based and multidimensional approach to classifying chronic pain conditions. *J. Pain* **2014**, *15*, 241–249. [CrossRef] [PubMed]
14. Chai, P.R.; Gale, J.Y.; Patton, M.E.; Schwartz, E.; Jambaulikar, G.D.; Taylor, S.W.; Edwards, R.R.; Boyer, E.W.; Schreiber, K.L. The impact of music on nociceptive processing. *Pain Med.* **2020**, *21*, 3047–3054. [CrossRef]
15. Yang, G.; Lin, Y. Using ECG Signal to Quantify Mental Workload Based on Wavelet Transform and Competitive Neural Network Techniques. *Biomed. Soft Comput. Hum. Sci.* **2009**, *14*, 17–25.
16. Liang, B.; Lin, Y. Using Physiological and Behavioral Measurements in a Picture-Based Road Hazard Perception Experiment to Classify Risky and Safe Drivers. *Transp. Res. Part F Traffic Psychol. Behav.* **2018**, *58*, 93–105. [CrossRef]
17. Yang, G.; Lin, Y.; Bhattacharya, P. A driver fatigue recognition model based on information fusion and dynamic Bayesian network. *Inf. Sci.* **2010**, *180*, 1942–1954. [CrossRef]
18. Wu, S.W.; Wang, Y.-C.; Hsieh, P.-C.; Tseng, M.-T.; Chiang, M.-C.; Chu, C.-P.; Feng, F.-P.; Lin, Y.-H.; Hsieh, S.-T.; Chao, C.-C. Biomarkers of neuropathic pain in skin nerve degeneration neuropathy: Contact heat-evoked potentials as a physiological signature. *Pain* **2017**, *158*, 516–525. [CrossRef]
19. Jang, E.H.; Park, B.J.; Park, M.S.; Kim, S.H.; Sohn, J.H. Analysis of physiological signals for recognition of boredom, pain, and surprise emotions. *J. Physiol. Anthr.* **2015**, *34*, 25. [CrossRef]

20. Johnson, A.; Yang, F.; Gollarahalli, S.; Banerjee, T.; Abrams, D.; Jonassaint, J.; Jonassaint, C.; Shah, N. Use of mobile health apps and wearable technology to assess changes and predict pain during treatment of acute pain in sickle cell disease: Feasibility study. *JMIR Mhealth Uhealth* **2019**, *7*, e13671. [CrossRef]
21. Lin, Y.; Xiao, Y.; Wang, L.; Guo, Y.; Zhu, W.; Dalip, B.; Kamarthi, S.; Schreiber, K.L.; Edwards, R.R.; Urman, R.D. Experimental Exploration of Objective Human Pain Assessment Using Multimodal Sensing Signals. *Front. Neurosci.* **2022**, *16*, 831627. [CrossRef] [PubMed]
22. Misra, G.; Wang, W.E.; Archer, D.B.; Roy, A.; Coombes, S.A. Automated classification of pain perception using high-density electroencephalography data. *J. Neurophysiol.* **2017**, *117*, 786–795. [CrossRef] [PubMed]
23. Elsayed, M.; Sim, K.S.; Tan, S.C. A novel approach to objectively quantify the subjective perception of pain through electroencephalogram signal analysis. *IEEE Access* **2020**, *8*, 199920–199930. [CrossRef]
24. Olesen, A.E.; Andresen, T.; Staahl, C.; Drewes, A.M. Human experimental pain models for assessing the therapeutic efficacy of analgesic drugs. *Pharmacol. Rev.* **2012**, *64*, 722–779. [CrossRef]
25. Zhu, W.; Xiao, Y.; Lin, Y. A Novel Labeling Method of Physiological-based Pressure Pain Assessment Among Patients with and Without Chronic Low Back Pain. *Proc. Human. Factors Ergon. Soc. Annu. Meet.* **2024**, *68*, 456–459. [CrossRef]
26. Dolgin, E. How a 'pain-o-meter' could improve treatments. *Nature* **2024**, *633*, S26–S27. [CrossRef]
27. Guo, Y.; Wang, L.; Xiao, Y.; Lin, Y. A Personalized Spatial-Temporal Cold Pain Intensity Estimation Model Based on Facial Expression. *IEEE J. Transl. Eng. Health Med.* **2021**, *9*, 4901008. [CrossRef]
28. Wang, L.; Guo, Y.; Dalip, B.; Xiao, Y.; Urman, R.D.; Lin, Y. An experimental study of objective pain measurement using pupillary response based on genetic algorithm and artificial neural network. *Appl. Intell.* **2021**, *52*, 1145–1156. [CrossRef]
29. Wang, L.; Xiao, Y.; Urman, R.D.; Lin, Y. Cold pressor pain assessment based on EEG power spectrum. *SN Appl. Sci.* **2020**, *2*, 1976. [CrossRef]
30. Chen, D.; Zhang, H.; Kavitha, P.T.; Loy, F.L.; Ng, S.H.; Wang, C.; Phua, K.S.; Tjan, S.Y.; Yang, S.-Y.; Guan, C. Scalp EEG-Based Pain Detection Using Convolutional Neural Network. *IEEE Trans. Neural Syst. Rehabil. Eng.* **2022**, *30*, 274–285. [CrossRef]
31. Zheng, J.; Lin, Y. Using Physiological Signals for Pain Assessment: An Evaluation of Deep Learning Models. In Proceedings of the 2024 30th International Conference on Mechatronics and Machine Vision in Practice (M2VIP), Leeds, UK, 3–5 October 2024; IEEE: Piscataway, NJ, USA, 2024; pp. 1–6. [CrossRef]
32. Zheng, J.; Lin, Y. An Objective Pain Measurement Machine Learning Model through Facial Expressions and Physiological Signals. In Proceedings of the 2022 28th International Conference on Mechatronics and Machine Vision in Practice (M2VIP), Nanjing, China, 16–18 November 2022; IEEE: Piscataway, NJ, USA, 2022; pp. 1–4. [CrossRef]
33. Khan, M.U.; Aziz, S.; Hirachan, N.; Joseph, C.; Li, J.; Fernandez-Rojas, R. Experimental Exploration of Multilevel Human Pain Assessment Using Blood Volume Pulse (BVP) Signals. *Sensors* **2023**, *23*, 3980. [CrossRef] [PubMed]
34. Pickering, G.; Achard, A.; Corriger, A.; Sickout-Arondo, S.; Macian, N.; Leray, V.; Lucchini, C.; Cardot, J.-M.; Pereira, B. Electrochemical Skin Conductance and Quantitative Sensory Testing on Fibromyalgia. *Pain Pract.* **2020**, *20*, 348–356. [CrossRef] [PubMed]
35. Gray, S.M.; Cuomo, A.M.; Proppe, C.E.; Traylor, M.K.; Hill, E.C.; Keller, J.L. Effects of Sex and Cuff Pressure on Physiological Responses during Blood Flow Restriction Resistance Exercise in Young Adults. *Med. Sci. Sports Exerc.* **2023**, *55*, 920–931. [CrossRef] [PubMed]
36. Tiemann, L.; Achard, A.; Corriger, A.; Sickout-Arondo, S.; Macian, N.; Leray, V.; Lucchini, C.; Cardot, J.; Pereira, B. Distinct patterns of brain activity mediate perceptual and motor and autonomic responses to noxious stimuli. *Nat. Commun.* **2018**, *9*, 4487. [CrossRef]
37. Nickel, M.M.; Hohn, V.D.; Dinh, S.T.; May, E.S.; Nickel, M.M.; Gross, J.; Ploner, M. Temporal–spectral signaling of sensory information and expectations in the cerebral processing of pain. *Proc. Natl. Acad. Sci. USA* **2022**, *119*, e2116616119. [CrossRef]
38. Tan, G.; Jensen, M.P.; Thornby, J.I.; Shanti, B.F. Validation of the brief pain inventory for chronic nonmalignant pain. *J. Pain* **2004**, *5*, 133–137. [CrossRef]
39. Zhu, W.; Kucyi, A.; Kramer, A.F.; Lin, Y. Multimodal Physiological Assessment of the Task-related Attentional States in a VR Driving Environment. In Proceedings of the 2022 28th International Conference on Mechatronics and Machine Vision in Practice (M2VIP), Nanjing, China, 16–18 November 2022; pp. 1–5.
40. Fernández, A.; García, S.; Herrera, F.; Chawla, N.V. SMOTE for Learning from Imbalanced Data: Progress and Challenges, Marking the 15-year Anniversary. *J. Artif. Intell. Res.* **2018**, *61*, 863–905.
41. Lin, Y.; Wang, L.; Xiao, Y.; Urman, R.D.; Dutton, R.; Ramsay, M. Objective Pain Measurement based on Physiological Signals. *Proc. Int. Symp. Hum. Factors Ergon. Health Care* **2018**, *7*, 240–247. [CrossRef]
42. Zhu, W.; Liu, C.; Yu, H.; Guo, Y.; Xiao, Y.; Lin, Y. COMPASS App: A Patient-centered Physiological based Pain Assessment System. *Proc. Hum. Factors Ergon. Soc. Annu. Meet.* **2023**, *67*, 1361–1367. [CrossRef]

43. Kent, M.L.; Tighe, P.J.; Belfer, I.; Brennan, T.J.; Bruehl, S.; Brummett, C.M.; Buckenmaier, C.C.; Buvanendran, A.; Cohen, R.I.; Desjardins, P.; et al. The ACTTION-APS-AAPM Pain Taxonomy (AAAPT) Multidimensional Approach to Classifying Acute Pain Conditions. *Pain Med.* **2017**, *18*, 947–958. [CrossRef]
44. Cerqueira, M.S.; Costa, E.C.; Oliveira, R.S.; Pereira, R.; Brito Vieira, W.H. Blood Flow Restriction Training: To Adjust or Not Adjust the Cuff Pressure Over an Intervention Period? *Front. Media* **2021**, *12*, 678407. [CrossRef]
45. Pouromran, F.; Lin, Y.; Kamarthi, S. Personalized Deep Bi-LSTM RNN Based Model for Pain Intensity Classification Using EDA Signal. *Sensors* **2022**, *22*, 8087. [CrossRef] [PubMed]
46. Lopez-Martinez, D.; Picard, R. Multi-task Neural Networks for Personalized Pain Recognition from Physiological Signals. August. *arXiv* **2017**, arXiv:1708.08755. [CrossRef]
47. Aguileta, A.A.; Brena, R.F.; Mayora, O.; Molino-Minero-Re, E.; Trejo, L.A. Multi-Sensor Fusion for Activity Recognition—A Survey. *Sensors* **2019**, *19*, 3808. [CrossRef]
48. Ivanov, P.C. The New Field of Network Physiology: Building the Human Physiolome. *Front. Netw. Physiol.* **2021**, *1*, 711778. [CrossRef]
49. Bashan, A.; Bartsch, R.P.; Kantelhardt, J.W.; Havlin, S.; Ivanov, P.C. Network physiology reveals relations between network topology and physiological function. *Nat. Commun.* **2012**, *3*, 702. [CrossRef]
50. Candia-Rivera, D.; Chavez, M.; De Vico Fallani, F. Measures of the coupling between fluctuating brain network organization and heartbeat dynamics. *Netw. Neurosci.* **2024**, *8*, 557–575. [CrossRef]
51. Bartsch, R.P.; Liu, K.K.L.; Bashan, A.; Ivanov, P.C. Network Physiology: How Organ Systems Dynamically Interact. *PLoS ONE* **2015**, *10*, e0142143. [CrossRef]
52. Rojas, R.F.; Brown, N.; Waddington, G.; Goecke, R. A systematic review of neurophysiological sensing for the assessment of acute pain. *NPJ Digit. Med.* **2023**, *6*, 76. [CrossRef]
53. Lewis, A.; Valla, V.; Charitou, P.; Karapatsia, A.; Koukoura, A.; Tzelepi, K.; Bergsteinsson, J., I; Ouzounelli, M.; Vassiliadis, E. Digital Health Technologies for Medical Devices—Real World Evidence Collection—Challenges and Solutions Towards Clinical Evidence. *Int. J. Digit. Health* **2022**, *2*, 8. [CrossRef]

MDPI AG
Grosspeteranlage 5
4052 Basel
Switzerland
Tel.: +41 61 683 77 34

Sensors Editorial Office
E-mail: sensors@mdpi.com
www.mdpi.com/journal/sensors

www.ingramcontent.com/pod-product-compliance
Lightning Source LLC
Chambersburg PA
CBHW061450210926
51850CB00008B/1666